Pigments in Fruits and Vegetables

Genomics and Dietetics

Chunxian Chen
Editor

Pigments in Fruits and Vegetables

Genomics and Dietetics

Springer

Editor
Chunxian Chen
Citrus Research and Education Center
University of Florida
Lake Alfred, Florida
USA

ISBN 978-1-4939-2355-7 ISBN 978-1-4939-2356-4 (eBook)
DOI 10.1007/978-1-4939-2356-4

Library of Congress Control Number: 2015933509

Springer New York Heidelberg Dordrecht London

Printed on acid-free paper

Springer is part of Springer Science+Business Media (www.springer.com)

Preface

Colors are ubiquitous in nature, particularly in living organisms ranging from bacteria and fungi to plants and animals. Many organisms have developed their own characteristic colors that vary by parts and developmental stage. These colors are not just visually decorative and attractive, but biologically essential in reproduction, coevolution, and ecosystem sustenance. Colors in plants, flowers, and fruits attract animals for pollination to produce seeds and for consumption to disperse seeds, which both help in species reproduction and diversification. Coloration-based camouflage in ecosystems to enhance survival is a good example of coevolution. The importance of colors in living organisms cannot be overstated. An old saying is apt: Colors can please the eye, gladden the heart, and nurture the mind. Biological pigments, the chemical components able to generate a full spectrum of visual colors in nature, are in fact much more important and valuable; they are biosynthesized behind the scene in living organisms and ultimately ingested in our daily diet.

Pigments produced in plants include four major classes: chlorophylls, carotenoids, flavonoids, and betalains. Chlorophylls are the primary green pigments for photosynthesis. The latter three are complementary nongreen pigments with diverse functions. Extensive research on the genetic mechanisms of their biosynthesis has yielded many exciting and insightful results over the last decades. On the other hand, many pigment-rich fruits and vegetables are consumed daily by human and animals. Potential nutritional and medicinal benefits from these pigments in fruits and vegetables have attracted nutritionists and clinical functional food researchers to study their health effects and encourage people to increase the daily consumption of these pigment-abundant foods.

Colorful fruits and vegetables attract visitors and eaters. Eating freshly harvested colorful vegetables while helping in my parents' vegetable garden remains among the most memorable moments in my childhood. Green cucumber and pea, red tomato and radish, and orange carrot and sweet potato, to name a few, are my favorites. My horticultural career might have started when I helped and wondered in the garden. Not only did the vegetables constantly attract me with their vibrant colors, but also ultimately nourished a future garden lover by their abundant tastes and nutrients. A time in the garden remained a joyful routine during every hometown visit. A small garden has been a must in my own family residence. If we believe there is a connection between an early childhood wonder and a later adulthood career, this book may give a casual explanation on it, and a delayed answer as well to my early curiosities about the distinct colors and tastes of the vegetables I ate in the garden.

This comprehensive treatise provides a systemic and insightful overview of current advances in the biosynthetic genomics/genetics and preventive dietetics of carotenoids, flavonoids, and betalains, from a general perspective, and in specific fruits and vegetables as well. Genomics/genetics focuses on what and how enzymatic and regulatory genes are involved in pigment biosynthesis. Dietetics emphasizes how these pigments contribute nutritional/medical benefits to health, prevent diseases, and act as potential nutraceuticals in the diet. The goal is to provide research scientists, nutrition specialists, healthy food advocates, students, and rainbow food (fruit and vegetable) lovers with an integrated resource on the biosynthetic and dietetic mechanisms of these pigments.

The book editor would like to acknowledge all the contributing authors, reviewers, and the entire Springer editorial staff. This book would not have been possible without their commitment and dedication. Particular thanks go to Hannah Smith, former plant sciences editor of Springer, for her initial enthusiastic endeavors, and to Michael D. Sova, developmental editor of Springer, for his persistent hard work.

Chunxian Chen

Contents

Contributors

Lílian Cristina Baldon Aizza Biologia Vegetal, Instituto de Biologia, Universidade Estadual de Campinas (UNICAMP), Campinas, SP, Brazil

Phyllis E. Bowen Kinesiology and Nutrition, University of Illinois at Chicago, Oak Park, IL, USA

Chunxian Chen Southeastern Fruit and Tree Nut Research Laboratory, USDA, ARS, Byron, Georgia, USA

Marcio Gilberto Cardoso Costa Biological Sciences Department, State University of Santa Cruz, Ilhéus, Bahia, Brazil

Veda Diwadkar-Navsariwala Bartlett, IL, USA

Marcelo Carnier Dornelas Biologia Vegetal, Universidade Estadual de Campinas (UNICAMP), Campinas, SP, Brazil

Andrea I. Doseff Department of Physiology and Cell Biology, Department of Molecular Genetics, Heart and Lung Research Institute, The Ohio State University, Columbus, OH, USA

Kevin M. Folta Horticultural Sciences, University of Florida, Gainesville, FL, USA

Julie Garden-Robinson Health, Nutrition, and Exercise Sciences, North Dakota State University, Fargo, ND, USA

Fred Gmitter Jr. Citrus Research and Education Center, University of Florida, Lake Alfred, FL, USA

Erich Grotewold Center for Applied Plant Sciences and Deptartment of Molecular Genetics, The Ohio State University, Columbus, OH, USA

Gregory J. Hatlestad Institute for Cell and Molecular Biology, The University of Texas at Austin, Austin, TX, USA

Maria A. Livrea Department of Stembio, University of Palermo, Palermo, Italy

Alan Lloyd Institute for Cell and Molecular Biology, The University of Texas at Austin, Austin, TX, USA

Angela Roberta Lo Piero Department of Agricultural and Food Science, University of Catania, Catania, Italy

Amanda Ferreira Da Silva Mendes Instituto Federal Da Bahia, Ilhéus, Bahia, Brazil

Christopher L. Owens Grape Genetics Research Unit, United States Department of Agriculture, Agricultural Research Service, Geneva, NY, USA

Arti Parihar Department of Biological Sciences, Government Post Graduate College, Center of Excellence, Ujjain, MP, India

Katia Petroni Dipartimento di Bioscienze, Università degli Studi di Milano, Milano, Italy

Jeremy Pillet Horticultural Sciences, University of Florida, Gainesville, FL, USA

Roberto Pilu Di.S.A.A.—Produzione, Territorio, Agroenergia, University of Milano, Milano, Italy

Manuel Rodriguez-Concepcion Centre for Research in Agriculture Genomics (CRAG) CSIC-IRTA-UAB-UB, Barcelona, Spain

Virgínia Lúcia Fontes Soares Biological Sciences Department, State University of Santa Cruz, Ilhéus, Bahia, Brazil

Maria Stacewicz-Sapuntzakis Chicago, IL, USA

Claudia Stange Department of Biology, University of Chile, Santiago, Chile

Luisa Tesoriere Department of Stembio, University of Palermo, Palermo, Italy

Chiara Tonelli Dipartimento di Bioscienze, Università degli Studi di Milano, Milano, Italy

Chapter 1
Overview of Plant Pigments

Chunxian Chen, PhD

Introduction

Pigments make nature colorful and likable. Plant pigments usually refer to four major well-known classes: chlorophylls, carotenoids, flavonoids, and betalains (Table 1.1). Each class may contain various numbers of chemical compounds that can be structurally categorized into distinct subgroups. Most pigments are colored. In general, the visible colors are the emission of certain wavelengths of light by colored pigments after they selectively absorb others specific to their molecules. The color spectra of pigments are illustrated in Fig. 1.1 with some examples of pigment-rich fruits and vegetables. Absorbed light may be captured by a few capable pigments as energy to fuel plant photosynthesis and biochemical reactions. These colored pigments not only visually attract animals for flower pollination and seed dispersal but also function in critical biological processes for plants and play essential coevolutionary roles in ecosystems [1–4]. The biological, ecological, and evolutionary importance of plant pigments and the derived colors cannot be overstated. On the other hand, many pigment-rich fruits and vegetables are critical in the human and animal diet. Some pigments are essential nutrients, and others may serve as nutraceuticals with additional medical benefits, including the prevention and treatment of certain diseases. Chlorophylls are the source of green in all land plants and green algae and function as the primary pigment to capture yellow and blue light for photosynthesis to power plant development and growth. Unlike chlorophylls, the other three are accessory pigments (generally with the absorbance spectrum complementary to chlorophylls) and secondary metabolites that possess much more diverse structures and functions in plants and offer more potential nutritional and

C. Chen (✉)
Citrus Research and Education Center, University of Florida,
Lake Alfred, Florida, USA
e-mail: chencx2k@gmail.com

C. Chen (ed.), *Pigments in Fruits and Vegetables,* DOI 10.1007/978-1-4939-2356-4_1

Table 1.1 Four major classes of plant pigments

Pigment class	Basic structure	Main subgroups	Typical colors	Examples[a]
Carotenoids	40-carbon polyene hydrocarbon chain	Carotenes and xanthophylls	Orange, yellow, pink, red	Citrus, banana, carrot, pepper, leaf veggies
Flavonoids	15-carbon benzo-γ-pyrone skeleton	Anthocyanins; flavonols; 7 others	Purple, blue, red, yellow	Blueberry, blackberry, eggplant, red cabbage
Betalains	Indole-derived glycoside	Betacyanins and betaxanthins	Red, violet, orange, yellow	Dragon fruit, cactus pear, beet, Swiss chard
Chlorophylls	Chlorine ring	a and b	Green	Any green plants

[a] All their scientific names are listed in Fig. 1.1 except cactus pear (*Opuntia ficus-indica*)

health benefits in the diet [3, 4]. In this context, attention is only given to carotenoids, flavonoids, and betalains, with emphasis on the basic biological attributes and dietary benefits of each class.

Carotenoids

Carotenoids are a large family of lipid-soluble tetraterpenoids with a basic 40-carbon polyene hydrocarbon chain structure. This family of over 600 members can be generally divided into two subgroups, carotenes ($C_{40}H_{56}$) and xanthophylls ($C_{40}H_{56}O_2$ or $C_{40}H_{56}O$, the oxygenated derivatives of carotenes), which differ in the terminal rings and oxygenation [5, 6]. For example, α-carotene, β-carotene, and lycopene are carotenes; lutein, zeaxanthin, and violaxanthin are xanthophylls. In plants, certain carotenoids function as complementary light-harvesting pigments to precisely absorb wavelengths of light not gathered by chlorophylls, the primary photosynthesis pigment. They also provide photo-protection against excess light damage to the photosynthetic reaction center by quenching excited species such as singlet oxygen and free radicals or by other carotenoid enabled mechanisms [7]. In addition, carotenoids can be potent antioxidants in lipid formations due to the linearly conjugated carbon bonds that provide high reduction potential. Apocarotenoid hormones, including abscisic acid (ABA) and strigolactone, are produced at the extended steps following the carotenoid biosynthesis pathway, and may play a regulatory or signaling role in the pathway [8, 9]. Most genes and enzymes in the plant carotenoid biosynthesis pathway have been well characterized, and the transcriptional and posttranscriptional regulations are being elucidated as well. This information facilitates more efficient selection of carotenoid-rich varieties through conventional breeding and large-scale biotechnological production of carotenoids via metabolic engineering in microorganisms [10].

A large number of fruits and vegetables are colored by carotenoids, such as citrus (*Citrus* spp.), banana (*Musa paradisiaca*), papaya (*Carica papaya*), tomato

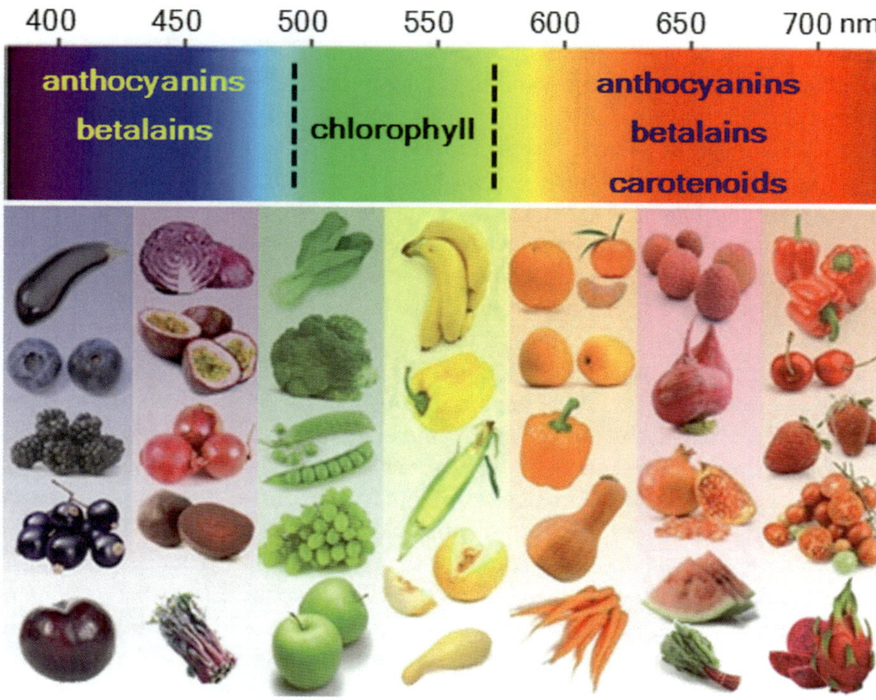

Fig. 1.1 Schematic color spectrum of carotenoids, anthocyanins, betalains and chlorophylls, and examples of pigment-rich fruits and vegetables. Each pigment class was marked in the zones with the approximately corresponding wavelengths (400-700 nm) of visible light and the examples given in the columns below with the symbolic colors. The two black dash lines flanking the green zone was to split and show the approximate color ranges of each pigment class. Some carotenoids are also found in green vegetables and many other colored fruits, however, the colors of these carotenoids at relatively low concentrations are usually masked by those of other predominant pigments. A for anthocyanins, B for betalains, and C for carotenoid were used to mark the primary pigment class in each colored fruit and vegetable, but no abbreviation for chlorophylls was used to mark the class in green fruits and vegetables. From left to right, column **1** from top to bottom are: eggplant (*Solanum melongena*, A), blueberry (*Vaccinium spp*., A), blackberry (*Rubus spp*., A), acai berry (*Euterpe oleracea*, A), plum (*Prunus domestica*, A); **2**: red cabbage (*Brassica oleracea* var. *capitata*, A), passion fruit (*Passiflora edulis*, A), red onion (*Allium cepa*, A), beet (*Beta vulgaris*, B), Zi cai tai ('purple shoot') (*Brassica campestris ssp. chinensis* var. *purpurea*, A); **3**: Shanghai bok choy (*Brassica rapa spp. chinensis*), broccoli (*Brassica oleracea*), pea (*Pisum sativum*), grape (*Vitis vinifera*), 'Granny Smith' apple (*Malus domestica*); **4**: banana (*Musa paradisiaca*, C), yellow bell pepper (*Capsicum annuum*, C), sweet corn (*Zea mays* var. *saccharata*, C), canary melon (*Cucumis melo*, C), squash (*Cucurbita pepo*, C); **5**: citrus (*Citrus spp*., C), persimmon (*Diospyros virginiana*, C), orange bell pepper (C), pumpkin (*Cucurbita maxima*, C), carrot (*Daucus carota*, C); **6**: litchi (*Litchi chinensis*, A), radish (*Raphanus sativus*, A), pomegranate (*Punica granatum*, A), watermelon (*Citrullus lanatus*, C), Swiss chard (*Beta vulgaris* spp. *cicla*, B); 7: red bell pepper (C), sweet cherry (*Prunus avium*, A), strawberry (*Fragaria ananassa*, A), tomato (*Solanum lycopersicum*, C), dragon fruit (*Hylocereus undatus*, B).

(*Solanum lycopersicum*), carrot (*Daucus carota*), pepper (*Capsicum annuum*), sweet corn (*Zea mays* var. *saccharata*), squash (*Cucurbita pepo*), pumpkin (*Cucurbita maxima*), canary melon (*Cucumis melo*), and watermelon (*Citrullus lana-*

tus) (Fig. 1.1). Some light-colored carotenoids are also found in green vegetables and many other colored fruits, but these carotenoid colors may be masked by green and other predominant colors. Depending on the concentrations and types, carotenoid-rich flowers, fruits, and other plant organs can show a wide spectrum of characteristic yellow, orange, or red colors (Fig. 1.1) [9]. For example, α-carotene, β-cryptoxanthin, and zeaxanthin primarily produce colors in the range of yellow, β-carotene, and lutein in the range of orange, and lycopene in the range of red. These six are also the most common carotenoids in the human diet. In contrast to lycopene, lutein and zeaxanthin that have no β-ionone ring at either terminal of the tetraterpenoid structure, α-carotene, β-carotene, and β-cryptoxanthin have at least one β-ionone ring. As a result, the latter three and the like are provitamin A carotenoids that can be converted to vitamin A in the human and animal body [11]. Therefore, intake of these provitamin A carotenoids can help the acquisition of vitamin A that is essential for human and animal vision, immune system function, normal development, and growth. For example, the immune system can be enhanced by β-carotene by strengthening T and B lymphocyte proliferative responses, stimulating effector T cell functions, and increasing the production of certain interleukins [11]. Lutein and zeaxanthin are the predominant carotenoids within the macula lutea; their physicochemical properties make them suitable candidates considered to act as photoprotectant preventing retinal degeneration. β-carotene and other carotenoids have also demonstrated antioxidant properties and singlet oxygen quenching capacities to prevent chronic disease in vitro and in animal models; the consumption of a diet rich in carotenoids has been epidemiologically correlated with a lower risk for several diseases [12]. A comparison showed cumulative incidence of liver cancer in β-cryptoxanthin-rich orange juice with carotenoids mixture capsules-treated group was statistically ($p=0.05$) lower than that in the control group [13]. There were also conflicting results from intervention studies on β-carotene to prevent certain cancers and cardiovascular diseases [12]. Data from many population-level intervention studies still suggest a daily intake of fruits and vegetables, which indeed is positively associated with reduced risks of obesity and related chronic diseases and with improved biomarkers of good health [14]. In addition, cooking oil and other food fat may facilitate dissolution of water-insoluble carotenoids and efficient absorption of the nutrient in the diet.

Flavonoids

Flavonoids are a huge family (over 9000 members) of water-soluble polyphenolic compounds with a basic 15-carbon benzo-γ-pyrone (C6–C3–C6) skeleton [15, 16]. Numerous combinations of several substitution groups diversify the chemical structures, properties, and biological functions, and result in at least nine major subgroups: anthocyanins (anthocyanidins), condensed tannins (proanthocyanidins), flavonols, flavones, flavandiols, isoflavonoids, chalcones, aurones, and phlobaphenes [17]. In plants, these colored and colorless flavonoids are synthesized in the complex flavonoid biosynthesis pathway and play much more diverse biological roles, compared to carotenoids. Only anthocyanins can provide a full spectrum of visible colors with the modification of substitution groups, change of pH values and metal ions [18]. As a result, they probably are the most studied flavonoids in plants [17, 19]. Anthocyanins, synthesized in the cytosol and stored in vacuoles, are also complementary pigments to help chlorophyll collect other available light for photosynthesis and to provide antioxidant protection for plant cells [4]. For example, in cotton plants, red cyanidin-3-O-β-glucoside (an anthocyanin, $C_{21}H_{21}ClO_{11}$) protects living cells from toxic effects of the plants' own phytoalexins by light filtering [20]. In addition to giving flowers and fruits attractive colors and providing the potent protection against oxidation and ultraviolet damage, some flavonoids can function as antimicrobial or defensive agents against biotic and abiotic stresses, and others secretively act as signaling molecules for the plant–microbe interaction in rhizosphere, for example, to establish the symbiosis between plants and nitrogen fixing rhizobia [21].

Flavonoids, particularly anthocyanins, are high in many colored fruits and vegetables, such as blueberry (*Vaccinium* spp.), blackberry (*Rubus* spp.), raspberry (*Rubus idaeus*), strawberry (*Fragaria ananassa*), grape (*Vitis vinifera*), sweet cherry (*Prunus avium*), plum (*Prunus domestica*), peach (*Prunus persica*), pomegranate (*Punica granatum*), red cabbage (*Brassica oleracea* var. *capitata*), and eggplant (*Solanum melongena*) (Fig. 1.1). In addition to the diverse structures and functions of flavonoids in plants, a wide range of biological and antioxidant activities are also observed among flavonoids in the human and animal diet against pathogens, allergens, carcinogens, and other agents causing inflammation. Diverse dietary benefits also include prevention and alleviation of some medical damaging conditions. Some of the activities are accomplished through interaction with host essential enzymes such as cytochromes P450 (CYPs). For example, the risk of some hormone-dependent breast and prostate cancers can be reduced, and menopausal symptoms prevented as well [22]. Accumulated knowledge of flavonoid biosynthesis and the structure–function relationship in plants and humans will facilitate production of targeted flavonoid compounds through metabolic engineering for use as more effective drug and/or chemopreventive agents [23].

Betalains

Betalains are a class of water-soluble indole-derived glycoside pigments that are found only in the order Caryophyllales (e.g., beets, cacti, and amaranths) and never coexist in plants with anthocyanins. In other words, betalains substitute for anthocyanins that are completely absent in the Caryophyllales plants [2, 24, 25]. Betalains differ from anthocyanins in the chemical structures and some properties, but share similarities to anthocyanins in the color spectra, biological functions, and other properties. For example, betalains contain nitrogen but anthocyanins do not. Similarly, betalains are also localized in vacuoles and reportedly offer potent antioxidant capacity and strong chemoprevention function [25]. Betalains can be structurally divided into betacyanins and betaxanthins that color flowers, fruits, and sometimes vegetative organs primarily into yellow, red, or violet. Betanin, with the molecular formula $C_{24}H_{27}N_2O_{13}$, is an important food colorant produced by beetroot. Compared to those many with carotenoids and flavonoids, betalains-containing fruits and vegetables are rather limited and less well known, but include beet (*Beta vulgaris*), Swiss chard (*Beta vulgaris* spp. *cicla*), dragon fruit (*Hylocereus undatus*), and cactus pear (*Opuntia ficus-indica*; Table 1.1; Fig. 1.1). The betalain biosynthesis pathway has not been elucidated as well as the other two pathways. Some main steps and key enzymes in the pathway were gradually uncovered in recent years, but many other questions remained to be answered [26].

In the end, it is worth noting that the United States Department of Agriculture (USDA) National Agricultural Library maintains a searchable National Nutrient Database (http://ndb.nal.usda.gov/) for over 8000 foods, which includes the main pigment compounds and other phytonutrients for almost all fruits and vegetables. For example, the database for the flavonoid content of selected foods (Release 3.1, December 2013) contains values for 506 food items and for five subgroups of flavonoids: flavonols, flavones, flavanones, flavan-3-ols, and anthocyanidins. These components, and benefits from potential functional foods and nutraceuticals, are also summarized to help people gain relevant knowledge and be encouraged to consume such foods on a regular basis. This book offers a more detailed coverage on the biosynthesis genomics and dietetics of these pigments and representative fruits/vegetables in the following chapters.

References

1. Moise AR, von Lintig J, Palczewski K (2005) Related enzymes solve evolutionarily recurrent problems in the metabolism of carotenoids. Trends Plant Sci 10(4):178–186
2. Stafford HA (1994) Anthocyanins and betalains—evolution of the mutually exclusive pathways. Plant Sci 101(2):91–98
3. Delgado-Vargas F, Jimenez AR, Paredes-Lopez O (2000) Natural pigments: carotenoids, anthocyanins, and betalains—characteristics, biosynthesis, processing, and stability. Crit Rev Food Sci Nutr 40(3):173–289
4. Tanaka Y, Sasaki N, Ohmiya A (2008) Biosynthesis of plant pigments: anthocyanins, betalains and carotenoids. Plant J 54(4):733–749
5. Hirschberg J (2001) Carotenoid biosynthesis in flowering plants. Curr Opin Plant Biol 4(3):210–218
6. Walter MH, Strack D (2011) Carotenoids and their cleavage products: biosynthesis and functions. Nat Prod Rep 28(4):663–692
7. Vershinin A (1999) Biological functions of carotenoids—diversity and evolution. Biofactors 10(2/3):99–104
8. Cazzonelli CI (2011) Carotenoids in nature: insights from plants and beyond. Funct Plant Biol 38(11):833–847
9. Ruiz-Sola MA, Rodriguez-Concepcion M (2012) Carotenoid biosynthesis in *Arabidopsis*: a colorful pathway. Arabidopsis Book 10:e0158

10. Lee PC, Schmidt-Dannert C (2002) Metabolic engineering towards biotechnological production of carotenoids in microorganisms. Appl Microbiol Biotechnol 60(1/2):1–11
11. Bendich A (1989) Carotenoids and the immune response. J Nutr 119(1):112–115
12. Stahl W, Sies H (2005) Bioactivity and protective effects of natural carotenoids. Biochim Biophys Acta 1740(2):101–107
13. Nishino H, Murakoshi M, Tokuda H, Satomi Y (2009) Cancer prevention by carotenoids. Arch Biochem Biophys 483(2):165–168
14. Woodside JV, Young IS, McKinley MC (2013) Fruits and vegetables: measuring intake and encouraging increased consumption. Proc Nutr Soc 72(2):236–245
15. Wang Y, Chen S, Yu O (2011) Metabolic engineering of flavonoids in plants and microorganisms. Appl Microbiol Biotechnol 91(4):949–956
16. Kandaswami C, Lee LT, Lee PP, Hwang JJ, Ke FC, Huang YT, Lee MT (2005) The antitumor activities of flavonoids. In Vivo 19(5):895–909
17. Winkel-Shirley B (2001) Flavonoid biosynthesis. a colorful model for genetics, biochemistry, cell biology, and biotechnology. Plant Physiol 126(2):485–493
18. Bjoroy O, Rayyan S, Fossen T, Andersen OM (2009) Structural properties of anthocyanins: rearrangement of C-glycosyl-3-deoxyanthocyanidins in acidic aqueous solutions. J Agric Food Chem 57(15):6668–6677
19. Holton TA, Cornish EC (1995) Genetics and biochemistry of anthocyanin biosynthesis. Plant Cell 7(7):1071–1083
20. Edwards WR, Hall JA, Rowlan AR, Schneider-Barfield T, Sun TJ, Patil MA, Pierce ML, Fulcher RG, Bell AA, Essenberg M (2008) Light filtering by epidermal flavonoids during the resistant response of cotton to *Xanthomonas* protects leaf tissue from light-dependent phytoalexin toxicity. Phytochemistry 69(12):2320–2328
21. Hassan S, Mathesius U (2012) The role of flavonoids in root-rhizosphere signalling: opportunities and challenges for improving plant-microbe interactions. J Exp Bot 63(9):3429–3444
22. Hodek P, Trefil P, Stiborova M (2002) Flavonoids-potent and versatile biologically active compounds interacting with cytochromes P450. Chem Biol Interact 139(1):1–21
23. Dixon RA, Steele CL (1999) Flavonoids and isoflavonoids—a gold mine for metabolic engineering. Trends Plant Sci 4(10):394–400
24. Brockington SF, Walker RH, Glover BJ, Soltis PS, Soltis DE (2011) Complex pigment evolution in the caryophyllales. New Phytol 190(4):854–864
25. Strack D, Vogt T, Schliemann W (2003) Recent advances in betalain research. Phytochemistry 62(3):247–269
26. Gandia-Herrero F, Garcia-Carmona F (2013) Biosynthesis of betalains: yellow and violet plant pigments. Trends Plant Sci 18(6):334–343

Chapter 2
Carotenoid Biosynthesis Genomics

Amanda Ferreira Da Silva Mendes, Virgínia Lúcia Fontes Soares, and Marcio Gilberto Cardoso Costa

Introduction

Carotenoids are a family of isoprenoid molecules that are widespread in nature. They are responsible for the typical yellow, orange, and red colors of most fruits, flowers, and vegetables, and for the characteristic colors of many birds, insects, fish, and crustaceous intaking carotenoids through the diet. The basic chemical structure of any carotenoid molecule is the long polyene chain, which may extend from 3 to 15 conjugated double bonds, acting as a chromophore that determines the absorption spectrum of the molecule, and hence its color [1]. This basic structure can be further modified in a number of ways, such as cyclization and oxygenation, to yield a family of more than 600 different carotenoids generally divided into two subgroups, carotenes (hydrocarbon carotenoids) and xanthophylls (oxygenated derivatives) [1]. Their oxidative breakdown products are called apocarotenoids.

Carotenoids perform a broad range of metabolic and ecological functions. Carotenoid pigments are essential components of the photosynthetic membranes in all photosynthetic organisms, including plants, algae, and cyanobacteria, protecting them against photooxidative damage [2, 3]. This protective function is crucial in oxygen-evolving photosynthetic organisms, since an impairment to produce cyclic carotenoids is eventually lethal [1]. In higher plants, carotenoids also act as accessory pigments in the light-harvesting antennae of chloroplasts, transferring energy to chlorophylls, and as precursors for biosynthesis of the phytohormones abscisic acid (ABA), which controls abiotic stress signaling pathways, and strigolactone, which controls lateral shoot growth [3–5]. An additional and important role of carotenoids in higher plants is as coloring agents in flowers and fruits to attract pollinators and

M. G. C. Costa (✉) · V. L. F. Soares
Biological Sciences Department, State University of Santa Cruz,
Rod. Jorge Amado, km 16, Ilhéus, Bahia, Brazil
e-mail: marciogc.costa@gmail.com

A. F. D. S. Mendes
Instituto Federal Da Bahia, Ilhéus, Bahia, Brazil

C. Chen (ed.), *Pigments in Fruits and Vegetables,* DOI 10.1007/978-1-4939-2356-4_2

agents of seed dispersal [2]. In these tissues, carotenoids accumulate in nonphotosynthetic chromoplasts, where they are found in association with lipid–protein complexes in plastoglobules and/or in carotenoid-accumulating structures of globular, crystalline, membranous, fibrillar, or tubular forms [6].

Since animals are unable to synthesize carotenoids de novo, they rely upon the diet as the source of these compounds. Dietary carotenoids contribute to animal health and behavior, because they stimulate the immune system and aid in the preferential selection by the sexual partner [7, 8]. In mammalians, including humans, carotenoid species containing a β-ring can be converted into retinal (the main visual pigment), retinol (vitamin A), and retinoic acid (a substance that controls morphogenesis) [9]. Additional beneficial effects of carotenoids in human health are attributed to their antioxidant and anti-inflammatory activities in vivo, which help to prevent certain cancers, cardiovascular diseases, light-induced erythema, and age-related diseases of eye such as cataract and macular degeneration [10, 11].

Here we present the current knowledge about the carotenoid biosynthesis gene families in higher plants from a genomic perspective. This includes information about the basic structure, function, and evolution of the genes and enzymes, as well as the molecular mechanisms regulating carotenoid biosynthesis in different plant tissues.

Carotenoid Biosynthetic Pathway

The biosynthetic pathways involved in carotenoid formation were elucidated during the second half of the twentieth century using both classical biochemical approaches and modern molecular biology techniques [12]. Nevertheless, the major advances in the identification of genes and enzymes of carotenoid biosynthesis occurred in the 1990s. Isolation of carotenoid-defective mutants in plants, and the information resources of the *Arabidopsis thaliana* EST database and the genome sequence of the cyanobacterium *Synechocystis* PCC6803 contributed to such advances [9, 12, 13]. Also important was the dissection of carotenoid biosynthesis pathway in bacterial systems, such as *Rhodobacter capsulatus*, *Erwinia uredovora*, and *Erwinia herbicola*. It allowed engineering strains of *Escherichia coli* accumulating a variety of carotenoid precursors for use as a simple and powerful in vivo system for the assay of enzyme function and substrate specificity [12]. In addition, the different colors exhibited by carotenoid-accumulating *E. coli* strains were exploited to visually screen complementary DNA (cDNA) and genomic libraries, in a procedure referred to as "color complementation," enabling the identification of a number of previously unidentified plant, algal, and cyanobacterial carotenogenic genes based on the visualization of color changes in *E. coli* colonies [14].

Carotenoids make a part of the plethora of chemical compounds that are produced via the general isoprenoid biosynthetic pathway (Fig. 2.1). As all other isoprenoids, carotenoids are built from the five-carbon (C_5) compound isopentenyl diphosphate (IPP) and its allylic isomer, dimethylallyl diphosphate (DMAPP)

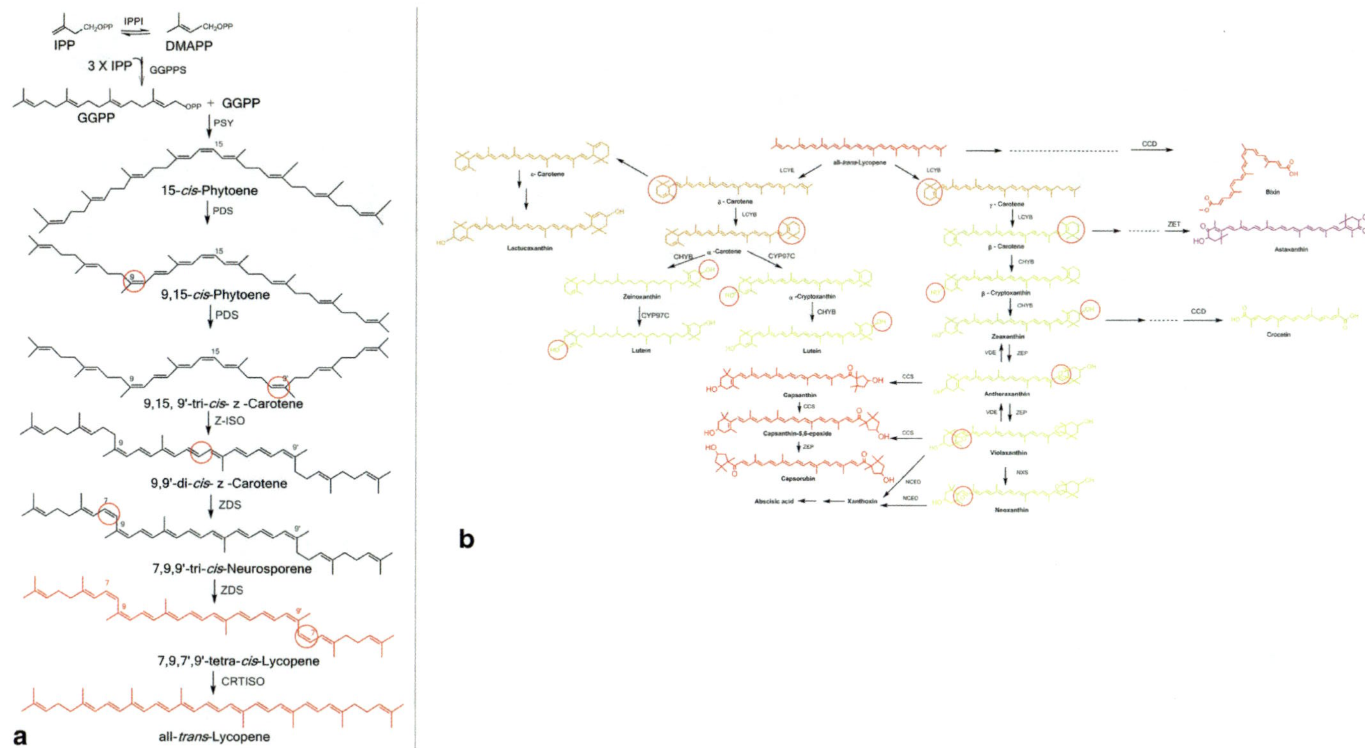

Fig. 2.1 The general isoprenoid carotenoid biosynthetic pathway in plants. **a** Formation of acyclic carotenoids from the precursor GGPP. **b** Formation of cyclic carotenoids from the precursor lycopene. Enzyme abbreviations are as follows: *IPPI* isopentenyl diphosphate isomerase, *GGPPS* geranylgeranyl diphosphate synthase, *PSY* phytoene synthase, *PDS* phytoene desaturase, *Z-ISO* ζ-carotene isomerase, *CrtISO* carotenoid isomerase, *ZDS* ζ-carotene desaturase, *LCYE* lycopene ε-cyclase, *LCYB* lycopene β-cyclase, *CHYB* β-ring hydroxylase, *CYP97C* ε-ring hydroxylase, *CCS* capsanthin/capsorubin synthase, *ZEP* zeaxanthin epoxidase, *VDE* violaxanthin de-epoxidase, *NXS* neoxanthin synthase, *CCD* carotenoid cleavage dioxygenase, *ZET* β-ring hydroxylase, *NCED* 9-cis-epoxycarotenoid dioxygenase, *IPP* isopentenyl diphosphate, *DMAPP* dimethylallyl diphosphate, *GGPP* geranylgeranyl diphosphate

[10, 13, 15]. Until recently, it was assumed that IPP was synthesized from acetyl-coenzyme A (CoA) via mevalonic acid (MVA) pathway [16]. However, in the early 1990s retro-biosynthetic studies established the presence of an alternative, MVA-independent pathway for the formation of IPP and DMAPP, termed 1-deoxy-D-xylulose-5-phosphate (DXP) or 2-C-methyl-D-erythritol-4-phosphate (MEP) pathway [17]. Several reports have indicated that eubacteria (*E. coli* and other pathogenic bacteria) and protozoans of apicomplexan phylum (*Plasmodium falciparum*) synthesize isoprenoids only via the MEP pathway, while archaebacteria, fungi, and animals (including humans) contain only the MVA pathway [18, 19]. The formation of isoprenoids in plants, however, can proceed from both MEP and MVA pathways [15, 18]. IPP is synthesized in plastids through the MEP pathway and in the cytosol through the MVA pathway. Thus, MEP pathway is a potential target for the development of new herbicides and anti-malarian and antimicrobial drugs that, besides the large spectrum of action, are not toxic for humans.

The MVA pathway involves a set of six reactions proceeding sequentially from acetyl-CoA to produce IPP and DMAPP [15, 16, 19]. Initially, two acetyl-CoA molecules, obtained through CO_2 fixation, are condensed to yield acetoacetyl-CoA, in a reaction catalyzed by acetyl-CoA *C*-acetyltransferase (AACT, EC 2.3.1.9). Then, a third acetyl-CoA molecule is condensed to acetoacetyl-CoA, forming 3-hydroxy-3-methylglutaryl-CoA (HMG-CoA) by the action of HMG-CoA synthetase (HMGS, EC 4.1.3.5). The nicotinamide adenine dinucleotide phosphate (NADP)-dependent HMG-CoA reductase converts the CoA-derived in (*R*)-MVA, which is phosphorylated to (*R*)-MVA 5-diphophate by the sequential action of mevalonate kinase (MK, EC 2.7.1.36) and diphosphomevalonate kinase (PMK, EC 2.7.4.2). (*R*)-MVA 5-diphophate is further decarboxylated by the mevalonate diphosphate decarboxylase (MDC, EC 4.1.1.33), producing a pool of IPP. Finally, IPP isomerase catalyzes the reversible conversion of IPP into DMAPP, maintaining the equilibrium between these two compounds [16, 20].

The MEP pathway consists of seven sequential reactions starting from the condensation of pyruvate and glyceraldehyde 3-phosphate (G3P) to yield 1-deoxy-$_D$-xylulose 5-phosphate (DXP) [10, 17, 18]. This transketolase reaction is catalyzed by DXP synthase (DXS, EC 4.1.3.37), an enzyme that requires thiamine pyrophosphate and a divalent cation (Mg^{2+}or Mn^{2+}) as cofactors [18, 21]. DXP is subsequently rearranged and reduced to MEP in a single step, in a reaction catalyzed by DXP reductoisomerase (DXR, EC 1.1.1.267). DXR requires NADPH and Mn^{2+}as cofactors [18, 21]. MEP is converted to IPP and DMAPP via 4-(cytidine 5′-diphospho)-2-*C*-methyl-D-erythritol (CDP-ME), 2-phospho-4-(cytidine 5′-diphospho)-2-C-methyl-D-erythritol (CDP-MEP), 2-C-methyl-D-erythritol-2,4-cyclodiphosphate (MCP), and 1-hydroxy-2-methyl-2-(*E*)-butenyl 4-phosphate (HMBPP). The enzymes responsible for these reactions are, respectively, MEP cytidyl transferase (MCT, EC 2.7.7.60), CDP-ME kinase (CMK, EC 2.7.1.148), MCP synthase (MCS, EC 4.6.1.12), and HMBPP synthase (HDS) [18, 21]. IPP/DMAPP synthase (IDS) is responsible for the conversion of HMBPP to a 5:1 mixture of IPP and DMAPP [10, 20, 21].

IPP and DMAPP are subsequently used as blocks in a modular assembly process that produces compounds of 5, 10, 15, 20, or more carbons (in multiples of 5), allowing the biosynthesis of the basic skeletons for the various isoprenoids, including carotenoids, with a relatively small number of basic reaction steps [10, 15, 19]. For instance, the C_{20} geranylgeranyl diphosphate (GGPP), which serves as the immediate precursor for carotenoids, is formed by the sequential and linear addition of three molecules of IPP to one molecule of DMAPP. The enzyme that catalyzes these reactions, the GGPP synthase (GGPS; EC 2.5.1.29), is encoded by a multigene family of 12 members in the *Arabidopsis* genome, suggesting the involvement of different isozymes in the production of specific groups of isoprenoids [10, 13].

Basic Structure, Function, and Evolution of Carotenoid Biosynthesis Genes and Enzymes

The C_{40} skeleton of all plant carotenoids is assembled from the condensation of two molecules of GGPP, which then suffer a series of enzymatic reactions of desaturation, cyclization, and oxidation [10, 19]. Genetic and molecular evidences indicate that all enzymes of carotenoid biosynthetic pathway in plants are encoded by nuclear genes and post-translationally imported to plastids [12, 22]. Here, the genes and enzymes of carotenoid biosynthesis are discussed sequentially in their order within the pathway, giving specific details for one or more examples of each in *Arabidopsis* and other higher plants whenever possible.

Phytoene Synthase

Phytoene synthase (PSY, EC 2.5.1.32) catalyzes the first committed step in the formation of carotenoids, by the condensation of two GGPP molecules to produce 15-*cis* phytoene (Fig. 2.1a) [10, 12]. Detailed biochemical characterization of tomato and pepper PSYs has demonstrated that they can be either thylacoid membrane associated, but not integral, or stroma-localized proteins [10, 22]. The catalytic site of PSY, at the carboxy terminus, contains a large central cavity, formed by antiparallel alpha-helices, with two aspartate-rich motifs (DELVD and DVGED) that are positioned on opposite walls of the central cavity [19, 23]. The high degree of sequence conservation of these motifs suggests that they are required for the interaction of enzyme with upstream products [23]. PSY also contains an active site (YAKTF) at the amino terminus and a squalene synthase (SQS) domain type located between the catalytic and active sites (Fig. 2.2) [19, 23]. There is a low sequence similarity at the amino terminus among PSYs of different plant species, partially due to the existence of plastid transit peptide sequences that are known to show a low degree of sequence conservation.

Arabidopsis possesses only one *PSY* gene, while tomato and tobacco have two *PSY*s and plants belonging to the Poaceae have three *PSY*s [24–29]. Scenarios of gene duplication and sub-functionalization can be invoked for the evolution of *PSY* genes and enzymes from an ancestral *PSY* gene prototype. The *PSY* paralogs are

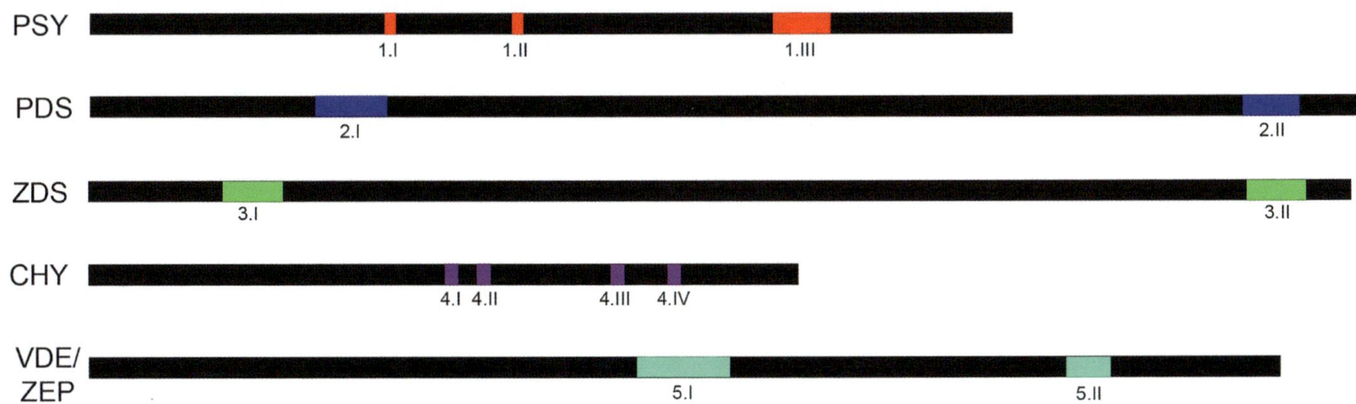

Fig. 2.2 Major functional domains present in some carotenogenic enzymes. 1.I, Active site of PSY; 1.II, squalene synthase domain type; 1.III, catalytic site of PSY. 2.I, Dinucleotide (FAD/NADP)-binding site domain of PDS; 2.II, carotenoid-binding domain of PDS. 3.I, Dinucleotide (FAD/NADP)-binding site domain of ZDS; 3.II, carotenoid-binding domain of ZDS. 4.I, Histidine-binding motif HXXXXXH type of β-CHY; 4.II, histidine-binding motif HXXHH type of β-CHY; 4.III, histidine-binding motif HXXXXXH type of β-CHY; 4.IV, histidine-binding motif HXXHH type of β-CHY. 5.I, Cystein-rich region of VDE and ZEP; 5.II, glutamate-rich region of VDE and ZEP. See text for further details. *PSY* phytoene synthase, *FAD/NADP* flavin adenine dinucleotide/nicotinamide adenine dinucleotide phosphate, *PDS* phytoene desaturase, *ZDS* ζ-carotene desaturase, *VDE* violaxanthin de-epoxidase, *ZEP* zeaxanthin epoxidase

involved with the carotenoid synthesis in different plant tissues. For instance, *PSY1* encodes a fruit- and flower-specific isoform in tomato, whereas *PSY2* encodes an isoform that predominates in photosynthetic tissues [25, 27]. In maize, *PSY1* and *PSY2* are required for endosperm carotenoid accumulation and photomorphogenesis in photosynthetic tissues, while *PSY3* is associated with root carotenogenesis and necessary for drought and salt stress-induced production of ABA [28, 30].

A comparison of the gene structures of *PSY*s among different plant species shows that the *Arabidopsis PSY* contains seven exons, as well as *Vitis vinifera*, while rice and maize *PSY*s show a loss of exon 1 (Fig. 2.3a). The gene structures at the

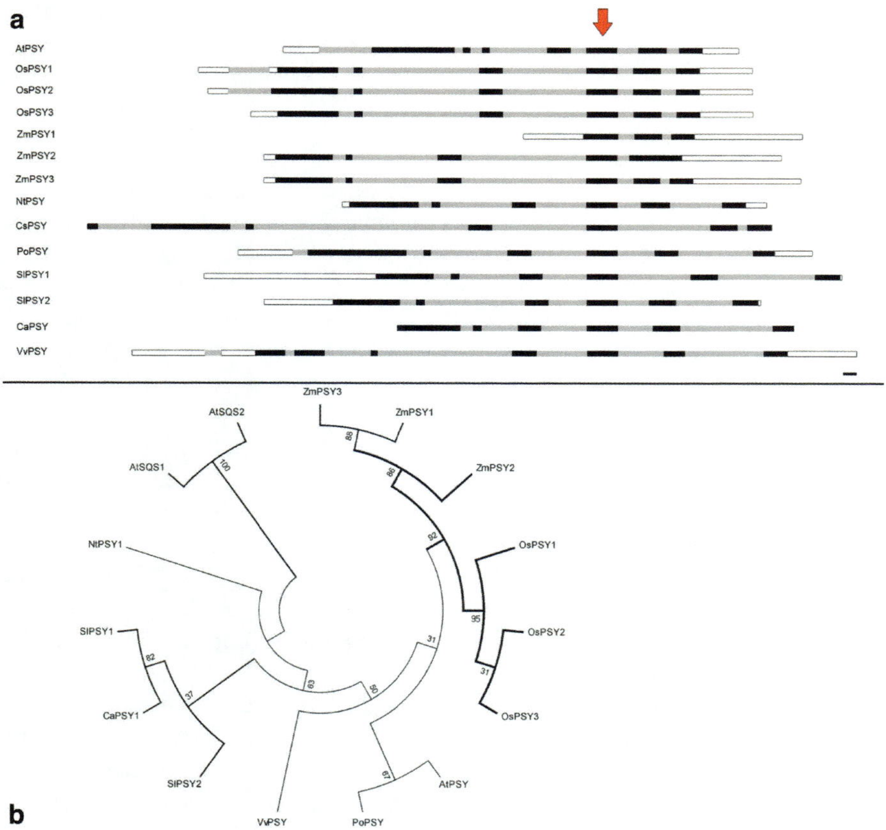

Fig. 2.3 *PSY* gene structure and phylogeny. **a** Exon/intron structure of PSY in different plant species. *Red arrow* denotes the catalytic site of PSY. *White, gray,* and *black thin bars* indicate UTR, intron and exon regions, respectively. **b** Similarity dendrogram of plant PSYs and squalene synthases (SQSs). Amino acid sequences were aligned using ClustalW and a neighbor-joining tree was constructed with a 1000-bootstrap replication support. Abbreviations for the name of plant species are as follows: At *Arabidopsis thaliana*, Os *Oryza sativa*, Zm *Zea mays*, Nt *Nicotiana tabacum*, Cs *Citrus sinensis*, Po *Populus trichocarpa*, Sl *Solanum lycopersicum*, Ca *Capsicum annum*, Vv *Vitis vinifera*, *PSY* phytoene synthase, *UTR* untranslated region

5′-untranslated region (UTR) of all *PSY* genes analyzed show differences in length and sequence, even among the different genes within a species, such as *OsPSY1* (227 bp), *OsPSY2* (147 bp), and *OsPSY3* (193 bp; Fig. 2.3a). In contrast, the lengths of exons 3, 4, 5, and 6 are comparable among the different plant species, with 45–51, 173, 236, and 193–211 bp, respectively. Such conservation may be associated to the presence of nucleotide sequences encoding conserved domains of enzyme, such as the catalytic site located at the exon 5 of *Arabidopsis*. The dendrogram of similarity reveals that PSYs of rice and maize clustered together, in a separated clade from PSYs of dicots, which includes the tomato PSY1 and PSY2. It suggests that the duplication event of *PSY* occurred separately in Poaceae and Solanaceae (Fig. 2.3b). Furthermore, the proximity between AtPSY and OsPSY1 represents the existence of an ancient PSY constituting a common ancestor of monocots and dicots [29].

Desaturases

Colorless phytoene undergoes a series of four desaturation reactions in plants that results in the formation of the red-colored carotenoid lycopene (Fig. 2.1a). These reactions are catalyzed by two related enzymes in plants: phytoene desaturase (PDS, EC 1.3.5.5) and ζ-carotene desaturase (ZDS, EC 1.3.5.6). PDS converts phytoene in phytofluene and then in ζ-carotene, while ZDS converts ζ-carotene in neurosporene and then in lycopene [10, 19]. In contrast with plants, bacteria and fungi contain only a single desaturase, *CrtI*, which catalyzes the four desaturation steps [19].

PDS and ZDS are found to be associated with other enzymes of carotenoid biosynthetic pathway, forming multimeric complexes of about 350 kDa [12, 31]. It has been proposed that two molecules of each PDS and ZDS, associated with one molecule of lycopene β-cyclase (β-LCY) and another of lycopene ε-cyclase (ε-LCY), form a multienzymatic complex responsible for the synthesis of α-carotene [12, 32]. A similar association of PDS and ZDS with two molecules of β-LCY would be responsible for the synthesis of β-carotene [12, 32]. PDS may be also associated with chloroplastic chaperonins (Cpn60) or heat-shock proteins (Hsp70) when located in stroma [31, 33, 34].

The active form of PDS is tightly bound to thylacoid membranes, whereas the stroma free form is inactive [31, 33]. Besides the association with membranes, the desaturases require cofactors for their complete activity. The removal of two hydrogen atoms during each desaturation step suggests the involvement of an electron transport chain for the regeneration of reductants [10]. A plastid terminal oxidase (PTOX) was identified in *Arabidopsis* mutants as one component of this electron transport chain required for the desaturation of phytoene [35]. PTOX is a plastoquinone oxidoreductase that regenerates the reduced plastoquinone formed during the desaturation of phytoene and ζ-carotene, using oxygen (O_2) as a terminal acceptor. Thus, PTOX and O_2 are considered as the main cofactors involved in the desaturation of phytoene, in both photosynthetic and nonphotosynthetic tissues [35].

The genomes of higher plants apparently contain only one copy of desaturase genes, *PDS* and *ZDS* [12, 26, 33, 36, 37]. A unique exception has been recently discovered in sweet orange (*Citrus sinensis*), in which 2 *PDS* and 12 *ZDS* members were found to be clustered, respectively, at one and three loci [38]. A high degree of similarity is observed in the deduced amino acid sequences of all plant desaturases [10]. All contain a conserved dinucleotide (FAD/NADP)-binding site domain at the amino terminus (Fig. 2.2). The carboxy terminus contains another conserved region, the carotenoid-binding domain.

Isomerases

In higher plants, phytoene occurs predominantly as the 15-*cis* isomer, while the predominant isomer of lycopene is all-*trans*, suggesting the existence of an enzymatic step mediating the *cis–trans* isomerization of lycopene precursors (Fig. 2.1a). Map-based cloning of the gene responsible for the *tangerine* tomato fruit phenotype, which accumulates 7,9,7′,9′-tetra-*cis* lycopene (prolycopene) and traces of its poly-*cis* precursors, resulted in the isolation of carotenoid isomerase (*CrtISO*). *CrtISO* encodes a carotenoid isomerase (CrtISO, EC 5.2.1.13) catalyzing the isomerization of prolycopene to all-*trans* lycopene [39]. A *CrtISO* homolog termed *Ccr-2* was also isolated in *Arabidopsis* [40].

Both tomato and *Arabidopsis* isomerases contain a dinucleotide (FAD/NADP)-binding domain, like PDS and ZDS. However, the isomerases show more identities to bacterial phytoene desaturases (CrtI) than the plant desaturases [39, 40]. In fact, the bacterial desaturase *CrtI* also possesses the function of isomerization in combination with that of desaturation, converting 15-*cis* phytoene to all-*trans* lycopene [10].

Cyclases

Cyclization of the linear carotenoid lycopene marks an important branching point in the carotenoid pathway: one branch leads to β-carotene and its derivative xanthophylls, whereas the other leads to α-carotene and lutein (Fig. 2.1b). These carotenoids differ in the type of cyclic end group that is added. It can be a ε- or β-ionone ring, depending on the position of a double bond within the cyclohexane ring. Carotenoids with two β-rings, such as β-carotene and zeaxanthin, are primarily involved in protection against photooxidative damage and dissipation of the excess of light energy in the photosynthetic membranes [12]. Carotenoids with one β-ring and one ε-ring, such as lutein, act as accessory pigments in light-harvesting antennae of the chloroplasts [12]. Carotenoids with two ε-rings, such as lactucaxanthin in lettuce, are rare [41].

The type of end group produced depends on the nature of cyclase enzyme. In higher plants, there are two major cyclases: β-LCY (EC 5.5.1.19), which introduces β-rings, and ε-LCY (EC 5.5.1.18) that introduces ε-rings. The formation of β-carotene requires the introduction of two β-rings by β-LCY, whereas α-carotene requires the interaction of both β-LCY and ε-LCY (Fig. 2.1b) [32]. In contrast with β-LCY, ε-LCY is able to incorporate only one ε-ring to the symmetrical lycopene, forming δ-carotene [32]. Lettuce ε-LCY is the only example of cyclase that can introduce two ε-rings to the lycopene molecule [41]. All lycopene cyclases, irrespective of class, proceed via a carbocationic mechanism [19].

A membrane-associated multienzymatic complex involving the association of β-LCY and ε-LCY with PDS and ZDS is postulated to act in the synthesis of α- and β-carotene (see the desaturases section). These carotenogenic complexes possibly are associated to other enzymes and cofactors that regulate their catalytic activity [12]. Flux directing towards the β,β- or β,ε-branch of the pathway seems to be determined by the relative amounts of enzymatic activity and/or substrate specificity of β-LCY and ε-LCY [32, 42].

The lycopene cyclases contain a dinucleotide (FAD/NADP)-binding site domain, apparently involved in allosteric activation, and two characteristic conserved motifs: cyclase I and cyclase II (Fig. 2.4) [43, 44]. The FAD/NADP-binding site domain is composed by a typical secondary structure (β-sheet/α-helice/β-sheet) present in all plant enzymes with lycopene cyclase activity [44]. The cyclization reaction seems to be a simple rearrangement that does not involve any change in the oxidation level of lycopene molecule [19]. Thus, the involvement NAD(P)H in the reaction is not expected. NAD(P)H seems to have an indirect action, participating in the enzymatic reaction as an allosteric activator [44].

The first searches for homology carried out with plant ε-LCYs revealed the existence of a conserved region (VQMQQ), which was termed "ε-cyclase conserved region" (Fig. 2.4) [32]. Similarly, β-LCYs contain a conserved region (PLYD) that has been identified as "β-cyclase conserved region." All lycopene cyclases also contain a conserved region that shows similarity to motifs of β-cyclase [32]. Since this region is present in cyclases that introduce β-, ε- or κ-ring, it has been termed "cyclase activity region" (Fig. 2.4).

Only a copy of *ε-LCY* gene has been identified in the genome of *Arabidopsis* and tomato [32, 45]. *Arabidopsis* also contains a copy of *β-LCY* [32], but two *β-LCY* copies, *Crtl-B* and *Cyc-B*, were identified in tomato [46, 47]. *Crtl-B* is active in photosynthetic tissues, whereas *Cyc-B* functions only in chromoplast-containing tissues [22]. The presence of two *β-LCY* genes, one with a chromoplast-specific expression, has been also reported in carotenogenic fruits other than tomato, including watermelon [48], orange, and grapefruit [49, 50].

Plant lycopene cyclases are also related to two other carotenoid cyclase enzymes: the capsanthin–capsorubin synthase (CCS, EC 5.3.99.8) of pepper [32] and the neoxanthin synthase (NSY, EC 5.3.99.9) of tomato [51] and potato [52]. CCS catalyzes the formation of the unusual five-carbon κ-ring [53], converting antheraxanthin or violaxanthin to capsanthin or capsorubin, respectively. In addition, CCS exhibits a β-LCY activity when lycopene is provided as a substrate [43]. NSY also modifies violaxanthin to the allenic product via a carbocation with a structure similar to the intermediate in the CCS-catalyzed reaction [54]. Although NSY operates mechanistically like CCS, its cryptic LCY activity has not been demonstrated [19].

Conservation of amino-acid sequences and their similar mechanisms of catalysis suggest that all plant cyclases, including CCS and NSY, have evolved from a common ancestor, most probably the cyanobacterial CrtL [55]. Since cyanobacteria do not

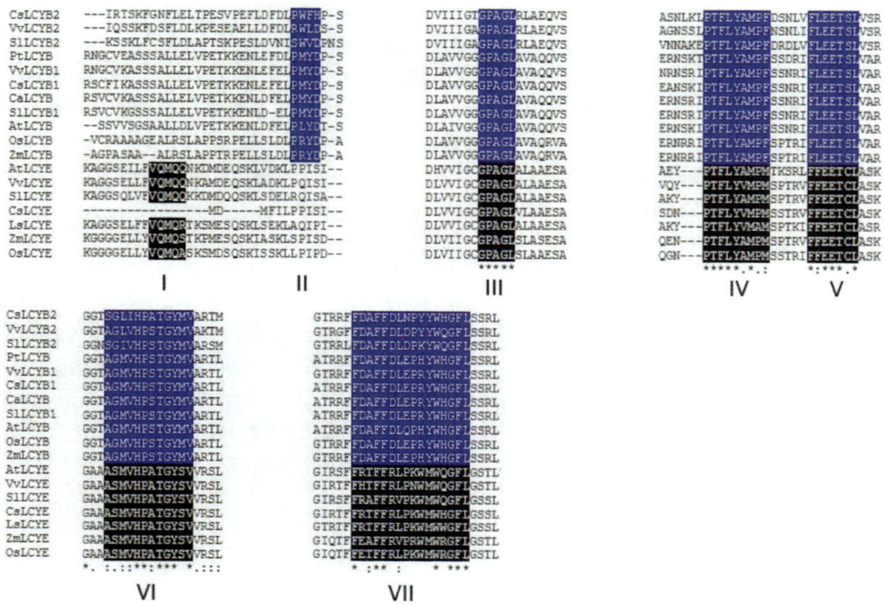

Fig. 2.4 Alignment of the partial amino acid sequences of β-LCY and ε-LCY from different plant species. Conserved amino acid sequences of β-LCY and ε-LCY in a given position are in white text on a blue and black background, respectively. *Dashes* denote a gap in the amino acid sequence. The conserved regions are plant ε-cyclase conserved region (I), plant β-cyclase conserved region (II), dinucleotide (FAD/NADP)-binding site domain (III), cyclase I motif (IV and V), cyclase II motif (VI) and cyclase activity region (VII). *Asterisk* (*), *colon* (:) and *dot* (.) symbols denote identical, conserved and similar amino acids, respectively. Abbreviations for the species names are as follows: At *Arabidopsis thaliana*, Ca *Capsicum annum*, Cs *Citrus sinensis*, Ls *Lactuca sativa*, Os *Oryza sativa*, Pt *Populus trichocarpa*, Sl *Solanum lycopersicum*, Vv *Vitis vinifera*, Zm *Zea mays*, *FAD/NADP* flavin adenine dinucleotide/nicotinamide adenine dinucleotide phosphate, *LCY* lycopene cyclase

contain any ε-LCY activity, it is presumed that this enzyme evolved by gene duplication of a β-LCY in prochlorophytes, where carotenoids with an ε-end group appear. In contrast with higher plants, *Prochlorococcus marinus* can synthesize α-carotene from lycopene by a single enzyme [55]. Thus, the *P. marinus* ε-LCY can be regarded as a premature form of the plant ε-LCY, which has not yet lost its β-cyclase activity [55].

Hydroxylases

Xanthophylls are oxidation products of α- and β-carotene in higher plants [10]. Hydroxylation of the number three carbon of each ring (C-3 and C-3′) of α-carotene and β-carotene results in the formation of lutein and zeaxanthin via α-cryptoxanthin and β-cryptoxanthin, respectively (Fig. 2.1b). These reactions are carried out by two types of enzymes, one specific for ε-rings (ε-hydroxylase, ε-CHY, EC 1.14.99.45) and the other for β-rings (β-CHY, EC 1.14.13.129) [22].

Two genes encoding β-CHY (*CHB1*, *CHB2*) were first cloned from *Arabidopsis* [56] and pepper [57]. The deduced amino acid sequences of β-CHY indicate the presence of transmembrane helices, suggesting that they are integral membrane enzymes. Pepper β-CHY displays histidine motifs, HXXXXH and HXXHH (Fig. 2.2) that play a vital role in coordinating the Fe ions in the active site [58]. Thus, β-CHY is a nonheme-di-iron monooxygenase that uses activated molecular oxygen to break the C–H bond, with the concomitant formation of unsaturated bonds and the retention of hydroxyl groups [57].

Although the β-ring of α-carotene is hydroxylated by a nonheme-di-iron monooxygenase, β-CHY, the ε-ring is not. Molecular analysis of the *Arabidopsis* mutant *lut1*, which shows up to a 95 % decrease in lutein content, led to the identification of a cytochrome P450-type oxygenase (CYP), CYP97C1, involved in the hydroxylation of the ε-ring of lutein [59]. Thus, the introduction of hydroxy group into the β- and ε-rings of α-carotene, to form lutein, is catalyzed by structurally unrelated enzymes such as nonheme-di-iron monooxygenases and cytochrome P450-type oxygenases.

Epoxidases and De-epoxidases

Zeaxanthin is converted into antheraxanthin and violaxanthin by the introduction of, respectively, one or two 5,6-epoxygroups into the 3-hydroxy-β-rings (Fig. 2.1b). This reaction is catalyzed by zeaxanthin epoxidase (ZEP, EC 1.14.13.90) [10]. The resulting antheraxanthin and violaxanthin are subjected to de-epoxidation by violaxanthin de-epoxidase (VDE, EC 1.10.99.3), which converts violaxanthin and antheraxanthin back to zeaxanthin [10]. These reactions of interconversion are known as the xanthophyll cycle, which has a key role in protecting the photosynthetic membranes against excess light by dissipating light energy as heat [19]. Zeaxanthin is converted into antheraxanthin and violaxanthin under light conditions that do not saturate photosynthesis or in the darkness [60]. Under light conditions that exceed the leaf photosynthetic capacity, violaxanthin is converted back to zeaxanthin [60].

ZEP and VDE are members of a group of proteins known as lipocalins, which bind and transport small hydrophobic molecules [61]. Lipocalins show a conserved secondary structure containing eight antiparallel β-sheets, which encompass three major motifs: motif I, first of the eight β-sheets, preceded by a stretch of α-helice; motif II, segments of sixth and seventh β-sheets, along with a loop between the sheets; and motif III, part of eighth β-sheet along with a carboxy terminal α-helice fragment and a loop between these two structures [61]. ZEP differs from VDE on the number of amino acid between the motifs I and II [62].

In addition, ZEP and VDE also contain a cystein-rich amino terminal region and a glutamate-rich carboxy terminal region (Fig. 2.2), probably involved in the formation of α-helices [62]. ZEP also contains a dinucleotide (FAD/NADP)-binding motif, like those in the carotenoid desaturases and cyclases. ZEP activity is controlled by the availability of β,β-xanthophyll precursors and requires the presence of FAD and ferredoxin [19]. VDE activity depends on the availability of violaxanthin, acidic pH of the lumen and protonated ascorbate [19]. During the light period, VDE binds to thylakoid membranes due to the acidification of lumen. In the dark, pH is neutral and VDE is released from thylakoid membranes.

VDE seems to be encoded by single copy genes in plants, while small gene families encode ZEP in plants such as maize, rice, and *Vitis* [26, 63, 64].

Neoxanthin Synthase

The conversion of violaxanthin into neoxanthin, by the rearrangement of a single epoxy group to 5-hydroxy, is the final step in the β,β-branch of carotenoid biosynthetic pathway (Fig. 2.1b). This reaction is catalyzed by NSY [10]. Neoxanthin can be cleaved to form a C_{15} precursor of ABA [4].

NSY has been characterized in potato and tomato [52, 54]. Interestingly, the amino acid sequence of NSY is homolog to CYC-B, the chromoplast-targeted isoform of β-LCY in tomato [47], and to CCS [43], which converts antheraxanthin and violaxanthin to capsanthin and capsorubin, respectively, in pepper. However, in contrast to β-LCY, which carries out the cyclization of end groups in the linear carotenoid lycopene, NSY acts by means of epoxidation, carbocation neutralization, and intramolecular rearrangement of the violaxanthin precursor [19]. NSY activity is similar to that of CCS, which also catalyzes an epoxidation reaction in the 5,6-epoxy cyclohexenyl end groups [19].

NSY is encoded by a single-copy gene in potato and tomato [52, 54]. Surprisingly, no *NSY* homologous gene has been found in the *Arabidopsis* genome yet. Its extensive homology with CYC-B has led to speculation that the NSY activity may be played by bifunctional β-LCYs, capable of converting both lycopene to β-carotene and violaxanthin to neoxanthin [65].

Regulation of Carotenoid Biosynthesis in Higher Plants

Since carotenoids play a central role in plant development and adaptation, their synthesis must be coordinated with the developmental processes, such as plastid differentiation, flowering and fruit development, and related metabolic pathways. Several mechanisms are used to regulate the carotenoid biosynthesis in higher plants, defining the types and amounts of the diverse carotenoids that will be synthesized in the different tissues [66]. Transcriptional regulation of carotenoid biosynthesis genes appears to be the major regulatory mechanism controlling carotenoid accumulation in the different plant tissues. However, other regulatory mechanisms have been also observed to control carotenoid biosynthesis, including posttranscriptional regulation, natural genetic variation and carotenoid degradation, turnover, and sequestration.

A key regulatory mechanism during the formation of isoprenoids is the control of precursor partitioning into the branches of the pathway. This has led to the concept of metabolic channeling among each branch of the pathway [10]. Considering the precursor supply as a limiting factor for carotenoid biosynthesis in plants, the enzymatic steps of MEP pathway involved in the synthesis of IPP and DMAPP would be the most restrictive ones to the flux into the pathway. Consistent with the idea of rate-limiting steps in the MEP pathway, changes in levels of DXS messenger RNA (mRNA) expression result in alterations in the contents of several isoprenoids, such as chlorophylls, tocopherols, carotenoids, and ABA [66]. For instance, expression analysis of *DXS* showed a developmental regulation during tomato fruit ripening, which correlated with an increase in *PSY* mRNA transcripts and carotenoid accumulation [67]. Furthermore, transgenic *Arabidopsis* seedlings overexpressing *DXS* and *DXR* showed increased levels of several isoprenoids, while antisense silencing of *DXS* significantly reduced isoprenoid formation [68, 69].

PSY catalyzes the most important rate-limiting step in the carotenoid biosynthetic pathway. *PSY* genes respond transcriptionally to ABA, high light, photoperiod, development cues, abiotic stresses, and posttranscriptional feedback regulation [66]. Some plant species possess more than one PSY, which show tissue-specific expression and different responses to environmental stimuli. For example, the mRNA expression of a flower- and fruit-specific *PSY* (*PSY-1*) significantly increases during tomato fruit ripening [45], contributing to the accumulation of lycopene. The induction of *PSY3* transcription in maize and rice under salt and drought stresses correlates with increased carotenoid flux and ABA synthesis in their roots [28, 29]. Allelic variation in *PSY* has been also described as a mechanism that may change PSY enzymatic activity. Alternative splicing of *PSY-A1* allele was considered to be a major quantitative trait locus (QTL) determinant of flour color in bread wheat [70]. The accumulation of phytoene has been demonstrated to involve negative feedback regulation of upstream genes, including *PSY*, as well as downstream genes such as *ZDS* and *β-LCY*, which were all downregulated in the *pds3* mutant [71].

CrtISO has recently emerged as a regulatory node in the carotenoid biosynthetic pathway. CRTISO mutants, such as *ccr2* and *tangerine*, accumulates *cis*-carotenes like 7,7′9,9′-tetra-*cis*-lycopene in etioplasts (dark-grown plastids) of seedlings and chromoplasts of fruit [39, 40]. In chloroplasts of the mutants, the biosynthetic pathway proceeds via photoisomerization, but there is delayed greening and substantial reduction in lutein in *Arabidopsis* and varying degrees of chlorosis in tomato and rice [39, 40, 72].

The branching point in the carotenoid pathway proceeding after lycopene has a major regulatory role in modulating the partitioning of β,β- and β,ε-ring carotenoids. Flux through the two branches can be controlled by ε-LCY and β-LCY [66]. The massive accumulation of lycopene during tomato fruit development is a result of developmentally controlled downregulation of mRNA expression of β-LCY and ε-LCY, besides the upregulation of PSY [45]. A similar mechanism of carotenoid gene expression regulation during fruit development has been found in other plant species. In citrus, such as oranges, lemons, mandarins, and grapefruit, there is a decrease in the lutein content after the green stage of fruit development and color appears in various tones, from light yellow to deep orange, due to the massive accumulation of β,β-xanthophylls in the flavedo and juice sacs [73, 74]. During this process, the mRNA levels of PSY, PDS, ZDS, β-LCY2, and β-CHY increase while that of ε-LCY disappears [49, 50, 73–75]. Natural genetic variation at the ε-LCY locus in maize was shown to be another mechanism regulating the flux down β,ε- or β,β-branch of the carotenoid pathway [76]. Four natural ε-LCY polymorphisms explained 58 % of the variation in lutein and β,β-carotenoids [76].

Several studies have demonstrated that a pool of carotenoids is also regulated, in part, by the rate of degradation by carotenoid cleavage dioxygenases (CCDs) [66]. Members of this family not only synthesize important apocarotenoid molecules, such as the phytohormones ABA and strigolactone, but also aid in the maintenance of suitable carotenoid levels in different plant tissues. In *Arabidopsis*, the high carotenoid levels observed in maturing seeds have been attributed to a loss of CCD activity [77]. In strawberry, there was a correlation between the increase in *FaCCD1* expression levels during ripening and a decrease in lutein content [78]. Petal color in chrysanthemum can also be regulated by CCD activity. Elevated transcript levels of *CmCCD4a* were observed in white petals, which correlated with a breakdown of the yellow carotenoid pigments [79].

Plastid biogenesis is another important regulatory mechanism providing a sink for carotenoid accumulation. The high-pigment-2 (*hp-2*) and *hp-3* tomato mutants contain higher fruit pigmentation caused by an enlargement of the plastid compartment size, enabling greater pigment biosynthesis and storage capacity for carotenoids in mature fruits [80, 81]. The cauliflower *Orange* (*Or*) mutant accumulates a high level of β-carotene in various normally white tissues of the plant, turning them orange, since the *Or* gene triggers the differentiation of proplastids and other colorless plastids into chromoplasts, creating a metabolic sink for carotenoid accumulation [82].

The sequestration of carotenoids within the various plastid types is another important form of regulation. Chloroplasts and chromoplasts appear to differ considerably in the way they sequestrate end-product carotenoids. In chloroplasts, end-product carotenoids are associated with light-harvesting complexes [10]. The unbound carotenoids are likely to be associated with specific proteins as small plastoglobuli. For instance, end-product carotenoids are typically esterified and associated with the fibrillin protein as fibrils in chromoplasts of pepper [83]. Tomato chromoplasts appear to sequestrate lycopene as crystals [84]. The plastid ultrastructure may be altered in response to changes in carotenoid content [84, 85]. Esterification of carotenoids appears to be an effective mechanism used by flowers of sunflower, daffodil, and marigold for a high-level carotenoid accumulation [10].

Transcriptional Co-regulation of Carotenoid Biosynthesis Genes

Transcriptional co-regulation has shown to play a central role in coordinating the cellular responses, which involve multiple genes and their products. Some studies have shown that many genes encoding metabolic enzymes that function within the same or functionally related pathways form co-expression modules [86, 87]. Thus, it is believed that the synthesis of functionally related isoprenoid molecules may be mediated by their transcriptional co-regulation. However, it is important to emphasize that changes in gene transcription do not necessarily reflect in altered protein abundance and functional activity due to the existence of posttranscriptional regulatory mechanisms of gene expression. Nevertheless, as the transcriptional regulation is the first level of protein synthesis regulation, changes in gene transcription in response to specific stimuli can be considered a primary regulatory response that reflects a change in requirement for specific proteins at a specific time period.

An expression correlation analysis using *PSY* and *β-LCY* as driver genes was carried out in order to determine the level of co-expression that they share with all of the other genes represented on the *Arabidopsis* co-expression tool (ACT)

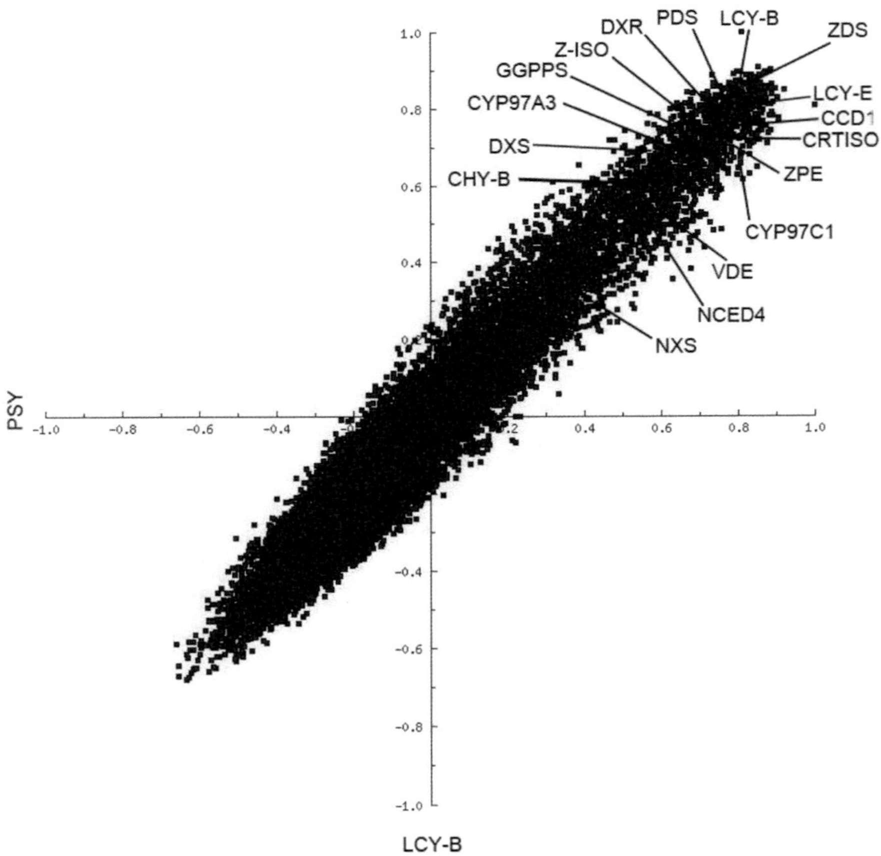

Fig. 2.5 Co-correlation scatter plot representing the level of co-expression of all *Arabidopsis* genes (*blue squares*) relative to PSY (*x*-axis) and β-LCY (LCY-B) (*y*-axis). Carotenogenesis-related genes are highlighted in *red*. Data were obtained using the *Arabidopsis* coexpression tool (ACT) (http://web.archive.org/20051222113921/www.arabidopsis.leeds.ac.uk/act/coexpanalyser.php). See text and figure legends for abbreviations. *PSY* phytoene synthase, *LCY* lycopene cyclase

(http://web.archive.org/20051222113921/www.arabidopsis.leeds.ac.uk/act/coexpanalyser.php). *PSY* and *β-LCY* were chosen as driver genes for this analysis as their transcriptions are known to be positively correlated with carotenoid accumulation, as discussed above. It is believed that genes that are highly co-expressed with *PSY* and *β-LCY* will have closely associated functional roles. A co-correlation scatterplot between *PSY* and *β-LCY* reveals that both genes have a high level of co-expression with other genes related to photosynthesis, including those encoding the photosystem II protein PsbP and chlorophyll a/b-binding proteins. The scatterplot also shows that expression of *PSY* and *β-LCY* is highly correlated with the expression of many genes involved in MEP and carotenoid biosynthetic pathways, with Pearson correlation coefficient (*r*-value) ranging from 0.52 to 0.87 (Fig. 2.5). These findings provide additional evidence that the coordinated transcriptional regulation of these biosynthesis genes is critical in regulating and coordinating the synthesis of functionally related carotenoids.

Concluding Comments

The carotenoid biosynthetic pathway is a wonderful example of successful interdisciplinary approach applied to clarify the fundamental reaction steps of the pathway and their associated genes and enzymes. The biochemical reaction sequences involved in the biosynthesis of carotenoids are now known, as are all the encoding genes. A significant progress has also been made in understanding the regulation of carotenoid biosynthesis accumulation in higher plants. Although most of the carotenoid-related research has focused on model species such as *Arabidopsis*, tomato, maize and rice, other exciting reports have emerged from vegetables and fruits such as orange. These studies along with more in-depth research on the field of genomics will strengthen knowledge about the carotenoid pathway and its regulation in the future. The recent progress in the technological capacities in genomics, including the emergence of next-generation sequencing technologies, can lead to more fascinating discoveries.

References

1. Britton G (1995) Structure and properties of carotenoids in relation to function. Faseb J 9(15):1551–1558
2. Goodwin TW (1986) Metabolism, nutrition, and function of carotenoids. Annu Rev Nutr 6:273–297
3. Li Z, Wakao S, Fischer BB, Niyogi KK (2009) Sensing and responding to excess light. Annu Rev Plant Biol 60:239–260
4. Nambara E, Marion-Poll A (2005) Abscisic acid biosynthesis and catabolism. Ann Rev of Plant Biol 56:165–185
5. Umehara M, Hanada A, Yoshida S, Akiyama K, Arite T, Takeda-Kamiya N et al (2008) Inhibition of shoot branching by new terpenoid plant hormones. Nature 455(7210):195–200

6. Vishnevetsky M, Ovadis M, Vainstein A (1999) Carotenoid sequestration in plants: the role of carotenoid-associated proteins. Trends Plant Sci 4(6):232–235
7. Toomey MB, McGraw KJ (2012) Mate choice for a male carotenoid-based ornament is linked to female dietary carotenoid intake and accumulation. BMC Evol Biol 12:3
8. Butler MW, McGraw KJ (2012) Differential effects of early- and late-life access to carotenoids on adult immune function and ornamentation in mallard ducks (*Anas platyrhynchos*). PLoS One 7(5):e38043
9. DellaPenna D, Pogson BJ (2006) Vitamin synthesis in plants: tocopherols and carotenoids. Annu Rev Plant Biol 57:711–738
10. Fraser PD, Bramley PM (2004) The biosynthesis and nutritional uses of carotenoids. Prog Lipid Res 43(3):228–265
11. Krinsky NI, Johnson EJ (2005) Carotenoid actions and their relation to health and disease. Mol Aspects Med 26(6):459–516
12. Cunningham FX, Gantt E (1998) Genes and enzymes of carotenoid biosynthesis in plants. Annu Rev Plant Phys 49:557–583
13. Ruiz-Sola MA, Rodriguez-Concepcion M (2012) Carotenoid biosynthesis in *Arabidopsis*: a colorful pathway. Arabidopsis Book 10:e0158
14. Cunningham FX Jr, Gantt E (2007) A portfolio of plasmids for identification and analysis of carotenoid pathway enzymes: *Adonis aestivalis* as a case study. Photosynth Res 92(2):245–259
15. Rodriguez-Concepcion M (2010) Supply of precursors for carotenoid biosynthesis in plants. Arch Biochem Biophys 504(1):118–122
16. McGarvey DJ, Croteau R (1995) Terpenoid metabolism. Plant Cell 7(7):1015–1026
17. Eisenreich W, Rohdich F, Bacher A (2001) Deoxyxylulose phosphate pathway to terpenoids. Trends Plant Sci 6(2):78–84
18. Kuzuyama T, Seto H (2012) Two distinct pathways for essential metabolic precursors for isoprenoid biosynthesis. Proc Jpn Acad Ser B Phys Biol Sci 88(3):41–52
19. Bouvier F, Rahier A, Camara B (2005) Biogenesis, molecular regulation and function of plant isoprenoids. Prog Lipid Res 44(6):357–429
20. Kirby J, Keasling JD (2009) Biosynthesis of plant isoprenoids: perspectives for microbial engineering. Annu Rev Plant Biol 60:335–355
21. Hunter WN (2007) The non-mevalonate pathway of isoprenoid precursor biosynthesis. J Biol Chem 282(30):21573–21577
22. Hirschberg J (2001) Carotenoid biosynthesis in flowering plants. Curr Opin Plant Biol 4(3):210–218
23. Fu Z, Yan J, Zheng Y, Warburton ML, Crouch JH, Li JS (2010) Nucleotide diversity and molecular evolution of the PSY1 gene in *Zea mays* compared to some other grass species. Theor Appl Genet 120(4):709–720
24. Bartley GE, Scolnik PA (1993) cDNA cloning, expression during development, and genome mapping of PSY2, a second tomato gene encoding phytoene synthase. J Biol Chem 268(34):25718–25721
25. Fraser PD, Kiano JW, Truesdale MR, Schuch W, Bramley PM (1999) Phytoene synthase-2 enzyme activity in tomato does not contribute to carotenoid synthesis in ripening fruit. Plant Mol Biol 40(4):687–698
26. Lange BM, Ghassemian M (2003) Genome organization in *Arabidopsis thaliana*: a survey for genes involved in isoprenoid and chlorophyll metabolism. Plant Mol Biol 51(6):925–948
27. Giorio G, Stigliani AL, D'Ambrosio C (2008) Phytoene synthase genes in tomato (*Solanum lycopersicum* L.)—new data on the structures, the deduced amino acid sequences and the expression patterns. FEBS J 275(3):527–535
28. Li FQ, Vallabhaneni R, Wurtzel ET (2008) PSY3, a new member of the phytoene synthase gene family conserved in the Poaceae and regulator of abiotic stress-induced root carotenogenesis. Plant Physiol 146(3):1333–1345
29. Welsch R, Wust F, Bar C, Al-Babili S, Beyer P (2008) A third phytoene synthase is devoted to abiotic stress-induced abscisic acid formation in rice and defines functional diversification of phytoene synthase genes. Plant Physiol 147(1):367–380

30. Li FQ, Vallabhaneni R, Yu J, Rocheford T, Wurtzel ET (2008) The maize phytoene synthase gene family: overlapping roles for carotenogenesis in endosperm, photomorphogenesis, and thermal stress tolerance. Plant Physiol 147(3):1334–1346
31. Lopez AB, Yang Y, Thannhauser TW, Li L (2008) Phytoene desaturase is present in a large protein complex in the plastid membrane. Physiol Plant 133(2):190–198
32. Cunningham FX Jr, Pogson B, Sun Z, McDonald KA, DellaPenna D, Gantt E (1996) Functional analysis of the beta and epsilon lycopene cyclase enzymes of *Arabidopsis* reveals a mechanism for control of cyclic carotenoid formation. Plant Cell 8(9):1613–1626
33. Al-Babili S, von Lintig J, Haubruck H, Beyer P (1996) A novel, soluble form of phytoene desaturase from *Narcissus pseudonarcissus* chromoplasts is Hsp70-complexed and competent for flavinylation, membrane association and enzymatic activation. Plant J 9(5):601–612
34. Bonk M, Hoffmann B, Von Lintig J, Schledz M, Al-Babili S, Hobeika E et al (1997) Chloroplast import of four carotenoid biosynthetic enzymes in vitro reveals differential fates prior to membrane binding and oligomeric assembly. Eur J Biochem 247(3):942–950
35. Carol P, Kuntz M (2001) A plastid terminal oxidase comes to light: implications for carotenoid biosynthesis and chlororespiration. Trends Plant Sci 6(1):31–36
36. Romer S, Hugueney P, Bouvier F, Camara B, Kuntz M (1993) Expression of the genes encoding the early carotenoid biosynthetic enzymes in capsicum annuum. Biochem Biophys Res Commun 196(3):1414–1421
37. Matthews PD, Luo R, Wurtzel ET (2003) Maize phytoene desaturase and zeta-carotene desaturase catalyse a poly-Z desaturation pathway: implications for genetic engineering of carotenoid content among cereal crops. J Exp Bot 54(391):2215–2230
38. Chen CX, Costa MGC, Yu QB, Moore GA, Gmitter FG (2010) Identification of novel members in sweet orange carotenoid biosynthesis gene families. Tree Genet Genomes 6(6):905–914
39. Isaacson T, Ronen G, Zamir D, Hirschberg J (2002) Cloning of tangerine from tomato reveals a carotenoid isomerase essential for the production of beta-carotene and xanthophylls in plants. Plant Cell 14(2):333–342
40. Park H, Kreunen SS, Cuttriss AJ, DellaPenna D, Pogson BJ (2002) Identification of the carotenoid isomerase provides insight into carotenoid biosynthesis, prolamellar body formation, and photomorphogenesis. Plant Cell 14(2):321–332
41. Cunningham FX Jr, Gantt E (2001) One ring or two? Determination of ring number in carotenoids by lycopene epsilon-cyclases. Proc Natl Acad Sci U S A 98(5):2905–2910
42. Bramley PM (2002) Regulation of carotenoid formation during tomato fruit ripening and development. J Exp Bot 53(377):2107–2113
43. Hugueney P, Badillo A, Chen HC, Klein A, Hirschberg J, Camara B et al (1995) Metabolism of cyclic carotenoids: a model for the alteration of this biosynthetic pathway in *Capsicum annuum* chromoplasts. Plant J 8(3):417–424
44. Hornero-Mendez D, Britton G (2002) Involvement of NADPH in the cyclisation reaction of carotenoid biosynthesis. FEBS Lett 515(1–3):133–136
45. Ronen G, Cohen M, Zamir D, Hirschberg J (1999) Regulation of carotenoid biosynthesis during tomato fruit development: expression of the gene for lycopene epsilon-cyclase is down-regulated during ripening and is elevated in the mutant delta. Plant J 17(4):341–351
46. Pecker I, Gabbay R, Cunningham FX Jr, Hirschberg J (1996) Cloning and characterization of the cDNA for lycopene beta-cyclase from tomato reveals decrease in its expression during fruit ripening. Plant Mol Biol 30(4):807–819
47. Ronen G, Carmel-Goren L, Zamir D, Hirschberg J (2000) An alternative pathway to beta-carotene formation in plant chromoplasts discovered by map-based cloning of beta and old-gold color mutations in tomato. Proc Natl Acad Sci U S A 97(20):11102–11107
48. Lewinsohn E, Sitrit Y, Bar E, Azulay Y, Meir A, Zamir D et al (2005) Carotenoid pigmentation affects the volatile composition of tomato and watermelon fruits, as revealed by comparative genetic analyses. J Agric Food Chem 53(8):3142–3148
49. Alquezar B, Zacarias L, Rodrigo MJ (2009) Molecular and functional characterization of a novel chromoplast-specific lycopene beta-cyclase from *Citrus* and its relation to lycopene accumulation. J Exp Bot 60(6):1783–1797

50. Mendes AF, Chen C, Gmitter FG Jr, Moore GA, Costa MG (2011) Expression and phylogenetic analysis of two new lycopene beta-cyclases from *Citrus paradisi*. Physiol Plant 141(1):1–10
51. Bouvier F, D'Harlingue A, Backhaus RA, Kumagai MH, Camara B (2000) Identification of neoxanthin synthase as a carotenoid cyclase paralog. Eur J Biochem 267(21):6346–6352
52. Al-Babili S, Hugueney P, Schledz M, Welsch R, Frohnmeyer H, Laule O et al (2000) Identification of a novel gene coding for neoxanthin synthase from Solanum tuberosum. FEBS Lett 485(2/3):168–172
53. Bouvier F, Hugueney P, d'Harlingue A, Kuntz M, Camara B (1994) Xanthophyll biosynthesis in chromoplasts: isolation and molecular cloning of an enzyme catalyzing the conversion of 5,6-epoxycarotenoid into ketocarotenoid. Plant J 6(1):45–54
54. Bouvier F, D'Harlingue A, Backhaus RA, Kumagai MH, Camara B (2000) Identification of neoxanthin synthase as a carotenoid cyclase paralog. Eur J Biochem 267(21):6346–6352
55. Sandmann G (2002) Molecular evolution of carotenoid biosynthesis from bacteria to plants. Physiol Plant 116(4):431–440
56. Sun Z, Gantt E, Cunningham FX Jr (1996) Cloning and functional analysis of the beta-carotene hydroxylase of *Arabidopsis thaliana*. J Biol Chem 271(40):24349–24352
57. Bouvier F, Keller Y, d'Harlingue A, Camara B (1998) Xanthophyll biosynthesis: molecular and functional characterization of carotenoid hydroxylases from pepper fruits (*Capsicum annuum* L.). Biochim Biophys Acta 1391(3):320–328
58. Shanklin J, Cahoon EB (1998) Desaturation and related modifications of fatty acids1. Annu Rev Plant Physiol Plant Mol Biol 49:611–641
59. Tian L, Musetti V, Kim J, Magallanes-Lundback M, DellaPenna D (2004) The *Arabidopsis* LUT1 locus encodes a member of the cytochrome p450 family that is required for carotenoid epsilon-ring hydroxylation activity. Proc Natl Acad Sci U S A 101(1):402–407
60. Muller P, Li XP, Niyogi KK (2001) Non-photochemical quenching. A response to excess light energy. Plant Physiol 125(4):1558–1566
61. Flower DR (1996) The lipocalin protein family: structure and function. Biochem J 318 (Pt 1):1–14
62. Hieber AD, Bugos RC, Yamamoto HY (2000) Plant lipocalins: violaxanthin de-epoxidase and zeaxanthin epoxidase. Biochim Biophys Acta 1482(1/2):84–91
63. Vallabhaneni R, Wurtzel ET (2009) Timing and biosynthetic potential for carotenoid accumulation in genetically diverse germplasm of maize. Plant Physiol 150(2):562–572
64. Young PR, Lashbrooke JG, Alexandersson E, Jacobson D, Moser C, Velasco R et al (2012) The genes and enzymes of the carotenoid metabolic pathway in *Vitis vinifera* L. BMC Genomics 13(1):243
65. North HM, De Almeida A, Boutin JP, Frey A, To A, Botran L et al (2007) The *Arabidopsis* ABA-deficient mutant aba4 demonstrates that the major route for stress-induced ABA accumulation is via neoxanthin isomers. Plant J 50(5):810–824
66. Cazzonelli CI, Pogson BJ (2010) Source to sink: regulation of carotenoid biosynthesis in plants. Trends Plant Sci 15(5):266–274
67. Lois LM, Rodriguez-Concepcion M, Gallego F, Campos N, Boronat A (2000) Carotenoid biosynthesis during tomato fruit development: regulatory role of 1-deoxy-D-xylulose 5-phosphate synthase. Plant J 22(6):503–513
68. Estevez JM, Cantero A, Reindl A, Reichler S, Leon P (2001) 1-Deoxy-D-xylulose-5-phosphate synthase, a limiting enzyme for plastidic isoprenoid biosynthesis in plants. J Biol Chem 276(25):22901–22909
69. Carretero-Paulet L, Cairo A, Botella-Pavia P, Besumbes O, Campos N, Boronat A et al (2006) Enhanced flux through the methylerythritol 4-phosphate pathway in *Arabidopsis* plants overexpressing deoxyxylulose 5-phosphate reductoisomerase. Plant Mol Biol 62(4/5):683–695
70. Howitt CA, Cavanagh CR, Bowerman AF, Cazzonelli C, Rampling L, Mimica JL et al (2009) Alternative splicing, activation of cryptic exons and amino acid substitutions in carotenoid biosynthetic genes are associated with lutein accumulation in wheat endosperm. Funct Integr Genomics 9(3):363–376

71. Qin G, Gu H, Ma L, Peng Y, Deng XW, Chen Z et al (2007) Disruption of phytoene desaturase gene results in albino and dwarf phenotypes in *Arabidopsis* by impairing chlorophyll, carotenoid, and gibberellin biosynthesis. Cell Res 17(5):471–482
72. Fang J, Chai C, Qian Q, Li C, Tang J, Sun L et al (2008) Mutations of genes in synthesis of the carotenoid precursors of ABA lead to pre-harvest sprouting and photo-oxidation in rice. Plant J 54(2):177–189
73. Kato M (2004) Accumulation of carotenoids and expression of carotenoid biosynthetic genes during maturation in citrus fruit. Plant Physiol 134(2):824–837
74. Rodrigo MJ, Marcos JF, Zacarias L (2004) Biochemical and molecular analysis of carotenoid biosynthesis in flavedo of orange (*Citrus sinensis* L.) during fruit development and maturation. J Agric Food Chem 52(22):6724–6731
75. Costa MG, Moreira CD, Melton JR, Otoni WC, Moore GA (2012) Characterization and developmental expression of genes encoding the early carotenoid biosynthetic enzymes in *Citrus paradisi* macf. Mol Biol Rep 39(2):895–902
76. Harjes CE, Rocheford TR, Bai L, Brutnell TP, Kandianis CB, Sowinski SG et al (2008) Natural genetic variation in lycopene epsilon cyclase tapped for maize biofortification. Science 319(5861):330–333
77. Auldridge ME, Block A, Vogel JT, Dabney-Smith C, Mila I, Bouzayen M et al (2006) Characterization of three members of the *Arabidopsis* carotenoid cleavage dioxygenase family demonstrates the divergent roles of this multifunctional enzyme family. Plant J 45(6):982–993
78. Garcia-Limones C, Schnabele K, Blanco-Portales R, Luz Bellido M, Caballero JL, Schwab W et al (2008) Functional characterization of FaCCD1: a carotenoid cleavage dioxygenase from strawberry involved in lutein degradation during fruit ripening. J Agric Food Chem 56(19):9277–9285
79. Ohmiya A, Kishimoto S, Aida R, Yoshioka S, Sumitomo K (2006) Carotenoid cleavage dioxygenase (CmCCD4a) contributes to white color formation in chrysanthemum petals. Plant Physiol 142(3):1193–1201
80. Kolotilin I, Koltai H, Tadmor Y, Bar-Or C, Reuveni M, Meir A et al (2007) Transcriptional profiling of high pigment-2dg tomato mutant links early fruit plastid biogenesis with its overproduction of phytonutrients. Plant Physiol 145(2):389–401
81. Galpaz N, Wang Q, Menda N, Zamir D, Hirschberg J (2008) Abscisic acid deficiency in the tomato mutant high-pigment 3 leading to increased plastid number and higher fruit lycopene content. Plant J 53(5):717–730
82. Lu S, Van Eck J, Zhou X, Lopez AB, O'Halloran DM, Cosman KM et al (2006) The cauliflower or gene encodes a DnaJ cysteine-rich domain-containing protein that mediates high levels of beta-carotene accumulation. Plant Cell 18(12):3594–3605
83. Deruere J, Romer S, d'Harlingue A, Backhaus RA, Kuntz M, Camara B (1994) Fibril assembly and carotenoid overaccumulation in chromoplasts: a model for supramolecular lipoprotein structures. Plant Cell 6(1):119–133
84. Romer S, Fraser PD, Kiano JW, Shipton CA, Misawa N, Schuch W et al (2000) Elevation of the provitamin a content of transgenic tomato plants. Nat Biotechnol 18(6):666–669
85. Shewmaker CK, Sheehy JA, Daley M, Colburn S, Ke DY (1999) Seed-specific overexpression of phytoene synthase: increase in carotenoids and other metabolic effects. Plant J 20(4):401–412X
86. Ihmels J, Bergmann S, Barkai N (2004) Defining transcription modules using large-scale gene expression data. Bioinformatics 20(13):1993–2003
87. Wei H, Persson S, Mehta T, Srinivasasainagendra V, Chen L, Page GP et al (2006) Transcriptional coordination of the metabolic network in *Arabidopsis*. Plant Physiol 142(2):762–774

Chapter 3
Carotenoids in Human Nutrition

Phyllis E. Bowen, Maria Stacewicz-Sapuntzakis, and Veda Diwadkar-Navsariwala

Introduction

Although there are about 600 known carotenoids in nature, the human diet contains approximately 40, and only ~10 were found in human tissues and plasma [1]. Some abundant dietary carotenoids, like violaxanthin and antheraxanthin in green plant tissues, are poorly absorbed in the human intestinal tract. Nevertheless, humans and monkeys are able to absorb and utilize a remarkably wide variety of carotenoids, from the nonpolar hydrocarbons (carotenes) to polar hydroxycarotenoids (lutein, zeaxanthin) [2]. Carotenoids containing an unsubstituted β-ionone ring (β-carotene, α-carotene, γ-carotene, β-cryptoxanthin) are partially converted to vitamin A in the intestinal mucosa and other tissues. Other animals may absorb only carotenes (cows, felines) [3], or are practically unable to absorb any carotenoids unless fed with large pharmacological doses (carnivores) [4]. Some animals seem to absorb only provitamin A carotenoids and convert them so efficiently to vitamin A that none can be found in their blood or tissues (rats, mice). Birds utilize provitamin A carotenoids, as well as polar hydroxy- and keto-carotenoids, to produce pigments coloring their skin and feathers [5]. Because of these great differences in carotenoid utilization and metabolism, it is difficult to translate the data obtained from research with laboratory animals to human physiological response and dietary recommendations. However, genetic and proteomic investigations in animals and humans may reveal the causes of the differences between species and within human populations [6].

P. E. Bowen (✉)
Kinesiology and Nutrition, University of Illinois at Chicago,
150 LeMoyne Pkwy, Oak Park, IL 60302, USA
e-mail: pbowen@uic.edu

M. Stacewicz-Sapuntzakis
5825 North Forest Glen Ave, Chicago, IL 60646, USA

V. Diwadkar-Navsariwala
Bartlett, IL, USA

C. Chen (ed.), *Pigments in Fruits and Vegetables,* DOI 10.1007/978-1-4939-2356-4_3

This chapter considers the role of carotenoids in various aspects of human physiology, possible benefits of dietary intake in preventing and treating disease, as well as the controversial issues of supplementation. Despite a very substantial body of existing literature, we are still at the beginning in the understanding of these complex problems.

Food Sources and Bioavailability

Plants are the main source of carotenoids in our diet, since the bright colors of many fruits and vegetables are due to the high content of these pigments. Carrots, sweet potato, pumpkin, squash, mango, and apricots are rich in β-carotene, which is also abundant in the dark-green leaves of spinach, turnip greens, and collards [7]. Carrots, pumpkin, and squash also contain considerable amounts of α-carotene, while oranges, tangerines, red peppers, yellow papaya, and persimmons are the main sources of β-cryptoxanthin. Lycopene is provided mostly by tomatoes and tomato products, but is also present in watermelon, guava, pink grapefruit, and red papaya. Lutein and zeaxanthin are delivered by dark-green leaves, broccoli, and corn (maize).

Canned, cooked, or dried fruits and vegetables often contain more carotenoids per unit of weight than their fresh counterparts, due to dehydration, although some losses occur in processing at high temperatures. However, the bioavailability of carotenoids is greatly increased by processing (mincing, pureeing, cooking, canning) since it releases them from the food matrix. Thus, they may be easier to digest in the intestine, i.e., to incorporate into micelles of lipids and bile salts. Carotenoids are fat-soluble pigments, and therefore require the presence of fat in the same meal to be absorbed. Salad dressings should contain some fat to facilitate the absorption of carotenoids [8]. Mixed dishes containing carotenoids, like soups, stews, casseroles, and sauces, are better vehicles for carotenoid delivery than raw produce. For the same reason, animal sources of carotenoids are excellent providers of bioavailable β-carotene (milk, butter, cheese, and beef liver) or lutein and zeaxanthin (eggs, chicken fat, liver, and skin).

Major contributors of individual carotenoids, as well as of total provitamin A carotenoids, in the diet of adult Americans are listed in Table 3.1 [9]. The average daily intake of carotenoids by adults >19 years of age is close to 10 mg/day, including 2.63 mg provitamin A carotenoids (2.11 mg β-carotene, 0.39 mg α-carotene, 0.13 mg β-cryptoxanthin), 5.52 mg lycopene, and 1.50 mg lutein and zeaxanthin [10]. However, it was estimated from the National Health and Nutrition Examination Survey (NHANES) 2003–2006 data that only 5 % of adult males and 7 % of adult females meet dietary guidance recommendations for fruit and vegetable intake. Those individuals have a two- to threefold higher intake of carotenoids than the rest of the US population. Table 3.2 provides a list of mixed foods that contribute to carotenoid intake and are well accepted even by people who do not eat many fruits and vegetables.

Table 3.1 Major contributors of carotenoids in diets of US adults > 19 years old (%)

Total provitamin A carotenoids[a]		α-Carotene[b]		β-Carotene[b]		β-Cryptoxanthin[b]		Lycopene[b]		Lutein + zeaxanthin[b]	
Carrots	17.1	Carrots	72.5	Carrots	30.8	Oranges, orange juice, tangerines	67.5	Tomatoes and tomato products	81.2	Spinach	27.8
Vegetable soups	8.2	Vegetable mixtures	7.2	Spinach	9.9	Yellow corn	5.1	Vegetable mixtures	6.1	Lettuce	5.6
Collard, mustard, turnip greens	7.0	Tomatoes and tomato products	5.5	Sweet potatoes	9.1	Watermelon	3.7	Watermelon	5.2	Eggs	5.2
Spinach	5.6	Soups and stews without tomatoes	3.4	Tomatoes and tomato products	8.7	Cucumbers	3.7	Soups and stews with tomatoes	3.8	Collards	5.2
Oranges, orange juice, tangerines	5.5	Pumpkin	2.3	Lettuce	5.5	–	–	–	–	Broccoli	5.0

[a] Adapted from [9]. Data were compiled from the Second National Health and Nutrition Examination Survey (NHANES II, 1976–1980)
[b] Adapted from [10]. Data were compiled from the National Health and Nutrition Examination Surveys (2003–2004 and 2005–2006)

Table 3.2 Carotenoid content of mixed dishes (µg/100 g serving). [Based on data from Food Composition Classics: USDA-NCC carotenoid database for US Foods, 1998 (http://www.ars.usda.gov/services/docs.htm?docid=9447). Accessed 14 Jan 2013]

Description	α-Carotene	β-Carotene	β-Cryptoxanthin	Lutein + zeaxanthin	Lycopene
Baked beans	147	408	–	–	–
Beef stew with vegetables (including potatoes, carrots), canned	700	1780	–	60	302
Chicken pot pie, with carrots, potatoes, peas, frozen	242	1048	27	105	–
Fruit cocktail, canned, drained, heavy syrup	–	138	52	112	–
Green peppers stuffed with beef and rice, in tomato sauce, frozen entrée, cooked	–	192	–	75	3092
Lasagna with meat and tomato sauce, frozen entrée, cooked	–	170	–	97	7750
Meatloaf with mashed potatoes and gravy, low-fat frozen entrée, cooked	–	110	–	–	930
Pasta in tomato sauce with cheese, canned	–	127	–	–	3162
Pasta with chicken and vegetables (carrots, peas, onions mushrooms) with oriental sauce, low-calorie frozen entrée, cooked	365	620	–	312	–
Pasta with shrimp and vegetables (broccoli, red peppers, yellow zucchini, onions) in lemon pepper sauce, low-fat frozen entrée, cooked	–	148	–	177	–
Pizza supreme with sausage and pepperoni, mushrooms, peppers, onions, cheese and sauce, thin crust, frozen	–	170	–	20	2071
Pizza with pepperoni, cheese and sauce, thin crust, frozen	–	264	–	15	4449
Sauce mole, prepared from a recipe	199	1282	828	–	–
Sauce for pasta, spaghetti/marinara, ready-to-serve	–	440	–	160	15,990
Soup, minestrone, canned, condensed, commercial	210	920	–	150	1480
Soup, tomato, canned, condensed, commercial	–	235	–	90	10,920
Soup, vegetable beef, canned, condensed, commercial	489	1618	–	92	364
Soup, vegetarian vegetable, canned, condensed, commercial	40	1500	–	160	1930
Spinach soufflé, frozen, cooked	–	1300	–	2727	–
Vegetable combination with butter sauce (broccoli, cauliflower, baby carrots)	333	450	–	142	–
Vegetables (white potatoes, sweet potatoes, rutabagas, green beans, onions) with beef and sauce, low-fat frozen entrée, cooked		352		70	285
Vegetable juice cocktail, canned	210	830	–	80	9660

Absorption of carotenoids in the intestine was long considered to occur passively together with dietary lipids [11]. After their release from plant tissues, they must be dissolved in fat, forming emulsions with the aid of bile acids. Carotenoid-containing lipid micelles are absorbed by the intestinal mucosal cells and incorporated in chylomicrons; however, this universal scheme does not explain considerable differences in carotenoid absorption among animal species. Recent cell culture and animal model studies indicate that carotenoid absorption is protein-mediated by a scavenger receptor class B type 1 (SR-B1), which also facilitates the uptake of α-tocopherol and cholesterol [6]. SR-B1 is regulated (attenuated) by the intestine-specific homeobox factor (ISX), which in turn is activated by retinoic acid. Further studies of the differences in activity and expression of these proteins may help to elucidate the diversity of carotenoid absorption among various species, as well as individual differences in the human population. It is worth mentioning that careful balance studies of carotenoids measured in the diet and feces provide some assessment of carotenoid absorption, although it may be overestimated due to bacterial degradation in the colon. The absorption of β-carotene from plant sources ranged from 5 to 65 % in adult humans [12] according to studies in the Netherlands, UK, and USA, a demonstration of the wide variation in carotenoid absorption within the human population.

Carotenoid Metabolism

The absorbed provitamin A carotenoids are centrally cleaved by β-carotene-15,15′-monooxygenase (BCMO1), producing retinal (vitamin A aldehyde) that is quickly reduced to retinol. Irreversible oxidation of retinol results in retinoic acid, which mediates the vitamin A functions in growth and development by binding to retinoic acid receptors (RARs). BCMO1, like the absorption facilitator SR-B1, is regulated (attenuated) by ISX, a transcription factor activated by RARs [6]. A heterozygotic mutation in BCMO1 causes elevated β-carotene and low retinol concentrations in blood [13]. Normally, about 40 % of absorbed β-carotene is not converted to vitamin A in the intestine, but incorporated in chylomicrons, entering circulation and taken up in various tissues by lipoprotein-specific receptors. Since BCMO1 is expressed in the liver and peripheral tissues (glandular cells of stomach, pancreas, colon, prostate, mammary tissue, steroidogenic cells of ovary, testis, and adrenals, as well as skin keratocytes, muscle cells, and retinal pigmented epithelium of the eyes), local production of retinoids may also occur.

The efficiency of provitamin A carotenoid conversion to vitamin A depends on bioavailability, absorption, and cleavage reactions that are obviously affected by many factors, like food preparation, health, and genetic variability in human subjects [12, 14]. Low vitamin A status may increase the conversion in the absence of parasites, anemia, and other diseases. On the other hand, large dietary intakes of provitamin A carotenoids may decrease the efficiency of conversion due to the negative feedback regulation by retinoic acid. The equivalency ratio (weight of β-carotene in food: estimated weight of retinol formed) was better for fruits (12:1) than for

Table 3.3 Vitamin A equivalency of β-carotene from plant sources, based on plasma retinol response. (Adapted from [12])

Plant source	Amount of β-carotene (mg)	Equivalency ratio (wt)	Range	CV (%)
Indian spinach (*Basella alba*)	4.5	10:1	–	–
Sweet potato	4.5	13:1	–	–
Spinach	11	21:1	10:1–47.1:1	43
Carrots	11	15:1	8:1–25:1	44
Maize (biofortified)	0.527	6.5:1	3.9:1–13.3:1	54
Golden Rice	1–1.53	3.8:1	1.9:1–6.4:1	45
Red palm oil	2.37	5.7:1	–	–
Spirulina	4.2	4.5:1	2.3:1–7.1	36

vegetables (26:1) in most studies, but the best was found for biofortified Golden Rice (3.8:1), biofortified maize (6.5:1), spirulina (4.5:1), and red palm oil (RPO; 5.7:1). However, the reported ranges within studies were very wide, as shown in Table 3.3. In 2001, the US Institute of Medicine proposed a new unit, the retinol activity equivalent (1 μg RAE = 1 μg retinol), and advised that 1 μg RAE = 12 μg β-carotene or 24 μg of other provitamin A carotenoids for calculations of vitamin A equivalency for plant-derived food [15].

Besides the cytoplasmic enzyme BCMO1, there is also a mitochondrial carotenoid-degrading enzyme, β-carotene-9′,10′-dioxygenase 2 (BCDO2), that cleaves a wide range of carotenoids at position 9, 10 or 9′,10′, forming apocarotenals, β-ionone, or acyclic fragments, depending on the substrate [16]. BCDO2 degrades provitamin A carotenoids, xanthophylls (lutein and zeaxanthin), and lycopene *cis* isomers. Inhibition of BCDO2 gene expression may help to explain the accumulation of xanthophylls in birds (yellow skin of chickens) or an unusual yellow-fat phenotype in sheep. Thus, the apparent lack of carotenoids in blood or tissues of various animal species may be the result of active degradation by BCDO2. Apocarotenals seem to be quickly metabolized, since they do not accumulate in animal tissues. However, minute amounts of β-apo-8′-carotenal, β-apo-10′-carotenal, and β-apo-12′-carotenal were found in mice and humans, derived from dietary apocarotenals or from β-carotene eccentric cleavage (enzymatic or nonenzymatic) [17]. Small amounts of β-apocarotenals (~1.5 % of β-carotene) were found in cantaloupe [18] and lycopenals (0.1 % of lycopene) were identified in tomatoes, tomato paste, watermelon, and red grapefruit [19].

Significance of Provitamin A Carotenoids in Developing Countries

Provitamin A carotenoid availability is of particular importance in developing countries where vitamin A deficiency (VAD) is a significant public health concern. The main underlying cause of VAD in low-income countries is a poor diet that is consistently insufficient in vitamin A, eventually leading to depleted stores that fail to achieve physiological needs [20, 21]. Persistent, severe deficiency can lead to xerophthalmia, a form of preventable, but irreversible, blindness in young children, and facilitates infectious diseases such as measles, diarrhea, and intestinal parasites, which increase infant mortality risks.

In such low-income populations, due to the poor availability of animal sources of preformed vitamin A, dietary carotenoids from plant sources, which need to be converted to vitamin A in the intestine, contribute to ~80% of daily vitamin A intake and become highly necessary [12]. However, intestinal conversion of provitamin A to vitamin A is often compromised due to intestinal parasitic infections in young children, further exacerbating the deficiency condition. Some success has been observed with improving vitamin A status through supplementation with low-dose β-carotene (1.2 mg daily) coupled with deworming for roundworms, such as *Ascaris lumbricoides*, in young Bangladeshi children with subclinical VAD [22].

World Health Organization (WHO; [21]) has reported that about 250 million preschool children are affected by VAD, and an estimated 250,000–500,000 children go blind every year. WHO has outlined three main community-intervention strategies to combat and reduce VAD in affected populations: (1) fortification, (2) supplementation, and (3) dietary diversification. The first approach involves increasing dietary intake through fortification of a staple food with vitamin A. While this approach has been implemented and successful in Central and South America where sugar has been fortified for many years [23], and also in high-income populations where fats, oils, and cereal products are enriched, its success in low-income countries is still limited. As mentioned earlier, vitamin A equivalency ratios appear to be lower for biofortified foods that are staples in populations at risk for VAD in developing countries. However, since these ratios were estimated in healthy US adults, it remains to be seen whether these biofortified foods will be as favorable in these high-risk areas where nutritional deficiencies and intestinal parasites are highly prevalent [12].

In high-risk populations, periodic supplementation with 200,000 IU vitamin A in preschool children (<5 years) has been shown to reduce the risk of xerophthalmia by 90% and mortality by 23% [20]. A recent meta-analysis evaluating the efficacy of vitamin A supplementation in reducing mortality and morbidity in children aged 6 months to 5 years found that out of the 43 trials included, 17 trials reported a 24% reduction in the all-cause mortality rate. Seven trials reported a 28% reduction in mortality associated with diarrhea, and vitamin A supplementation was associated with a reduced incidence of diarrhea and measles and a reduced prevalence of vision problems including night blindness and xerophthalmia [24]. Although effective, supplementation targets an immediate need, but is not regarded as a sustainable and economically viable approach.

Dietary diversification has been promoted in these regions in order to enhance the overall nutritional status of a community through nutrition education, and encouraging the consumption of vitamin A- and provitamin A-rich foods. The importance of home gardens to improve availability has also been emphasized. However, supplementation with plant sources of β-carotene in children and pregnant/lactating women have had variable effects [12]. In Mozambique, the promotion of production and consumption of orange-fleshed sweet potato (OSP) resulted in increased serum retinol concentrations in preschool children from the intervention area compared with children in the control area, following a 2-year period [25]. Similar results were obtained in Uganda, where introduction of OSP to farming households led to a decrease in the prevalence of inadequate vitamin A intake in both children and women, and a 9.5 % reduction in prevalence of serum retinol < 1.05 μmol/L in children [26]. In poor Bangladeshi women, however, 60 days of daily consumption of OSP increased plasma β-carotene concentrations from 0.1 ± 00 μmol/L to 0.18 ± 0.09 μmol/L ($P < 0.0001$), but the vitamin A body pool size was not significantly affected [27]. In Gambian children, the consumption of dried mango for 4 months increased serum retinol concentrations, compared to the control group who received a single placebo capsule of 40 mg α-tocopherol, followed by no intervention for 4 months. Dark-green leafy vegetables or orange-colored fruits, provided to Indonesian schoolchildren for 9 weeks, increased serum retinol concentrations compared to the control group that received a diet low in provitamin A. Filipino children also showed a positive response to a 9-week intervention with β-carotene-rich vegetables [12]. While these studies indicated that consumption of β-carotene through fruits and vegetables can increase the vitamin A status of at risk children, it is argued that the extent of vitamin A status improvement is difficult to quantify, because serum retinol is under homeostatic control and assessing these levels alone cannot provide information on the magnitude of change in vitamin A status [12].

Another food widely promoted for controlling VAD is RPO, the richest naturally occurring source of β-carotene, which generally contains a total of 500–800 mg of provitamin A carotenoids/kg oil [28] and is traditionally used for cooking in tropical rain forest regions of West Africa [29]. Rice and Burns [30] have presented a comprehensive review on the efficacy of RPO in preventing VAD. In the reviewed studies, RPO was provided mainly to preschool-aged children and pregnant or lactating women, either as a daily supplement, an in-home fortificant (such as mixing with breakfast, foods, and regular meals), or fortified into other foods (such as biscuits or cassava flour). In general, most studies showed an improvement in serum retinol concentrations in the targeted populations. Carotenoid levels increased in breast milk and maternal serum, but did not increase milk retinol concentrations in early lactation. The authors observed that the distinctive odor, color, and taste of RPO could be a potential barrier in achieving intervention success in cultures and countries where the oil is not traditionally used for cooking. However, if incorporated successfully, there is considerable evidence supporting the efficacy of RPO in ameliorating VAD.

In recent years, genetically modified (GM) crops have been advocated as a safe and super-efficient mode to increase the provitamin A levels in staple crops that otherwise contain negligible amounts of carotenoids, and Golden Rice (GR) has become a frontrunner in this regard. By genetically engineering the components of the β-carotene biosynthetic pathway into the endosperm of rice [31], scientists have been able to produce GR containing up to 37 μg total provitamin A carotenoids (~30 μg β-carotene) in 1.0 g milled and uncooked rice [32]. This technological advancement has provided an opportunity to improve the availability of provitamin

A carotenoids to South Asian countries where rice is a main staple and VAD still prevalent. Tang et al. [33] compared the efficiency of isotope-labeled GR to spinach or pure β-carotene in oil in providing vitamin A to Chinese preschool children. The results indicated that GR was as effective as β-carotene in the oil capsule and better than spinach in contributing to the vitamin A intake in these children. Furthermore, the efficiency of conversion of β-carotene from GR to vitamin A was found to be better (2.3:1) in the Chinese children than that in US adults (3.8:1) [34]. If supplementation with GR produces similar outcomes in other high-risk countries, a considerable reduction in VAD can be expected due to the large consumption of this staple food. However, the economic cost of producing and supplying large amounts of GM crops to densely populated areas may be a challenge to consider. In the meantime, the consumption of carotenoids through homegrown vegetables and crops needs be continually encouraged in order to maintain a vitamin A status that would prevent fatal consequences.

Fertility and Reproductive Success

It is well known that carotenoids are crucial for the propagation of animal species, especially birds [35]. The bright colors of feathers or skin advertise gender (usually male), health and vigor, and the ability to procure a nutritious diet for offspring. Sexual attraction and choice of mate often depend, in birds, on the dietary supply of carotenoids. Carotenoids are accumulated in the egg yolk and provide a necessary supply of lutein and zeaxanthin for the retina of the developing young bird. Bird retina has more photoreceptors than human retina, and each cone cell contains an oil droplet with a high concentration of carotenoids that allows them to have excellent vision and to distinguish a great range of colors [36].

Recent simulation experiments indicated that subjects preferred carotenoid-related yellowish skin tones over suntan in potential mates, possibly because it advertises good health and may also signal robust fertility [37]. Dietary carotenoids normally found in the testes [38, 39] and seminal plasma [40] were often at lower concentrations in men suffering from infertility. The relationship between carotenoids and fertility was investigated in 30 men with idiopathic sperm quality issues [41] who were given 2 mg lycopene twice a day for 3 months. Lycopene administration significantly improved sperm count, motility, and morphology. This short study resulted in six pregnancies.

Carotenoids also accumulate in ovaries, especially the corpus luteum, which develops in female mammals from the ovarian follicle after release of the egg, during the luteal phase of the estrous or menstrual cycle. Bovine corpus luteum is so rich in β-carotene that it is often referred to as corpus rubrum, and it increases from 14 to 175 μg/g during the estrous cycle [42]. The corpus luteum synthesizes progesterone to prepare the uterine endometrium for implantation of the fertilized egg, thus maintaining pregnancy. When studied in bovine luteal cell culture, this steroidogenesis was found to require the replenishing of high-density lipoprotein (HDL) cholesterol (substrate) and β-carotene [43]. It is possible that β-carotene fulfills a dual function in corpus luteum, as an antioxidant protecting the tissue from the action of free radicals released during the synthesis of progesterone [44] and as an endogenous source of vitamin A produced there by the enzyme BCMO1. Reproductive performance of cows seems to be enhanced by the intake of carotenoids from fresh hay and/or supplemental β-carotene [45].

Human corpus luteum is the primary source of progesterone for 4–5 weeks after implantation [46], until placental production takes over. It is quite obvious that the concentration of carotenoids in human corpus luteum is higher than in the rest of the ovary, but the quantitative data are missing due to the difficulty of obtaining specimens from different phases of the menstrual cycle and pregnancy. The ovaries contain all the dietary carotenoids found in human plasma (β-carotene, lutein, lycopene) [38, 39]. It is tempting to speculate that an increased intake of dietary carotenoids may improve conception rates and maintenance of early pregnancy in some women with unexplained infertility.

Maternal deficiency of vitamin A increases the risk of pregnancy-related mortality and impairs infant survival. Although not of concern in high-income populations, where preformed vitamin A-rich and fortified foods are readily available, in low-income, high-risk countries, supplemental vitamin A is recommended during pregnancy to prevent night blindness and postnatal infant mortality.

A randomized, double blind, placebo-controlled trial in Nepal administered 7 mg retinol, 42 mg β-carotene, or placebo to pregnant women on a weekly basis [47]. Both vitamin A and β-carotene significantly reduced maternal mortality during pregnancy and 12 weeks after birth. However, no such effect was seen in a similar trial conducted in rural Bangladesh, where VAD was much less prevalent and the diet included more protein and fat [48]. Infant mortality (till 12th week post partum) and the rate of stillbirth were not reduced by the weekly dose of retinol or β-carotene.

In a meta-analysis of 17 trials using vitamin A or β-carotene supplementation during pregnancy, no significant overall effect on birth weight, preterm birth, stillbirth, miscarriage, or fetal loss was found [49]. Supplementation was protective in HIV positive women against low birth weight (<2.5 kg; RR=0.79, CI=0.64, 0.99) but not for preterm delivery or small-for-gestational-age infants. However, some evidence indicated that concurrent supplementation with vitamin A (5000 IU) and β-carotene (30 mg) was associated with an increase in HIV transmission from mother to child [50], thereby warranting caution against supplementation programs in high HIV prevalence areas. These authors concluded that the evidence in favor of vitamin A/β-carotene supplementation on maternal and infant mortality was lacking.

However, the WHO [51] has suggested a vitamin A supplementation scheme in areas with severe public health problems of night blindness and infant mortality. A daily dose of 10,000 IU or 25,000 IU weekly vitamin A in the form of an oral liquid or oil-based preparation of retinyl palmitate or retinyl acetate is recommended. Supplementation should be considered for a minimum of 12 weeks during pregnancy until delivery where prevalence of night blindness is ≥5% in pregnant women and children aged 24–59 months. No recommendation for provitamin A carotenoids was given. Supplementation of provitamin A carotenoids or vitamin A during the last trimester of pregnancy may improve levels of retinol in breast milk [51], but a comprehensive study has not been performed. The initial postpartum breast secretion, colostrum, contains more carotenoids than later milk, especially in women who have lactated after earlier pregnancies [52]. The variability is great, but the carotenoids likely benefit the health of the newborn child. A multinational study of breast milk of healthy mothers revealed that the provitamin A carotenoids accounted for >50 % of milk carotenoids and were highest in Japanese mothers [53]. US mothers had the lowest concentrations of total carotenoids.

Few studies have explored the associations between carotenoids and preeclampsia in pregnant women. Significantly lower plasma β-carotene levels were associated with the subsequent development of preeclampsia in pregnant women with diabetes [54]. Both placenta and plasma from preeclamptic women were significantly lower in β-carotene and lycopene compared to normal pregnant women [55]. Since oxidative stress characterizes the pathophysiology of preeclampsia [56], it is conceivable that consumption of carotenoids during pregnancy may favor an environment that minimizes oxidative damage. In any case, since pregnancy is a period of rapid growth and development, a carotenoid-rich diet may help to sustain a pregnancy free of complications and promote a healthy outcome.

Prevention of Oxidative Stress and Inflammation

The carotenes and xanthophylls are some of nature's most efficient quenchers of singlet oxygen. Singlet oxygen arises from exposure of chromophores such as porphyrins, chlorophylls, and riboflavin to sunlight. These activated molecules can damage DNA, proteins, and lipids. Carotenoids absorb the excess energy of singlet oxygen and dissipate it as heat [57]. However, the idea that carotenoids are important classical antioxidants, which play a role in major diseases where oxidative stress and inflammation are causative factors, is less well established. Small amounts of carotenoids can be found in almost all the lipid membranes of the body. In the animal kingdom, they are often associated with specific proteins. The xanthophylls, such as lutein and zeaxanthin, having hydrophilic hydroxyl groups, orient themselves across membranes. The carotenes, such as β-carotene and lycopene, are oriented within the bilipid layers and can disturb the phospholipid structure to a small extent, allowing for greater penetration of small molecules [58]. Their location and orientation may play a role in their ability to act as classical antioxidants.

In vitro studies have shown that all the carotenoids accept the unpaired electron from a number of free radical species (sulfur-containing radicals, glutathione, nitric oxide, NO_2, –ONOO–, and peroxyl radicals), as well as superoxide. The resulting carotenoid radical must be converted back to its original state to be considered an antioxidant. The residence time of the specific carotenoid radical, oxygen partial pressure, and the availability of other antioxidants determine whether a carotenoid acts as a prooxidant or an antioxidant [59]. For example, protection of lymphocytes from nitrogen radical attack increases almost tenfold in the presence of vitamins C and E and β-carotene compared to β-carotene alone [60]. Carotenoids can also accept electrons from each other, with lycopene being the ultimate acceptor from all other carotenoids [61].

Oxidative stress and inflammation processes occur together. Several human studies have found an association between low dietary carotenoids and/or blood carotenoid levels and increased markers of oxidative stress and inflammation [62–66]. Such associations may be more related to antioxidant- and anti-inflammation-promoting dietary patterns that include fruit and vegetable consumption than to carotenoid intake alone. Cell culture and animal studies, using models of oxidative stress, have found redox-based regulation of proinflammatory pathways by lutein, β-carotene, and lycopene [67–69].

Supplementation studies have been equivocal and their outcomes may depend upon the level of oxidative stress and the antioxidant status of participants during the study. Lutein supplementation of preterm infants [70] and healthy adults [71], mixed carotenoids in postmenopausal women [72], and numerous lycopene supplementation studies [73] have found decreases in markers of oxidative stress or inflammation. The two studies that used supplement combinations of β-carotene, vitamins C and E, and selenium found no change in markers of inflammation [74, 75].

Photoprotection and Skin Health

Sunlight is an environmental hazard over the life of human skin. Not only UV-A and UV-B but also visible and infrared light are responsible for singlet oxygen and radical production, especially in the presence of natural photosensitizers such a porphyrins and riboflavin. This can result in photoaging (roughened or patchy skin, wrinkles), UV-induced erythema (sunburn), and skin cancer [76]. Given the role that carotenoids play in nature as photoprotectors, do they play the same role in human skin? A variety of carotenoids accumulate in various dermal layers of the skin. In fact, their presence in skin due to higher consumption of fruits and vegetables or carotenoid supplements can be visibly detected by ordinary observers [77, 78]. As noted before, the resulting facial skin tones are favored and are considered a sign of greater health and vigor [79, 80]. Various case reports document carotenodermia in individuals who have gone overboard in consuming carrot or tomato juice. The extreme yellow-to-orange pigmentation that concentrates in the palms of the hands disappears harmlessly with the termination of intake [81, 82]. Canthaxanthin was sold as tanning capsules in European countries in the 1980s until it was discovered that golden crystals formed in the paramacular region of the eye [83].

The first medical use of β-carotene was developed by Micheline Mathews-Roth and her collaborators. People suffering from erythropoietic protoporphyria (EPP) accumulate large quantities of protoporphyrin IX, a precursor of hemoglobin synthesis. Protoporphyrin is a photosensitizer in the visible light range, so sunscreens developed against UV light do not prevent the burning sensation and edema that occur when sufferers are exposed to even a small amount of sunlight. Mathews-Roth made the connection between carotenoid protection for chlorophyll in plants in the wavelength range of 380–560 nm and the similar structure of protoporphyrin. Clinical trials were very effective, with doses of 180 mg/day of β-carotene providing 84 % of patients a threefold greater ability to tolerate sunlight, especially for children who could now play outside. In 1975, the US Food and Drug Administration approved the use of β-carotene for the treatment of EPP [84]. The availability of β-carotene for human use motivated its employment in subsequent trials for cancer prevention.

A number of studies have evaluated dietary supplementation with β-carotene, lycopene, or mixed carotenoids, as protection from UV-radiation caused erythema. More positive results were observed for tomato extracts compared to lycopene alone. Phytoene and phytofluene (found in tomatoes) are precursors in the lycopene synthetic pathway and, unlike lycopene, absorb UV radiation. Greater efficacy was found in trials lasting ≥7 weeks with carotenoid doses ≥12 mg/day, possibly in combination with vitamin E [85, 86]. A meta-analysis of β-carotene supplementation trials found a protective effect regardless of dose, but increased in effectiveness with time ≥10 weeks. The sun protection factor (SPF) was 4 compared to sunscreens with SPFs between 10 and 40 [87]. Topical application of β-carotene in lotion may have benefits, providing protection from infrared light that has also been shown to produce radicals from distressed mitochondria [88].

Skin ages in light-exposed areas (extrinsic or photoaging) as well as covered areas (intrinsic aging). Reactive oxygen species (ROS) formation via mutations in mitochondrial DNA are important for both processes, but UVA-light exposure increases the mutation rate by 40 %. UV light accounts for 80 % of facial skin aging, with visible wrinkles and rough skin explained by the breakdown of collagen, degradation of elastin fibers, and 50 % slower renewal of the epidermis by the age of 80 years [89]. There is a general impression that a nonsmoking, healthy lifestyle leads to a more youthful appearance, but only a few studies have explored the association of carotenoid status to fewer wrinkles. A cross-sectional study of people >70 years in a range of countries found that back-of-the-hand skin wrinkling was negatively associated with higher intakes of vegetables, olive oil, and legumes, with healthy diet explaining 34 % of the variance [90]. Forehead skin roughness (an early stage of wrinkling) was associated with dermal lycopene ($R^2=0.7$) concentration (measured by resonance Raman spectroscopy) but not with aging in a small study with a narrow age range (40–56 years) [91]. Photoaging nude mice models have shown Anti-wrinkling effects for β-carotene and lutein + zeaxanthin. A variety of carotenoids are associated with a lower incidence of skin cancer [92].

Vision and Diseases of the Eye

Aside from the conversion of some of the carotenoids to vitamin A compounds, the strongest evidence of the human need for carotenoids comes from the study of the eye and its diseases associated with aging. The human macula lutea, an indented area located roughly in the center of the retina, occupying a diameter of 5–6 mm and accounting for the central 15°–16° of vision, accumulates lutein and zeaxanthin at roughly 1000-fold the concentrations found in plasma [93]. The result is a visibly yellow hue referred to as macular pigment. A binding or transport protein that might explain this accumulation has not yet been unequivocally identified. These xanthophylls are organized with zeaxanthin predominating in the central region to about 2.5 mm from the center, with lutein exceeding zeaxanthin 2:1 at the periphery, and *meso*-zeaxanthin (thought to be an intermediate product) appearing throughout [94]. The relative location of lutein to zeaxanthin in the macula varies greatly from person to person with some people having concentric rings of greater pigmentation [95]. Lutein and zeaxanthin are localized in the Henle fiber layer that covers the photoreceptors, so light must be filtered through these pigments. An arrangement perpendicular to the Henle fibers is consistent with maximal absorption of blue-wavelength light, but their specific organization (protein-bound or not) is not known. The orientation of zeaxanthin's β-ionone rings provides a more orderly arrangement compared to lutein [96].

The obvious function of macular pigment is to filter blue light (the most energetic in the light spectrum), thus protecting the photoreceptors, retinal pigment epithelium (RPE), and the underlying choriocapillaries during 70+ years of light exposure. The reduction in blue light intensity can range between 40 and 90 %, which is sufficient to account for the observed reduction in risk for age-related macular degeneration (AMD) in some epidemiological studies [96]. In addition, a role as *in situ* antioxidants has been proposed. Zeaxanthin and, to a lesser extent, lutein are effective quenchers of singlet oxygen [$^1O_2$] coming from UV exposure [97, 98]. This quenching is a physical process with the excess energy dissipated as heat and does not destroy the xanthophylls; so macular pigment levels are maintained long after xanthophyll intake has ceased [99]. They may also act as classical antioxidants because they are present in the perifoveal and peripheral regions of the retina that are exposed to high oxygen tensions and have high rates of metabolism. Several carotenoid derivatives are found in the retina including meso-zeaxanthin, (3 S,3′S)-zeaxanthin, and epilutein that may have been formed by redox reactions [96].

Whether there is an association between lutein and zeaxanthin status and the risk of AMD would seem to be a logical question. Late-stage AMD is the leading cause of legal blindness in people >65 years old in industrialized countries, affecting more than 10 million people in the USA and about 200,000 in the UK [100]. Non-Hispanic black persons have a lower prevalence [101]. Early-stage AMD starts in the macula with soft drusen (yellowish deposits containing no xanthophylls) and other pigmentation abnormalities in the RPE. It continues on to later stages characterized by atrophy of the photoreceptors and the RPE (dry AMD), often progressing further to choroidal neovascularization, retinal hemorrhage (wet AMD), detachment of the RPE, and retinal scarring [102]. The nature of this progressive disease complicates the interpretation of epidemiological studies and clinical trials. A recent meta-analysis of six longitudinal cohort studies found that the pooled relative risk (RR) of early- stage AMD was not related to baseline dietary intake of lutein and zeaxanthin, but late-stage AMD (RR = 0.74) and neovascular AMD (RR = 0.68) were risks for those in the lowest intake category [100].

Since dietary lutein and zeaxanthin are related to fruit, vegetable, and egg intake, does lutein or zeaxanthin supplementation increase macular pigment optical density (MPOD) in people with healthy sight and those suffering from various stages of AMD? Is increased MPOD related to visual function and the development or reversal of AMD? Lutein, lutein esters, zeaxanthin plus lutein, or these xanthophylls plus various antioxidant mixes do increase MPOD in college students [103], older men with healthy sight [104, 105], those with early AMD [106], and men with mild to moderately advanced AMD [107, 108]. There appeared to be few differences between these formulations (10–20 mg/day) in their effect on MPOD over the 6–12 month period in these studies. Furthermore, the ring structures of MPOD seen in some individuals, and thought to be protective, were neither attenuated nor generated de novo with xanthophylls supplementation [95]. In addition, those with lower baseline MPOD tended to have the greater increase in MPOD with xanthophyll supplementation. Lutein or zeaxanthin supplementation also improved visual acuity, foveal shape discrimination, subjective glare recovery, and contrast sensitivity. These improvements were correlated with the increases seen in MPOD in several of the studies cited above. A small study, using 6 mg/day of lutein plus a vitamin/mineral mixture for 9 or 18 months, failed to find improvements in the visual performance parameters listed above [105].

The Age-Related Eye Disease Study 1 (AREDS1), a national study using a daily supplement containing vitamins C and E, β-carotene (15 mg/d), and the minerals zinc and copper, reduced the 5-year risk of developing advanced AMD by 25% in eyes with intermediate AMD, but had no effect on the development or risk of cataract [109, 110]. A recent Cochrane review evaluating four high-quality randomized placebo-controlled studies that included 62,520 people, using vitamin E and β-carotene in their supplement formulations, failed to prevent the onset of AMD [111]. The AREDS1 study was included in this review. At the time these studies commenced, neither lutein nor zeaxanthin were yet available.

The goal of AREDS2 was to determine whether lutein (10 mg/d) plus zeaxanthin (2 mg/d), or the omega-3 long-chain fatty acids (eicosapentaenoic acid (EPA) 650 mg/d plus docosahexanoic acid (DHA), 350 mg/d) in addition to the original AREDS1 formulation, could reduce the 5-year risk of progression to advanced AMD in those with intermediate AMD [112]. More than 4000 men and women were randomized to one of four treatments. A secondary randomization of the underlying AREDS formulation was made to answer the concern for the safety of β-carotene or high-dose zinc supplementation. There were no differences in the progression to advanced AMD for any of the primary treatments, although lutein+zeaxanthin appeared to have some efficacy for the reduction of neovascular AMD (wet AMD; HR=0.89, $P=0.05$) compared to geographic atrophy (dry AMD). Further subgroup analysis showed that those in the lowest quintile of lutein+zeaxanthin intake (0.1–1.4 mg/d) had a lower risk of AMD progression (HR=0.74, $P=0.01$) when supplemented compared to low-intake groups not receiving the lutein+zeaxanthin supplement, whereas there was no difference among the higher quintiles of intake. Smokers and recent quitters had a higher risk of lung cancer with β-carotene supplementation and there was evidence for competition for absorption between β-carotene and lutein + zeaxanthin. The authors were hesitant to make any conclusions based on these subgroup analyses but suggested that substitution of lutein + zeaxanthin in the AREDS1 formulation might be a safer choice for smokers.

Due to the macular location of lutein and zeaxanthin, and given their unique optical properties, they may enhance visual performance for everyone. Abundant macular pigment may (1) enhance visual acuity by reducing chromic aberration (like filters on a camera), (2) reduce visual discomfort by attenuating glare and dazzle, and (3) facilitate enhancement of details and visual contrast by the absorption of "blue haze," (e.g., how mountains appear blue in the distance) [113, 114]. Studies enrolling younger, normal-sighted individuals have found positive relationships between macular pigment and contrast sensitivity, visual acuity, glare reduction, photo-stress recovery, and time discrimination of changing light [115–118]. Clinical trials using lutein or zeaxanthin supplements (10–20 mg/day) for 6–12 months found improved visual acuity, contrast sensitivity, glare attenuation, and light/dark adaptation [119–122]. Another supplementation study (6 mg/day lutein plus vitamins and minerals for 9 months) found no effect on visual performance parameters [123]. All of these studies had small numbers of subjects and none found a correlation between macular pigment density and improved visual function. Since a number of subjects may have already achieved optimal macular pigment before supplementation, any association with visual performance may have been attenuated.

The lens is clear, and along with the cornea, focuses a sharp image onto the retina. It is composed of tightly ordered fibrous cells that lose their nuclei, so their major proteins cannot be regenerated. Over our lifetime, new cells accumulate at the outer surface of the lens like the layers of an onion. The central tightly packed denucleated cells are subject to the so-called nuclear cataract formation. Cortical cataracts are formed in outer cells laid down after birth. The prevalence of age-related cataract varies widely throughout the world in people >70 years old, with cortical cataract being more prevalent in some populations (30–45 %) and nuclear cataract more prevalent in others (10–80 %). Cataracts are observed as opacities of the lens that range from white to yellow to dark brown. The crystalline proteins (especially β) are chopped and cross-linked, forming large aggregates. This process is accompanied by the loss of reduced glutathione (the major antioxidant in the lens) and the enzyme, glutathione reductase, that regenerates it. There is also a loss of sulfhydryl groups in sulfur-containing amino acids of the lens proteins. Therefore, oxidative stress is thought to be a major factor in cataract formation [124, 125]. Do dietary antioxidants such as vitamins C and E and the carotenoids reduce the risk of cataract? The human lens contains substantial amounts of lutein plus zeaxanthin and vitamins A and E, but no hydrocarbon carotenoids, such as β-carotene and lycopene. Lutein and zeaxanthin are more concentrated in the cortical regions of the lens [126] and can be substantially increased throughout the lens with long-term supplementation [127]. A number of large population studies have found reduced risk for nuclear cataract (less so for cortical cataract) with higher intakes of lutein and zeaxanthin or higher plasma levels [128]. These associative studies may also be related to other substances in fruits and vegetables, and other lifestyle factors associated with healthy diets. For example, the Healthy Eating Index score was the strongest modifiable predictor for low prevalence of nuclear cataract (reduced by 37%) in the Women's Health Initiative Observational Study [129]. Furthermore, a number of population studies and clinical trials have found significant risk reductions for nuclear cataract with long-term use of Centrum and other multivitamin–mineral supplements [130]. Centrum has included small amounts of lutein (0.25 mg/day) for several years. Xanthophyll supplementation studies have focused on slowing visual decline in those with developing cataract, and visual performance was improved in two trials [131, 132]. Despite a plausible physiological rationale and largely consistent circumstantial evidence, the lack of xanthophyll-based, placebo-controlled clinical trials for the prevention of cataract or macular degeneration prompted the US Food and Drug Administration to determine that no credible evidence existed for a health claim about the intake of lutein or zeaxanthin and the risk of AMD or cataracts in 2006 [133].

Retinitis pigmentosa is an inherited disease affecting 1.5 million people worldwide. It starts with night vision problems in adolescence, progressing through the loss of peripheral vision, with final loss of central vision after the age of 60 years. The cause is the progressive loss of rod and cone photoreceptors [134]. The authors of a major 4-year clinical trial of 12 mg/day of lutein plus vitamin A (15,000 IU/day) versus vitamin A alone, in 225 nonsmoking retinitis pigmentosa patients, concluded that the lutein supplement slowed the loss of midperipheral visual field, but only in those achieving the highest serum level of lutein or those with the highest MPOD as a result of treatment. There was no difference in the primary endpoints [135]. This invoked a letter from their data safety-monitoring committee indicating that despite sound study design, the committee disapproved the use of subgroup analysis as the basis of the authors' main conclusion [136]. Two small clinical trials have also found positive results using lutein supplementation [137, 138], but one 6-month study (20 mg lutein/day) found that only 50% of their participants had any increase in MPOD resulting in no detectable change in central vision function during the intervention [139]. Whether lutein supplementation is beneficial for those suffering from retinitis pigmentosa is likely to be complicated by variable accumulation of lutein in the retina. MPOD increase with lutein supplementation was inversely proportional to serum total cholesterol levels, and MPOD became higher in those with brown irises and those who had retained more photoreceptors [140].

The possible anti-inflammatory, antiapoptotic effect of lutein on other retinal structures is a new area of inquiry, especially in acute retinal ischemic/reperfusions that are experienced during operations for retinal reattachment and arterial blockades. Preliminary studies on animals hold some promise [141–143].

Cognitive Decline and Alzheimer's Disease

More than 16 carotenoids have been found in human brain tissue, with the xanthophylls (lutein and zeaxanthin) accounting for 66–77%. Their distribution is not homogeneous, with the frontal cortex being particularly rich [144]. Tissue concentrations in some sections of the brain appear to be correlated with macular pigment density. A study of xanthophyll concentrations in postmortem brain tissue of subjects enrolled in the Georgia Centenarian study found positive correlations with better age-adjusted performance on various cognitive function tests, especially retention, with higher tissue levels [145].

Cognitive impairment refers to the subclinical complaint concerning memory functioning in the elderly and is so common that it has come to be thought of as an inevitable feature of the aging process. Cognitive decline is evaluated in longitudinal studies by changes in performance over time, in one or two domains of a series of cognitive tests (memory, orientation, language, executive function, or praxis). Using functional tests, cognitive impairment is defined as: (1) at least one standard deviation below the mean for young adults on one or more tests or (2) greater-than-expected decline in score for a person's age and education level. Impairment is considered a risk factor for the development of Alzheimer's disease (AD) [146].

AD is a degenerative disorder of the brain causing memory loss, progressing to the inability to perform activities of daily living, followed by death. It accounts for 60–80 % of cases of dementia and its current US prevalence is estimated at 5.1 million. Mild cognitive impairment occurs at much higher rates. This disability constitutes a large share of the ballooning health-care costs. Postmortem, it is characterized by senile amyloid-β protein plaques and neurofibrillary tangles composed of phosphorylated tau proteins that first accumulate in cortical tissue. The most vulnerable cortical tissue is that which was laid down during the later stages of fetal and newborn development. It then spreads to the hippocampus. This pathology can be found also in those with mild cognitive impairment, which makes diagnosis difficult in prevention and treatment studies. Furthermore, coexistent Parkinson disease and the evidence of infarcts or hemorrhage have been found in 20–25 % of AD brains. Aging, genetics (ApoE4—a genetic variant of a cholesterol reverse transport protein) and oxidative stress (tobacco use, diabetes) have been identified as causative factors. Progress is being made with imaging techniques such as magnetic resonance imaging (MRI) and position emission tomography (PET) correlated with postmortem pathology, so that more precise endpoints can be identified for future clinical trials [147–149].

A number of longitudinal studies of middle-aged populations have explored the relationship between plasma or dietary levels of β-carotene and cognitive decline. These studies have been mixed, but are largely negative. A number of clinical trials, designed to evaluate supplements containing vitamins C and E and β-carotene for heart disease and/or cancer endpoints, have added tests for cognitive decline. The results for all cognitive endpoints have been disappointing. A systematic review of both study types found no evidence that any of these antioxidants prevented cognitive decline in later life [150]. This systematic review was used as the basis for a National Institutes of Health (NIH) State-of-the-Science Conference Statement that explored modifiable factors that might reduce the risk of AD and cognitive decline in older adults. The major conclusion was that there was currently insufficient evidence to identify any modifiable factors that might reduce risk for either. This was due to the limitations of studies currently available [151, 152], including short follow-up time, starting interventions too late in the disease process, and lack of baseline measurements in the clinical trials allowing no subgroup analysis of those subjects who had low baseline intakes (a group more likely to benefit from intervention) [153]. None of these studies evaluated the xanthophylls, which have a more stabilizing effect on neuronal membranes.

A small preliminary study, supplementing DHA (an omega-3 fatty acid) or lutein (12 mg/day) for only 4 months to mentally unimpaired elderly women found improvements in memory scores and rates of learning with either or both supplements [154]. Cerebral ischemia with reperfusion (I/R) is often used in animal studies to simulate the effects of a stroke. Lutein supplementation, given to mice shortly before or subsequent to reperfusion, produced better survival rates, better neurological scores, and smaller areas of infarct [155]. Lutein also appears to protect neurons in models of retinal I/R [156].

It would be unfortunate to discard the possibility that the long-term intake of a fruit and vegetable-rich diet, that includes the xanthophylls and other carotenoids, might slow the progression of age-related cognitive decline before there are sufficient data.

Cancer Prevention and Treatment

Cancer is an umbrella term encompassing a number of tissue diseases and a variety of mechanistic antecedents. The retinoids have potent anticancer effects against a wide variety of experimental cancers, but their toxic side effects may be inseparable from their mode of action [157]. When the first population studies were published, showing a reduced risk for cancer with higher blood levels of retinol and β-carotene [158, 159], the search was on to find associations with other cancers in various populations. Some of these studies found risk reductions for greater intakes or higher blood levels of β-carotene or lycopene and occasionally the xanthophylls, but many did not [160]. Such studies may rather indicate beneficial dietary patterns (including greater intake of fruits and vegetables) and lifestyles associated with such patterns [161–163]. With high hopes, several clinical trials supplemented their subjects with large doses of β-carotene, combined with vitamins A or E, and followed cancer incidence. Surprisingly, two studies found an 18–24 % increase in lung cancer and 8–17 % increase in mortality over a 4–8 year follow-up, especially in those who received only β-carotene [α-tocopherol, β-carotene (ATBC) study] [164, 165]. A follow-up to one of these (ATBC study) found that the excess risk for any cancer, including lung cancer, was no longer evident 4–6 years after the ending of the intervention [166]. The early excess risk due to β-carotene supplementation in smokers and asbestos workers in these trials was widely hypothesized to be due to components of cigarette smoke in the presence of relatively high oxygen partial pressure in the lung [167]. The other clinical trials using β-carotene found neither reduced nor increased risk for any type of cancer [168, 169]. A study using a population of vitamin- and mineral-deficient Chinese men and women found that a mixture of vitamin E, β-carotene, and selenium decreased total cancers by 13 % and deaths by 9 % during a 5-year follow-up [170].

A search for mechanism ensued with numerous cell culture studies using cancer cell lines and animal cancer models. Carotenoids, especially β-carotene or lycopene, were found to exhibit functions consistent with blocking the development or growth of various cancers through the following mechanisms: (1) modulation of nuclear receptor superfamilies, (2) decreasing angiogenesis, (3) increasing apoptosis, (4) restoring gap junction communication, (5) prooxidant and antioxidant effects, especially protection from DNA damage, (6) inhibition of cell proliferation, and (7) modulation of phase I and phase II enzymes [160, 171, 172]. Cell culture experiments often used concentrations that were higher than physiologically feasible, with long incubations at high temperatures and oxygen partial pressures that could produce oxidation products. Indeed, under these conditions, oxidation products have been identified for lycopene, but not for β-carotene [173]. Oxidation products or metabolites of lycopene have also been identified in human plasma and tissue. These lycopenoids, mostly lycopenals, have similar structures to 9-*cis* retinoic acid (a powerful ligand for several nuclear receptors) and are present in comparable concentrations in human samples [174]. Beside β-carotene, lycopene is a substrate for the eccentric cleavage enzyme, BCDO2, which cleaves the 5-*cis* and 13-*cis* isomers for lycopene, but not the all-*trans* isomer. Lycopene *cis* isomers are prevalent in circulation after ingestion of tomato products or lycopene supplements, and the BCDO2-mediated conversion to apo-10′-lycopenal and then to apo-10′-lycopenoic acid which may be a substitute ligand for 9-*cis* retinoic acid. This hypothesis was investigated using the major retinoid receptors with only modest activity [173]. However, the lycopene degradation products have been shown to have a variety of anticarcinogenic activities and are often more potent than lycopene. Their actions include regulation of the cell cycle, apoptosis of cancer cells, and the induction of two important systems. The first is the electrophile response element/antioxidant response element (EpRE/ARE) system that mediates the induction of detoxifying and antioxidant enzymes responsible for inhibiting the mutagenic effects of carcinogens and oxygen radicals. The second is the nuclear factor kappa B (NF-κB) system which is responsible for the normal functioning of the immune system, but is crucial for the deleterious inflammatory response that often is associated with cancer risk [173, 174].

Several epidemiological studies have identified risk reduction with lycopene exposure while finding no other carotenoid associations. The data are especially strong for cancers of the prostate [175]. A systematic review of eight clinical trials using 15–30 mg/day lycopene supplements found reductions in serum prostate-specific antigen (PSA), but only a modest reduction in benign prostate hyperplasia (BPH), which was not statistically significant [176].

Cancer patients may be tempted to add various carotenoids as dietary supplements to their standard care which may include chemotherapy. As chemotherapeutic agents generate oxidative stress, there has been concern that antioxidant therapy may blunt the effects of these agents. However, numerous studies have demonstrated that antioxidants do not inhibit but actually enhance the cytotoxic effect of antineoplastic drugs on cancer cells. Several β-carotene and lycopene cell culture and animals studies using various agents have found an enhancement of their effects [177, 178]. A systematic review of 19 trials of antioxidant plus chemotherapy (only one of which had β-carotene as part of a vitamin mixture) found no detrimental effects, while several reported increased survival times and tumor responses as well as fewer toxicities compared to controls [179]. In a small trial of 50 patients with high-grade gliomas treated with radiation and paclitaxel, there was a modestly improved response in the lycopene-treated group [180].

The early promise of a variety carotenoids, as benign agents, for lowering the risk of various cancers is somewhat tarnished, but the intriguing biological activities of these carotenoids and their metabolites raise hope that they may yet be found to be useful adjuncts in the disruption of various carcinogenic processes.

Metabolic Syndrome, Obesity, Cardiovascular Disease, and Diabetes

Metabolic syndrome (MetS) is a clustering of many conditions including abdominal obesity, high blood pressure (HBP), hyperglycemia, elevated fasting triglycerides, and low levels of HDL cholesterol [181]. These metabolic abnormalities increase the risk for cardiovascular disease and diabetes mellitus, as also does obesity which is correlated with cardiovascular mortality through hypothesized instigating factors, such as an increase in chronic inflammation and oxidative stress [182]. Over the last decade, a consistent rise in the rates of obesity and MetS has been observed in the USA. As reported by the Centers for Disease Control and Prevention (CDC), 35.7 % of adults are obese, and it is expected that by 2030 more than half of US adults will be obese. More disconcerting is the increase in childhood obesity (16.9 %) and, consequently, of preventable diseases in this age group.

Carotenoids, by virtue of their antioxidant properties, may help to prevent the progression of chronic conditions related to obesity and MetS, and thereby decrease related morbidity and mortality rates. Many previous observational studies have indicated an inverse association between serum/dietary carotenoid levels and MetS [183–186]. In a recent analysis of NHANES cross-sectional data, serum carotenoid concentrations were found to be consistently lower in US adults with MetS compared to those without MetS (0.057–0.863 vs. 1.62–10.114 μmol/L total carotenoids), with an inverse association observed with all MetS components [187]. Furthermore, adolescents with MetS also had lower serum carotenoid concentrations compared to their counterparts without MetS that were inversely related to the inflammatory C-reactive protein (CRP) in their serum [188]. In middle-aged and elderly men, higher dietary intakes of total carotenoids, β-carotene, α-carotene, and lycopene were associated with lower waist circumference and less visceral and subcutaneous fat, while lycopene intake alone was related to lower serum triglyceride concentrations [186]. Despite the associations observed in these studies, long-term supplementation with antioxidants including a combination of vitamins C and

E, β-carotene, zinc, and selenium did not prevent the incidence of MetS in adults free of the condition at baseline, although baseline serum β-carotene concentrations were negatively associated with the risk of MetS [189]. Due to the scarcity of similar randomized clinical trials, it is still unclear whether long-term carotenoid supplementation would prevent progression to MetS in susceptible individuals. Short-term supplementation studies in obese and overweight individuals have presented mixed results. Four weeks of supplementation with 30 mg/day lycopene in the form of tomato-derived Lyc-O-Mato did not affect the markers of inflammation and oxidative products in severely obese individuals [190]. However, in moderately overweight middle-aged individuals, 70 mg lycopene/week enhanced HDL functionality [191], as determined by increases in the activity of the antioxidant enzyme paraoxonase-1 and decreases in the cholesteryl ester transfer protein (CETP). Tomato juice consumption in overweight and obese females was also found to reduce systemic inflammation through decreases in inflammatory biomarkers such as interleukin (IL)-6, IL-8, and tumor necrosis factor (TNF)-α [192]. Most of these investigations have examined the effect on obesity-induced inflammation, since this is considered the therapeutic strategy to retard the progression to cardiovascular disease and diabetes mellitus, and among the carotenoids, lycopene has consistently demonstrated anti-inflammatory properties in humans, as well as in animal and cell culture models [193–195].

Low levels of carotenoids found in MetS and obesity possibly point to an unbalanced diet, leading to a deficiency in these micronutrients and/or an increased requirement due to a heightened state of oxidative stress. It is of interest that intake of a low caloric diet that provided more than adequate micronutrients [exceeding the dietary reference intakes (DRI) requirements] was unable to correct deficiencies of many micronutrients, including lycopene, in obese subjects undergoing a 3-month weight loss regimen, suggesting a larger need for antioxidants in obese individuals, especially during weight loss-induced stress [196].

Associations between carotenoids and cardiovascular disease (CVD) have been examined in epidemiological studies through the measurement of dietary intake, serum, and adipose tissue levels. Some early prospective studies showed that higher β-carotene status was related to a reduced risk of myocardial infarct and a modest decrease in the risk of stroke in men and women [197]. In a recent study, a twofold increase in the risk of sudden cardiac death was observed in men with lower serum β-carotene levels [198]. Many recent studies have focused on lycopene as a primary player in the prevention of CVD, likely due to its potency as a singlet oxygen quencher. Increased serum concentrations of lycopene were associated with decreased arterial stiffness and lower oxidized low-density lipoprotein (LDL) in healthy women [199]. When supplemented with 15 mg/d of lycopene for 8 weeks, healthy men showed a significant improvement in endothelial function with decreases in CRP and systolic blood pressure and an increase in the antioxidant enzyme superoxide dismutase [200]. A prospective study of Finnish men at risk for ischemic heart disease found that men in the highest quartile of serum lycopene had a 59 % lower risk of ischemic stroke and a 55 % lower risk of all strokes [201]. Moreover, in patients with heart failure, higher lycopene intake was associated with a significantly longer cardiac event-free survival, compared to those with lower lycopene intake [202]. These findings from human studies are corroborated by experimental studies using animal models, which suggest that lycopene's antiatherogenic effects occur through the reduction of atherosclerotic plaques, decreases in serum total cholesterol and LDL cholesterol, and increases in HDL cholesterol [203]. In elderly Finnish men, the concentrations of plasma β-cryptoxanthin, lycopene, and α-carotene were found to decrease linearly with increasing intima-media thickness of the common carotid artery [204]. Serum levels of lutein and zeaxanthin were also found to be lower in early atherosclerotic patients compared to healthy subjects [205].

Oxidative stress may contribute to the etiology of diabetes by inducing insulin resistance in the peripheral tissues and impairing secretion from the pancreatic cells [206], and the antioxidant capacity of carotenoids may protect against the development or progression of the disease. High levels of commonly measured metabolic parameters such as fasting blood glucose, 2-h glucose level in the glucose tolerance test, and glycosylated hemoglobin were inversely correlated with serum or dietary carotenoid levels in cross-sectional studies [207–209], but few prospective studies have investigated this association. It has been reported that higher dietary intake of β-cryptoxanthin, but not other carotenoids, significantly reduced the risk of type 2 diabetes [210]. Long-term β-carotene supplementation in a randomized controlled trial did not affect the incidence of type 2 diabetes [211]. Baseline plasma carotenoid concentrations failed to show an association with the risk of type 2 diabetes [212].

Since cardiovascular disease and diabetes mellitus are characterized by high levels of oxidative stress, it would be natural to assume that an increased consumption of antioxidants such as carotenoids through fruits, vegetables, and other sources in these chronic conditions would decrease further oxidative stress-related deterioration. Whether regular intake of carotenoids would prevent onset of disease is unclear from available evidence.

Toxicity of Carotenoids

Large doses of preformed vitamin A may cause acute poisoning, but high intake of provitamin A carotenoids does not result in hypervitaminosis. In patients with erythropoietic protoporphyria, therapeutic doses up to 150 mg β-carotene per day greatly increased tolerance to sunlight and did not have deleterious effects [213]. Similarly, the use of two richest sources of β-carotene, RPO [214] and *gac* fruit (*Momordica cochinchinensis*) [215] in cooking is not associated with vitamin A toxicity.

However, high doses of supplemental β-carotene are not advisable, especially in people under unusual oxidative stress. The previously noted randomized intervention trial of smokers in Finland found significantly increased incidence of lung cancer [216] on a daily dose of 20 mg β-carotene for 5–8 years. Similar results were found in the US asbestos workers and former smokers [165] who received 30 mg β-carotene per day for ~4 years. However, the US trial participants also received 25,000 IU retinyl palmitate, which is more than eightfold the recommended daily allowance (RDA) for men and 10.5-fold RDA for women. The excess risks were restricted primarily to women and former smokers. Another randomized trial followed patients after removal of colorectal adenoma who received placebo or 25 mg β-carotene/day for 4 years [217]. The risk of recurrent adenoma was doubled by taking β-carotene supplements, but only for those who smoked and also drank more than one alcoholic drink per day. The nonsmokers and nondrinkers actually had their risk of adenoma markedly decreased by β-carotene supplements [218].

It has been long suspected that carotenoids may have prooxidant properties when present in high concentration or at high oxygen pressure [219]. Medical literature has noted a few cases of very high concentration of carotenoids in human tissues. Habitual and very excessive intake of tomato products caused lycopene deposits in the liver, accompanied by liver enlargement and abdominal pain [220]. Golden yellow crystals were observed in the retina of people taking canthaxanthin "tanning pills," as well as in macaque monkeys treated with the same supplement [221]. The pills were banned, although they did not seem to damage the vision of the subjects.

Smoke exposure with concurrent β-carotene supplementation was studied in ferrets, a useful animal model for carotenoid absorption. High amounts of administered β-carotene, equivalent to 30 mg/day in humans, caused precancerous lung lesions in the smoke-exposed animals [222]. The lung extracts from these ferrets had enhanced β-carotene breakdown into apocarotenals *in vitro*. The ferret lung and other tissues express BCDO2 enzyme, which cleaves β-carotene to β-apo-10′-carotenal, while other β-apocarotenals may be formed in enzymatic or nonenzymatic reactions, especially at high oxygen pressures and the free radical-rich environment of smoker lungs. Recently, all possible β-apocarotenoids were investigated for their biological activity on retinoid receptors [17] using molecular modeling assays. One of them, β-apo-13-carotenone, was found to be an antagonist of the 9-*cis*-retinoic acid activation of retinoid X receptor (RXRα), as well as of three all-*trans*-retinoic acid receptors (RARα, RARβ, RARγ). Considering the importance of these nuclear hormone receptors in cell differentiation, the excessive production of antagonists may help to explain the negative effects of supplemental β-carotene trials.

Recent studies with human cell cultures (liver carcinoma HepG2) and BCDO2-deficient mice found that an excess accumulation of various carotenoids (β-carotene, lutein, zeaxanthin, lycopene) in mitochondria may produce oxidative stress [16]. It is hypothesized that gene polymorphism in the carotenoid splitting genes BCMO1 and BCDO2 may alter carotenoid and vitamin A homeostasis in some individuals and make them more susceptible to toxic effects of excess carotenoid supplementation.

Summary

Plants have evolved a vast number of carotenoids as essential compounds for their development and survival. Animals and humans, as plant consumers, have also evolved to take advantage of the unique properties of carotenoids. The mechanisms for the absorption and metabolism of carotenoids are well established. Vitamin A toxicity is efficiently averted. However, the great variability in these processes in human omnivores, whether they adapted as predominate plant eaters or hunter gatherers, raises the likelihood that carotenoids have no obligate dietary requirement but rather act as dietary enhancements. Human carnivores survive quite well on the preformed vitamin A obtained from animal flesh and dairy products. On the other hand, vegans must pay attention to the daily consumption of provitamin A carotenoids (rich sources, like carrots, sweet potato, winter squash, pumpkin, or RPO) in order to meet their vitamin A requirement [15]. Increased fruit and vegetable consumption along with carotenoid-enhanced foods such as Golden Rice is an important strategy for the elimination of the devastating effects of VAD that still is too prevalent in several countries.

The accumulation of the xanthophylls in the macula of the eye points to the utility of these carotenoids as blue light filters and may be the strongest evidence of a specific human function. To support eye health, the AREDS1 formulation, with the substitution of 10 mg/d of lutein and 2 mg/d of zeaxanthin for β-carotene (especially for smokers), may be our best interim estimate for beneficial effects [112]. The inclusion of a one half cup serving of a cooked dark-green leafy vegetable and/or egg yolk per day would be necessary to provide a similar dose of lutein + zeaxanthin from foods sources [7]. The accumulation of carotenoids in the corpus luteum and skin presents an intriguing possibility of their importance in fertility and sexual attraction, as it was found in many animals. The presence of xanthophylls in specific sections of the human brain should encourage further research into their role in protection from age-related cognitive decline and AD.

Carotenoids accumulate in lipid droplets and the lipids of various cell membranes and likely occupy a different niche than other exogenous antioxidants, such as vitamins C and E, or the endogenous antioxidants, such as glutathione and the antioxidant enzymes. Therefore, they can act as players among the cast of actors against oxidative stress and inflammation. Their ability to act as prooxidants in situations of high oxygen partial pressure and oxidative stress that was evidenced in cancer trials among smokers and asbestos workers argues against high levels of supplementation [167]. The array of population studies evaluating various cancers, heart disease, and diabetes, that have found reduced risk with higher intakes of dietary carotenoids, may be more a product of health-promoting dietary patterns, including bountiful intakes of carotenoid-containing fruits and vegetables.

In one metabolic abnormality, erythropoietic protoporphyria, massive doses of β-carotene far beyond what could be provided by diet alone were found to have great clinical benefit [84]. Other metabolic disorders, where fat absorption is compromised, such as cystic fibrosis, jeopardize the absorption of all fat-soluble vitamins including the carotenoids [223]. There is an indication that obesity and other circumstances that produce mild oxidative stress and inflammation may benefit from the increased consumption of carotenoids, since many of these conditions are accompanied by lower plasma levels of various carotenoids. Given that few individuals in the US population meet dietary advice for the consumption of fruits and vegetables, many of these conditions could be ameliorated by actually consuming at least five servings for fruits and vegetables per day.

The carotenoids are remarkable pigments. They color our foods and provide pleasure to our eating experience. They contribute to our vitamin A requirement and the reduction of oxidative stress. They may even contribute to seeing those pigmented foods more clearly, even if it is no longer in the bush but rather in the supermarket.

References

1. Bendich A (1993) Biological functions of carotenoids. In: Canfield LM et al (eds) Carotenoids in human health. New York Academy of Sciences, New York, pp 61–67
2. Davies BH (1991) Carotenoid metabolism as a preparation for function. Pure Appl Chem 63:131–140
3. Slifka KA, Bowen PE, Stacewicz-Sapuntzakis M, Crissey SD (1999) A survey of serum and dietary carotenoids in captive wild animals. J Nutr 129:380–390
4. Korytko PJ, Rodvold KA, Crowell JA, Stacewicz-Sapuntzakis M, Diwadkar-Navsariwala V, Bowen PE, Schalch W, Levine BS (2003) Pharmacokinetics and tissue distribution of orally administered lycopene in male dogs. J Nutr 133:2788–2792
5. McGraw KJ (2006) Mechanics of carotenoid-based coloration. In: Hill GE, McGraw KJ (eds) Bird coloration, vol 1, Mechanisms and measurements. Harvard University Press, Cambridge, pp 177–242
6. von Lintig J (2012) Provitamin A metabolism and functions in mammalian biology. Am J Clin Nutr 96(suppl):1234–1244
7. Holden JM, Eldridge AL, Beecher GR, Buzzard I, Bhagwat S, Davis CS, Douglass LW, Gebhardt S, Haytowitz D, Schakel S (1999) Carotenoid content of U.S. Foods: an update of database. J Food Comp Anal 12:169–196
8. Brown MJ, Ferruzzi MG, Nguyen ML, Cooper DA, Eldridge AL, Schwartz SJ, White WS (2004) Carotenoid bioavailability is higher from salads ingested with full-fat than with fat-reduced salad dressing as measured by electrochemical detection. Am J Clin Nutr 80:396–403
9. Block G (1994) Nutrient sources of provitamin A carotenoids in the American diet. Am J Epidemiol 139:290–293
10. Murphy MM, Barraj LM, Herman D, Bi X, Cheatham R, Randolph RK (2012) Phytonutrient intake by adults in the United States in relation to fruit and vegetable consumption. J Acad Nutr Diet 122:222–229
11. Kopsell DA, Kopsell DE (2006) Accumulation and bioavailability of dietary carotenoids in vegetable crops. Trends Plant Sci 11:499–507
12. Haskell MJ (2012) The challenge to reach nutritional adequacy for vitamin A: β-carotene bioavailability and conversion-evidence in humans. Am J Clin Nutr 96(suppl):1193S–203S
13. Lindqvist A, Sharvill J, Sharvill DE, Andersson S (2007) Loss-of-function mutation in carotenoid 15,15′-monooxygenase identified in a patient with hypercarotenemia and hypovitaminosis A. J Nutr 137:2346–2350
14. Tang G (2012) Techniques for measuring vitamin A activity from β-carotene. Am J Clin Nutr 96(suppl):1185S–1188S
15. Food and Nutrition Board, Institute of Medicine (2001) Dietary reference intakes for vitamin A, vitamin K, arsenic, boron, chromium, copper iodine, iron, manganese, molybdenum, nickel, silicon, vanadium, and zinc. National Academy Press, Washington, DC, pp 82–161
16. Amengual J, Lobo GP, Golczak M, Li HNM, Klimova T, Hoppel CL, Wyss A, Palczewski K, von Lintig J (2011) A mitochondrial enzyme degrades carotenoids and protects against oxidative stress. FASEB J 25: 948–959
17. Harrison EH, dela Sena C, Eroglu A, Fleshman MK (2012) The formation, occurrence, and function of β-apocarotenoids: β-carotene metabolites that may modulate nuclear receptor signaling. Am J Clin Nutr 96(suppl):1189S–1192S

18. Fleshman MK, Lester GE, Riedl KM, Kopec RE, Narayanasamy S, Curley RW, Schwartz SJ, Harrison EH (2011) Carotene and novel apocarotenoid concentrations in orange-fleshed *Cucumis melo* melons: determinations of β-carotene bioaccessibility and bioavailability. J Agric Food Chem 59:4448–4454
19. Kopec RE, Riedl KM, Harrison EH, Curley RW Jr, Hruszkewycz DP, Clinton CK, Schwartz SJ (2010) Identification and quantification of apolycopenals in fruits, vegetables and human plasma. J Agric Food Chem 58:3290–3296
20. World Health Organization (WHO) (2009) Global prevalence of vitamin A deficiency in populations at risk 1995–2005. WHO global database on vitamin A deficiency, pp 1–55. www.whqlibdoc.who.int/publications/2009/9789241598019_eng.pdf
21. WHO (2012) http://www.who.int/nutrition/topics/vad/en/index.html. Accessed 27 Dec 2012
22. Haque R, Ahmed T, Wahed MA, Mondal D, Hamidur Rahman ASM, Albert MJ (2010) Low dose β-carotene supplementation and deworming improve serum vitamin A and β-carotene concentrations in preschool children of Bangladesh. J Health Popul Nutr 28(3):230–237
23. Arroyave G, Mejia LA (2010) Five decades of vitamin A studies in the region of Central America and Panama. Food Nutr Bull 31:118–129
24. Mayo-Wilson E, Imdad A, Herzer K, Yakoob MY, Bhutta ZA (2011) Vitamin A supplements for preventing mortality, illness, and blindness in children aged under 5: systematic review and meta-analysis. BMJ 343:d5094. doi:10.1136/bmj.d5094. Pubmed PMID: 21868478
25. Low JW, Arimond M, Osman N, Cunguara B, Zano F, Tschirley D (2007) A food-based approach introducing orange-fleshed sweet potatoes increased vitamin A intake and serum retinol concentrations in young children in rural Mozambique. J Nutr 137:1320–1327
26. Hotz C, Loechl C, Lubowa A, Tumwine JK, Ndeezi G, Nandutu Masawi A, Baingana R, Carriquiry A, de Brauw A, Meenakshi JV, Gilligan DO (2012) Introduction of β-carotene-rich orange sweet potato in rural Uganda resulted in increased vitamin A intakes among children and women and improved vitamin A status among children. J Nutr 142:1871–1880
27. Jamil KM, Brown KH, Jamil M, Peerson JM, Keenan AH, Newman JW, Haskell MJ (2012) Daily consumption of orange-fleshed sweet potato for 60 days increased plasma β-carotene concentration but did not increase total body vitamin A pool size in Bangladeshi women. J Nutr 142:1896–18902
28. Sundram K, Sambanthamurthi R, Tan YA (2003) Palm fruit chemistry and nutrition. Asia Pac J Clin Nutr 12:355–362
29. Atinmo T, Bakre AT (2003) Palm fruit in traditional African food culture. Asia Pac J Clin Nutr 12:350–354
30. Rice AL, Burns JB (2010) Moving from efficacy to effectiveness: red palm oil's role in preventing vitamin A deficiency. J Am Coll Nutr 29:302S–313S
31. Ye X, Al-Babili S, Kloti A, Zhang J, Lucca P, Beyer P, Potrykus I (2000) Engineering the provitamin A β-carotene biosynthetic pathway into (carotenoid free) rice endosperm. Science 287:303–305
32. Paine JA, Shipton CA, Chaggar S, Howells RM, Kennedy MJ, Vernon G, Wright SY, Hinchliffe E, Adams JL, Silverstone AL, Drake R (2005) Improving the nutritional value of Golden Rice through increased provitamin A content. Nat Biotechnol 23:482–487
33. Tang G, Hu Y, Yin SA, Wang Y, Dallal GE, Grusak MA, Russell RM (2012) β-Carotene in Golden Rice is as good as β-carotene in oil at providing vitamin A to children. Am J Clin Nutr 96:658–664
34. Tang G, Qin J, Dolnikowski GG, Russell RM, Grusak MA (2009) Golden Rice is an effective source of vitamin A. Am J Clin Nutr 89:1776–1783
35. Blount JD, Houston DC, Sural PF, Pape Moller A (2004) Egg-laying capacity is limited by carotenoid pigment availability in wild gulls *Larus fuscus*. Proc R Soc Lond B 271:S79–S81
36. Vorobyev M, Osorio D, Bennett ATD, Marshall NJ, Cuthill IC (1998) Tetrachromancy, oil droplets and bird plumage colours. J Comp Physiol A 183:621–633
37. Stephen ID, Coetzee V, Perrett DI (2011) Carotenoid and melanin pigment coloration affect perceived human health. Evol Human Behav 32:216–227
38. Kaplan LA, Lau JM, Stein EA (1990) Carotenoid composition, concentrations, and relationships in various human organs. Clin Physiol Biochem 8:1–10

39. Stahl W, Schwarz W, Sundquist AR, Sies H (1992) Cis-trans isomers of lycopene and β-carotene in human serum and tissues. Archiv Biochem Biophys 294:173–177
40. Palan P, Naz R (1996) Changes in various antioxidant levels in human seminal plasma related to immunofertility. Arch Androl 36:139–143
41. Gupta NP, Kumar R (2002) Lycopene therapy in idiopathic male infertility: a preliminary report. Int J Urol Nephrol 34:369–372
42. Schweigert FJ (2003) Research note: changes in the concentration of β-carotene, α-tocopherol and retinol in the bovine corpus luteum during the ovarian cycle. Arch Anim Nutr 57:307–310
43. O'Shaughnessy PJ, Wathes DC (1998) Bovine luteal cell activity in culture. Maintenance of steroidogenesis by high density lipoprotein containing high or low beta-carotene concentrations. Anim Reprod Sci 17:165–176
44. Rapoport R, Sklan D, Wolfeson D, Shaham-Albancy A, Hanukoglu I (1998) Antioxidant capacity is correlated with steroidogenic status of the corpus luteum during the bovine estrous cycle. Biochem Biophys Acta 1380:133–140
45. Folman Y, Rosenberg M, Ascarelli I, Kaim M, Herz Z (1983) The effect of dietary and climatic factors on fertility, and on plasma progesterone and oestradiol-17 beta levels in dairy cows. J Steroid Biochem 19:863–868
46. Csapo AI, Pulkkinen M (1978) Indispensability of the human corpus luteum in the maintenance of early pregnancy: luteectomy evidence. Obstet Gynecol Surv 33:69–81
47. West KP Jr, Katz J, Khatry SK, LeClerq SC, Pradhan EK, Shrestha SR, Connor PB, Dali SM, Christian P, Pokhrel RP, Sommer A (1999) Double blind, cluster randomized trial of low dose supplementation with vitamin A or β-carotene on mortality related to pregnancy in Nepal. BMJ 318:570–575
48. West KP Jr, Christian P, Labrique AB, Rashid M, Shamin AA, Klemm RDW, Massie AB, Mehra S, Schulze KJ, Ali H, Ullah B, Wu LSF, Katz J, Banu H, Akhter HH, Sommer A (2011) Effects of vitamin A or beta carotene supplementation on pregnancy-related mortality and infant mortality in rural Bangladesh. A cluster randomized trial. JAMA 305:1986–1995
49. Thorne-Lyman A, Fawzi WW (2012) Vitamin A and carotenoids during pregnancy and maternal, neonatal and infant health outcomes: a systematic review and meta-analysis. Paediatric Perinatal Epidemiol 26(Suppl 1):36–54
50. Villamor E, Koulinska IN, Aboud S, Murrin C, Bosch RJ, Manji KP, Fawzi WW (2010) Effect of vitamin supplements on HIV shedding in breast milk. Am J Clin Nutr 92:881–886
51. WHO (2011) Guideline: vitamin A supplementation in pregnant women. http://www.who.int/elena/titles/vitamina_pregnancy/en/. Accessed 28 Dec 2012
52. Patton S, Canfield LM, Huston GE, Ferris AM, Jensen RG (1990) Carotenoids of human colostrums. Lipids 25:159–165
53. Canfield LM, Clandinin MT, Davies DP, Fernandez MC, Jackson J, Hawkes J, Goldman WJ, Pramuk K, Reyes H, Sablan B, Sonobe T, Bo X (2003) Multinational study of major breast milk carotenoids of healthy mothers. Eur J Nutr 42:133–141
54. Azar M, Basu A, Jenkins AJ, Nankervis AJ, Hanssen KF, Scholz H, Henriksen T, Garg SK, Hammad SM, Scardo JA, Aston CE, Lyons TJ (2011) Serum carotenoids and fat-soluble vitamins in women with type 1 diabetes and preeclampsia: a longitudinal study. Diabetes Care 34:1258–1264
55. Palan PR, Mikhail MS, Romney SL (2001) Placental and serum levels of carotenoids in preeclampsia. Obstet Gynecol 98:459–462
56. Agarwal A, Aponte-Mellado A, Premkumar BJ, Shaman A, Gupta S (2012) The effects of oxidative stress on female reproduction: a review. Reprod Biol Endocrinol 10:49. doi:10.1186/1477-7827-10-49. Pubmed PMID: 22748101
57. Bohm F, Edge R, Truscott G (2012) Interactions of dietary carotenoids with activated (singlet) oxygen and free radicals: potential effects for human health. Mol Nutr Food Res 56:205–216
58. Gruszecki W (2010) Carotenoids in lipid membranes. In: Landrum J (ed) Carotenoids: physical, chemical and biological functions and properties. CRC Press, Boca Raton, pp 19–30

59. Edge R, Truscott G (2010) Properties of carotenoid radicals and excited states and their potential role in biological systems. In: Landrum J (ed) Carotenoids: physical, chemical and biological functions and properties. CRC Press, Boca Raton, pp 283–308
60. Bohm F, Edge R, McGarvey D, Truscott T (1998) β-carotene with vitamins E and C offers synergistic cell protection against NO_x. FEBS Lett 436:387–389
61. Bohm F, Edge R, Truscott G (2012) Interactions of dietary carotenoids with singlet oxygen ($^1O_2$) and free radicals: potential effects for human health. Acta Biochim Pol 59:27–30
62. Walston J, Xue Q, Semba R, Ferrucci L, Cappola A, Ricks M, Guraink J, Fried L (2006) Serum antioxidants, inflammation, and total mortality in older women. Am J Epidemiol 163:18–26
63. Holt E, Steffen L, Moran A, Basu S, Steinberger J, Ross J, Hong C, Sinaiko A (2009) Fruit and vegetable consumption and its relation to markers of inflammation and oxidative stress in adolescents. J Am Diet Assoc 109:414–421
64. Thomson C, Stendell-Hollis N, Rock C, Cussler E, Flatt S, Pierce J (2007) Plasma and dietary carotenoids are associated with reduced oxidative stress in women previously treated for breast cancer. Cancer Epidemiol Biomarkers Prev 16:2006–2015
65. Hozawa A, Jacobs D, Steffes M, Gross M, Steffen L, Duk-Hee L (2007) Relationships of circulating carotenoid concentrations with several markers of inflammation, oxidative stress, and endothelial dysfunction: The Coronary Artery Risk Development in Young Adults (CARDIA)/Young Adult Longitudinal Trends in Antioxidants (YALTA) study. Clinic Chem 53:447–455
66. D'Adamo C, Miller R, Shardell M, Orwig D, Hochberg M, Ferrucci L, Semba R, Yu-Yahiro J, Magaziner J, Kicks G (2012) Higher serum concentrations of dietary antioxidants are associated with lower levels of inflammatory biomarkers during the year after hip fracture. Clin Nutr 31:659–665
67. Kim J-H, Na H-J, Kim C-K, Kim J-Y, Ha K-S, Lee H, Chung H-T, Kwon H, Kwaon Y-G, Kim Y-M (2008) The non-provitamin A carotenoid, lutein, inhibits NF-κB-inducing kinase pathways: role of H_2O_2 in NF-κB activation. Free Radic Biol Med 45(6):885–896
68. Izumi-Nagai K, Nagai N, Ohgami K, Satfuka S, Ozawa Y, Tsubota K, Umezawa K, Ohno S, Oike Y, Ishida S (2007) Macular pigment lutein is anti-inflammatory in preventing choroidal neovascularization. Aterioscler Thromb Vasc Biol 27:2555–2562
69. Di Tomo P, Canali R, Ciavardelli D, Di Silvestre S, De Marco A, Giardinelli A, Pipino C, Di Pietro N, Virgili F, Pandolfi A (2012) β-carotene and lycopene affect endothelial response to TNF-α reducing nitro-oxidative stress and interaction with monocytes. Mol Nutr Food Res 56:217–227
70. Rubin L, Chan G, Barrett-Reis B, Fulton A, Hansen R, Ashmeade T, Oliver R, Mackey A, Dimmit R, Hartmann E, Adamkin D (2012) Effect of carotenoid supplementation on plasma carotenoids, inflammation and visual development in preterm infants. J Perinatol 32:418–424
71. Graydon R, Hogg RE, Chakravarthy U, Young I, Woodside J (2012) The effect of lutein- and zeaxanthin-rich foods vs supplements on macular pigment level and serological markers of endothelial activation, inflammation and oxidation: pilot studies of healthy volunteers. Br J Nutr 108:334–342
72. Zhao X, Aldini G, Johnson E, Rasmussen H, Kraemer K, Woolf H, Musaeus N, Krinsky N, Russell R, Yeum K (2006) Modification of lymphocyte DNA damage by carotenoid supplementation in postmenopausal women. Am J Clin Nutr 83:163–169
73. Basu A, Imrhan V (2007) Tomatoes versus lycopene in oxidative stress and carcinogenesis: conclusions from clinical trials. European J Clin Nutr 61:295–303
74. Dunstan J, Breckier L, Hale J, Lehmann H, Franklin P, Lyons G, Ching S, Mori T, Barden A, Prescott S (2007) Supplementation with vitamins C, E, beta-carotene and selenium has no effect on anti-oxidant status and immune responses in allergic adults: a randomized controlled trial. Clin Exp Allergy 37:180–187
75. Rytter E, Vessby B, Asgard R, Ersson C, Moussavian S, Sjodin A, Abramsson-Zetterberg L, Moller L, Basu S (2010) Supplementation with a combination of antioxidants does not affect glycaemic control, oxidative stress or inflammation in type 2 diabetes subjects. Free Radic Res 44:1445–1453

76. Meinke M, Friedrich A, Tscherch K, Haag S, Darvin M, Vollert H, Groth N, Lademann J, Rohn S (2013) Influence of dietary carotenoids on radical scavenging capacity of the skin and skin lipids. Eur J Pharmaceut Biopharm Dec 13; 84:365–373
77. Mayne S, Cartmel B, Scarmo S, Lin H, Leffell D, Welch E, Ermakov I, Bhosale P, Bernstein P, Getterman W (2010) Noninvasive assessment of dermal carotenoids as a biomarker of fruit and vegetable intake. Am J Clin Nutr 92:794–800
78. Alaluf S, Heinrich U, Stahl W, Tronnier H, Wiseman S (2002) Dietary carotenoids contribute to normal human skin color and UV photosensitivity. J Nutr 132:399–403
79. Stephen I, Smith M, Stirrat M, Perret D (2009) Facial skin coloration affects perceived health of human faces. Int J Primatol 30:845–857
80. Whitehead R, Ozakinci G, Stephen I, Perrett D (2012) Appealing to vanity: could potential appearance improvement motivate fruit and vegetable consumption? Am J Pub Health 102:207–211
81. Karthik S, Campbell-Davis D, Isherwood D (2006) Carotenemia in infancy and is association with prevalent feeding practices. Ped Dermatol 23:571–573
82. Takita Y, Ichimiya M, Hamamoto Y, Muto M (2006) A case of carotenemia associated with ingestion of nutrient supplements. J Dermatol 2:132–134
83. Hueber A, Rosentreter A, Severin M (2011) Canthaxanthin retinopathy: long-term observations. Ophthalmic Res 46:103–106
84. Mathews-Roth M (1986) Beta-carotene therapy for erythropoetic protoporphyria and other photosensitivity diseases. Biochimie 68:875–884
85. Stahl W, Sies H (2012) β-carotene and other carotenoids in protection from sunlight. Am J Clin Nutr 96(suppl):1179–1184
86. Zussman J, Ahdout J, Kim J (2010) Vitamins and photoaging: do scientific data support their use? J Am Acad Dermatol 63:507–525
87. Kopcke W, Krutmann J (2008) Protection from sunburn with β-carotene: a meta-analysis. Photochem Photobiol 84:284–288
88. Darvin M, Fluhr J, Meinke M, Zastrow L, Sterry W, Lademann J (2011) Topical beta-carotene protects against infra-red-light-induced free radicals. Exper Dermatol 20:125–129
89. Kohl E, Steinbauer J, Landthaler M, Szeimies R (2011) Skin ageing. J Euro Acad of Dermatol Venereol (JEADV) 25:873–884
90. Purba M, Kouris-Blazos A, Wattanapenpalboon N, Lukito W, Rothenberg E, Steen B, Wahlqvist M (2001) Skin wrinkling: can food make a difference. J Am Coll Nutr 20:71–80
91. Darvin M, Patzelt A, Gehse S, Schanzer S, Benderoth C, Sterry W, Lademann J (2008) Cutaneous concentration of lycopene correlates significantly with the roughness of skin. Eur J Pharmaceut Biopharmaceut 69:943–947
92. Evans J, Johnson E (2010) The role of phytonutrients in skin health. Nutrients 2:903–928
93. Drasdo N, Fowler CW (1974) Non-linear projection of the retinal image in a wide-angle schematic eye. Br J Ophthalmol 58:709–714
94. Bone R, Landrum J, Fernandez L, Tarsis S (1988) Analysis of the macular pigment by HPLC: retinal distribution and age study. Invest Ophthalmol Vis Sci 29:843–949
95. Zeimer M, Dietzel M, Hense H, Helmes B, Austermann U, Pauleikhoff D (2012) Profiles of macular pigment optical density and their changes following supplemental lutein and zeaxanthin: new results from the LUNA study. Invest Ophthalmol Vis Sci 53:4852–4859
96. Krinsky N, Landrum J, Bone R (2003) Biologic mechanisms of the protective role of lutein and zeaxanthin in the eye. Ann Rev Nutr 23:171–201
97. Edge R, McGarvey D, Truscott G (1997) The carotenoids as antioxidants: a review. J Photochem Photobiol B Biol 41:189–200
98. Ahmed S, Lott M, Marcus D (2005) The macular xanthophylls. Surv Ophthalmol 50: 183–193
99. Bone R, Landrum J, Guerra L, Ruiz J (2003) Lutein and zeaxanthin dietary supplements raise macular pigment density and serum concentrations of these carotenoids in humans. J Nutr 133:992–998

100. Ma L, Dou H-L, Wu Y-Q, Huang Y-M, Huang Y-B, Xu X-R, Zou Z-Y, Lin X-M (2012) Lutein and zeaxanthin intake and the risk of age-related macular degeneration: a systematic review and meta-analysis. Br J Nutr 107:350–359
101. Klein R, Chou C-F, Klein B, Zhang X, Meuer S, Saddine J (2011) Prevalence of age-related macular degeneration in the US population. Arch Ophthalmol 129:75–80
102. Muni R, Altaweel M, Tennant M, Weaver B, Kertes P (2008) Agreement among Canadian retina specialists in the determination of treatment eligibility for photodynamic therapy in age-related macular degeneration. Retina 28:1421–1426
103. Bone R, Landrum J (2010) Dose-dependent response of serum lutein and macular pigment optical density to supplementation of lutein esters. Arch Biochem Biophys 504:50–55
104. Schalch W, Cohn W, Barker F, Kopcke W, Mellerio J, Bird A, Robson A, Fitzke F, van Kujk F (2007) Xanthophyll accumulation in the human retina during supplementation with lutein or zeaxanthin: the LUXEA (LUtein Xanthophyll Eye Accumulation) study. Arch Biochem Biophys 458:128–135
105. Barlett H and Eperjesi F (2008) A randomized controlled trial investigating the effect of lutein and antioxidant dietary supplementation on visual function in healthy eyes. Clinical Nutr 27:218–227
106. Ma L, Yan S-F, Huang Y-M, Lu X-R, Qian F, Pang H-L, Xu X-R, Zou Z-Y, Dong P-C, Xiao X, Wang X, Sun T-T, Dou H-L, Lin X-M (2012) Effect of lutein and zeaxanthin on macular pigment and visual function in patients with early age-related macular degeneration. Ophthalmol 119:2290–2297
107. Richer S, Stiles W, Statkute L, Pulido J, Frankowski J, Rudy D, Pei K, Tsipursky M, Nyland J (2004) Double-masked, placebo-controlled, randomized trial of lutein and antioxidant supplementation in the intervention of atrophic age-related macular degeneration: the veterans LAST study (Lutein Antioxidant Supplementation Trial). Optometry 75:216–230
108. Richer S, Stiles W, Graham-Hoffman K, Levin M, Ruskin D, Wrobel J, Park D-W, Thomas C (2011) Randomized, double-blind, placebo-controlled study of zeaxanthin and visual function in patients with atrophic age-related macular degeneration. Optometry 82:667–680
109. Age-Related Eye Disease Study Group (2001) A randomized, placebo-controls clinical trial of high-dose supplementation with vitamins C and E, beta carotene, and zinc for age-related macular degeneration and vision loss. AREDS report no. 8. Arch Ophthalmol 119:1417–1436
110. Age-Related Eye Disease Study Group (2001) A randomized, placebo-controlled, clinical trial of high-dose supplementation with vitamin C and E and beta-carotene for age-related cataract with vision loss: AREDS report no. 9. Arch Ophthalmol 119:1439–1452
111. Evans J, Lawrenson J (2012) Antioxidant vitamin and mineral supplements for preventing age-related macular degeneration (Review) The Cochrane Collaboration Cochrane Library 2012, Issue 6 www.thecochranelibrary.com John Wiley
112. The Age-Related Eye Disease Study 2 (AREDS2) Research Group (2013) Lutein + Zeaxanthin and omega-3 fatty acids for age-related macular degeneration. The age-related eye disease study 2 (AREDS2) randomized clinical trial. JAMA 309:2005–2015
113. Loughman J, Davidson P, Molan J, Akkali M, Beatty S (2010) Macular pigment and its contribution to visual performance and experience. J Optometry 3:74–90
114. Hammond B, Fletcher L (2012) Influence of the dietary carotenoids lutein and zeaxanthin on visual performance: application to baseball. Am J Clin Nutr 96(Suppl):1207–1300
115. Stringham J, Hammond B (2007) The glare hypothesis of macular pigment function. Optometry Vis Sci 84:859–864
116. Stringham J, Garcia P, Smith P, McLin L, Foutch B (2011) Macular pigment and visual performance in glare: benefits for photostress recovery, disability glare and visual discomfort. Ophthalmol Vis Sci 52:7406–7415
117. Loughman J, Akkali M, Beatty S, Scanlon G, Davidson P, O'Dwyer V, Cantwell T, Major P, Stack J, Nolan J (2010) The relationship between macular pigment and visual performance. Vis Res 50:1249–1256

118. Renzi L, Hammond B (2010) The relation between the macular carotenoids, lutein and zeaxanthin, and temporal vision. Ophthal Physiol Opt 30:351–357
119. Ma L, Lin X-M, Zou Z-Y, Xu X-R, Li Y, Xu R (2009) A 12-week lutein supplementation improved visual function in Chinese people with long-term computer display light exposure. Br J Nutr 102:186–190
120. Kvansakul J, Rodriguez-Camona M, Edgar D, Barker F, Kopcke W, Schalch W, Barbur J (2006) Supplmentation with the carotenoids lutein and zeaxanthin improves human visual performance. Ophthal Physiol Opt 26:362–371
121. Nolan J, Loughman J, Akkali M, Stack J, Scanlon G, Davidson P, Beatty S (2011) The impact of macular pigment augmentation on visual performance in normal subjects: COMPASS. Vis Res 51:459–469
122. Loughman J, Nolan J, Howard A, Connolly E, Meagher K, Beatty S (2012) The impact of macular pigment augmentation on visual performance using different carotenoid formulations. Invest Ophthalmol Vis Sci 53:7871–7880
123. Barlett H, Eperjesi F (2007) Effect of lutein and antioxidant dietary supplementation on contrast sensitivity in age-related macular disease: a randomized controlled trial. Eur J Clin Nutr 61:1121–1127
124. Michael R, Bron A (2011) The ageing lens and cataract: a model of normal and pathological ageing. Phil Trans R Soc B 366:1278–1292
125. Truscott R (2005) Age-related nuclear cataract: oxidation is the key. Exper Eye Res 80:709–725
126. Yeum K-J, Shang F, Schalch W, Russell R, Taylor A (1999) Fat-soluble nutrient concentrations in different layers of human cataractous lens. Curr Eye Res 19:502–505
127. Bhosale P, Zhao D-Y, Bernstein P (2007) HPLC measurement of ocular carotenoid levels in human donor eyes in the lutein supplementation era. Invest Ophthalmol Vis Sci 48:543–549
128. Barker F (2010) Dietary supplementation: effects on visual performance and occurrence of AMD and cataracts. Curr Med Res Opin 26:2011–2023
129. Mares J, Voland R, Adler R, Tinker L, Millen A, Moeller S, Blodi B, Gehrs K, Wallace R, Chappell R, Meuhouser M, Sarto G (2010) Healthy diets and the subsequent prevalence of nuclear cataract in women. Arch Ophthalmol 128:738–749
130. Milton R, Sperduto R, Clemons T, Ferris F (2006) Centrum use and progression of age-related cataract in the age-related eye disease study: a propensity score approach. AREDS report No. 21. Ophthalmology 113:1264–1270
131. Olmedilla B, Granado F, Blanca I, Vaquero M, Cajigal C (2001) Lutein in patients with cataracts and age-related macular degeneration: a long term supplementation study. J Sci Food Agric 81:904–909
132. Olmedilla B, Granado F, Blanco I, Vaquero M (2003) Lutein, but not alpha tocopherol supplementation improves visual function in patients with age-related cataracts: a two-year double-blind, placebo-controlled pilot study. Nutrition 19:21–24
133. Trumbo P, Ellwood K (2006) Lutein and zeaxanthin intakes and risk of age-related macular degeneration and cataracts: an evaluation using the Food and Drug Administration's evidence-based review system for health claims. Am J Clin Nutr 84:971–974
134. Berson E (1993) Retinitis pigmentosa. The Friedenwald Lecture. Invest Ophthalmol Vis Sci 34:1659–1676
135. Berson E, Rosner B, Sandberg M, Weigel-DeFranco C, Brockhurst R, Hayes C, Johnson E, Anderson E, Johnson C, Guadio A, Willet W, Schaefer E (2010) Clinical trial of lutein in patients with retinitis pigmentosa receiving vitamin A. Arch Ophthalmol 128:403–411
136. Wittes J, Gorin M, Mayne S, McCarthy C, Sternberg P, Wall M (2011) Letter from the DSMC regarding a clinical trial of lutein in patients with retinitis pigmentosa. Arch Ophthalmol 129:675
137. Dagnelie G, Zorge I, McDonald T (2000) Lutein improves visual function in some patients with retinal degeneration: a pilot study via the internet. Optometry 71:147–164
138. Bahrami H, Melia M, Dagnelie G (2006) Lutein supplementation in retinitis pigmentosia: PC-based vision assessment in a randomized double-masked placebo-controlled clinical trial. BMC Ophthalmol 6:23. www.biochemicalcontrol.com/1471-2415-6-23

139. Aleman T, Duncan J, Bieber M, de Castro E, Marks D, Gardner L et al (2001) Macular pigment and lutein supplementation in retinitis pigmentosa and Usher syndrome. Invest Ophthalmol Vis Sci 42:1873–1881
140. Sandberg M, Johnson E, Berson E (2010) The relationship of macular pigment optical density to serum lutein in retinitis pigmentosa. Invest Ophthalmol Vis Sci 51:1086–1091
141. Li S-Y, Fung F, Zhong J-F, Wong D, Chan H, Lo A (2012) Anti-inflammatory effects of lutein in retinal ischemic/hypoxic injury: in vivo and in vitro studies. Invest Opth={thalmol Vis Sci 53:5976–5984
142. Li S-Y, Fu Z-J, Ma H, Jang W-C, So K-F, Wong D, Lo A (2009) Effect of lutein on retinal neurons and oxidative stress in a model of acute retinal ischemia/reperfusion. Invest Ophthalmol Vis Sci 50:836–843
143. Woo T, Li S-Y, Lai W, Wong D, Lo A (2013) Neuroprotective effects of lutein in a rat model of retinal detachment. Graefes Arch Clin Exp Ophthalmol 251:41–51
144. Craft N, Haitema T, Garnett K, Fitch K, Dorey C (2004) Carotenoid, tocopherol and retinol concentrations in elderly human brain. J Nutr Health Aging 8:156–162
145. Johnson E (2012) The possible role for lutein and zeaxanthin in cognitive function in the elderly. Am J Clin Nutr 96:1161S–1165S
146. Ritchie K, Touchon J (2000) Mild cognition impairment: conceptual basis and current neurological status. Lancet 355:225–228
147. Lahiri D, Greig N (2004) Lethal weapon: amyloid β-peptide, role in the oxidative stress and neurodegeneration of Alzheimer's disease. Neurobiol Aging 25:581–587
148. Farooqui T, Farooqui A (2009) Aging: an important factor for the pathogenesis of neurodegenerative diseases. Mech Aging Dev 130:203–215
149. Thompson P, Vinters H (2012) Pathologic lesions in neurodegenerative diseases. Prog Mol Biol Trans Sci. 107:1–40
150. Plasman B, Williams J, Burke J, Holsinger T, Benjamin S (2010). Systematic review: factors associated with risk for and possible prevention of cognitive decline in later life. Ann Intern Med 153:182–193
151. Daviglus M, Bell C, Berrettini W, Bowen P, Connolly S, Cox N, Dunbar-Jacob J, Granieri E, Hunt G, McGarry K, Patel D, Potosky A, Sanders-Bush E, Silberbert D, Trevisan M (2010) National Institutes of Health State-of-the-Science Conference statement: preventing Alzheimer's disease and cognitive decline. Ann Intern Med 153:176–181
152. Daviglus M, Plassman B, Pirzuda A, Bell C, Bowen P, Burke J, Connolly S, Dunbar-Jacob J, Granieri E, McGarry K, Patel D, Trevisan M, Williams J (2011) Risk factors and preventive interventions for Alzheimer's disease. Arch Neurol 68:1185–1190
153. Yaffe K (2007) Antioxidants and prevention of cognitive decline: does duration of use matter? Arch Intern Med 167:2167–2168
154. Johnson E, McDonald K, Caldarella S, Chung H, Troen A, Snodderly D (2008) Cognitive findings of an exploratory trial of docosahexaenoic acid and lutein supplementation in older women. Nutr Neurosci 11:75–83
155. Li S-Y, Fu Z-J, Woo T, Wong D, Lo A (2012) Lutein enhances survival and reduces neuronal damage in a mouse model of ischemic stroke. Neurobiol Dis 45:624–632
156. Li S-Y, Fu Z-J, Lo A (2012) Hypoxia-induced oxidative stress in ischemic retinopathy. Oxid Med Cell Longev 2012:426769. doi:10:1155/2012/426769
157. Greenberg E (1993) Retinoids or carotenoids: is there another choice? Prev Med 22: 723–727
158. Peto R, Doll R, Buckley J, Sporn M (1981) Can dietary beta-carotene materially reduce human cancer rates? Nature 290:201–208
159. Wald N (1987) Retinol, beta-carotene and cancer. Cancer Surv 6:635–651
160. Tanaka J, Shnimizu M, Moriwaki H (2012) Cancer prevention by carotenoids. Molecules 17:3202–3242
161. Vainio H, Weiderpass E (2006) Fruit and vegetables in cancer prevention. Nutr Cancer 54:111–142
162. Marques-Vidal P, Ravasco P, Camilo M (2006) Foodstuffs and colorectal cancer risk: a review. Clin Nutr 25:14–36

163. Flood A, Rasotogi T, Wirfalt E, Mitrou P, Reedy J, Subar A, Kipnis V, Mouw T, Hollenbeck A, Leitzmann M, Schatzkin A (2008) Dietary patterns as identified by factor analysis and colorectal cancer among middle-age Americans. Am J Clin Nutr 88:176–184
164. Heinonen O, Albanes D (1994) The effect of vitamin E and β-carotene on the incidence of lung cancer and other cancers in male smokers. The Alpha-Tocopherol, Beta-Carotene Cancer Prevention Study Group. N Engl J Med 330:1029–1035
165. Omenn G, Goodman G, Thornquist M, Balmes J, Cullen M, Glass A, Koegh J, Meyskens F, Valanis B, Williams H et al (1996) Effects of a combination of β-carotene and vitamin A on lung cancer and cardiovascular disease. N Engl J Med 334:1150–1155
166. The ATBC Study Group (2003) Incidence of cancer and mortality following α-tocopherol and β-carotene supplementation. JAMA 290:476–485
167. Mayne S, Handelman G, Beecher G (1996) β-carotene and lung cancer promotion in heavy smokers: a plausible relationship. J Natl Cancer Inst 88:1513–1515
168. Hennekens C, Buring J, Manson J, Stampfer M, Rosner B, Cook N, Belanger C, LaMotte F, Gaziano J, Ridker P et al (1996) Effects of a combination of β-carotene and vitamin A on lung cancer and cardiovascular disease. N Eng J Med 334:1150–1155
169. Kelloff G, Boone C, Crowell J, Steele V, Lubet R, Doody L, Malone W, Hawk E, Sigman C (1996) New agents for cancer prevention. J Cell Biochem Suppl 26:1–28.
170. Ziegler R, Mayne S, Swanson C (1996) Nutrition and lung cancer. Cancer Causes Control 7:157–177
171. Chatterjee M, Roy K, Janarthan M, Das S, Chatterjee M (2012) Biological activity of carotenoids: its implications in cancer risk and prevention. Curr Pharmaceut Biotech 13: 180–190
172. Kelkel M, Schumacher M, Dicato M, Diederich M (2011) Antioxidant and anti-proliferative properties of lycopene. Free Rad Res 45:925–940
173. Sharoni Y, Linnewiel-Hermoni K, Khanin M, Salman H, Veprik A, Danilenko M, Levy J (2012) Carotenoids and apocarotenoids in cellular signaling related to cancer: a review. Mol Nutr Food Res 56:259–269
174. Ford N, Erdman J (2012) Are lycopene metabolites metabolically active? Acta Biochim Pol 59:1–4
175. Etminan M, Takkouche B, Caamano-Isoma F (2004) The role of tomato products and lycopene in the prevention of prostate cancer: a meta-analysis of observational studies. Cancer Epidemiol Biomark Prev 13:340–345
176. Ilic D, Misso M (2012) Lycopene for the prevention and treatment of benign prostatic hyperplasia and prostate cancer: a systematic review. Maturitas 72:269–276
177. Conklin K (2003) Dietary antioxidants during cancer therapy: impact on chemotherapeutic effectiveness and development of side effects. Nutr Cancer 37:1–18
178. Sahin K, Sahin N, Kucuk O (2010) Lycopene and chemotherapy toxicity. Nutr Cancer 62:988–995
179. Block K, Koch A, Mead M, Tothy P, Newman R, Gyllenhaal C (2007) Impact of antioxidant supplementation on chemotherapeutic efficacy: a systematic review of the evidence from randomized controlled trials. Cancer Treatment Rev 33:407–418
180. Puri T, Goyal S, Julka K, S, Nair O, Sharma D, Roth G (2010) Lycopene in treatment of high-grade gliomas: a pilot study. Neurol India 58(1):20–23
181. Meigs JB (2002) Epidemiology of the metabolic syndrome, 2002. Am J Manag Care 8(11 Suppl):283S–292S
182. Roberts CK, Sindhu KK (2009) Oxidative stress and metabolic syndrome. Life Sci 84: 705–712
183. Ford ES, Mokdad AH, Giles WH, Brown DW (2003) The metabolic syndrome and antioxidant concentrations: findings from the Third National Health and Nutrition Examination Survey. Diabetes 52:2346–2352
184. Sugiura M, Nakamura M, Ogawa K, Ikoma Y, Matsumoto H, Ando F, Shimokata H, Yano M (2008) Associations of serum carotenoid concentrations with the metabolic syndrome: interaction with smoking. Br J Nutr 100:1297–1306

185. Coyne T, Ibiebele TI, Baade PD, McClintock CS, Shaw JE (2009) Metabolic syndrome and serum carotenoids: findings of a cross-sectional study in Queensland, Aust Br J Nutr 102:1668–1677
186. Sluijs I, Beulens JW, Grobbee DE, van der Schouw YT (2009) Dietary carotenoid intake is associated with lower prevalence of metabolic syndrome in middle-aged and elderly men. J Nutr 139:9879–9892
187. Beydoun MA, Shroff MR, Chen X, Beydoun HA, Wang W, Zonderman AB (2011) Serum antioxidant status is associated with metabolic syndrome among U.S. adults in recent national surveys. J Nutr 141:903–913
188. Beydoun MA, Canas AJ, Beydoun HA, Chen X, Shroff MR, Zonderman AB (2012) Serum antioxidant concentrations and metabolic syndrome are associated among U.S. adolescents in recent national surveys. J Nutr 142:1693–1704
189. Czernichow S, Vergnaud AC, Galan P, Arnaud J, Favier A, Faure H, Huxley R, Hercberg S, Ahluwalia N (2009) Effects of long-term antioxidant supplementation and association of serum antioxidant concentration with risk of metabolic syndrome in adults. Am J Clin Nutr 90:329–335
190. Markovits N, Amotz AB, Levy Y (2009) The effect of tomato derived lycopene on low carotenoids and enhanced systemic inflammation and oxidation in severe obesity. IMAJ 11:598–601
191. McEneny J, Wade L, Young IS, Masson L, Duthie G, McGinty A, McMaster C, Thies F (2013) Lycopene intervention reduced inflammation and improves HDL functionality in moderately overweight middle-aged individuals. J Nutr Biochem 24:163–168. doi:10.1186/1475-2891-11-34.
192. Ghavipour M, Saedisomeolia A, Djalali M, Sotoudeh G, Eshraghyan MR, Moghadam AM, Wood LG (2013) Tomato juice consumption reduces systemic inflammation in overweight and obese females. Br J Nutr 109:2031-2035
193. Riso P, Visioli F, Grand S, Guarnieri S, Gardana C, Simonetti P, Porrini M (2006) Effect of tomato based drink on markers of inflammation, immunomodulation, and oxidative stress. J Agric Food Chem 54:2563–2566
194. Palozza P, Simone R, Catalano A, Monego G, Barini A, Mele MC, Parrone N, Trombino S, Picci N, Ranelletti FO (2011) Lycopene prevention of oxysterol-induced proinflammatory cytokine cascade in human macrophages: inhibition of NF-κB nuclear binding and increase in PPARγ expression. J Nutr Biochem 22(3):259–268.
195. Gouranton E, Thabuis C, Riollet C, Malezet-Desmoulins C, El Yazidi C, Amiot MJ, Borel P, Landrier JF (2011) Lycopene inhibits proinflammatory cytokine and chemokine expression in adipose tissue. J Nutr Biochem 22:642–648
196. Damms-Machado A, Weser G, Bischoff S (2012) Micronutrient deficiency in obese subjects undergoing low calorie diet. Nutr J 11:34. doi:10.1186/1475-2891-11-34. Pubmed PMID: 22657586
197. Voutilainen S, Nurmi T, Mursu J, Rissanen TH (2006) Carotenoids and cardiovascular health. Am J Clin Nutr 83:1265–1271
198. Karppi J, Laukkanen JA, Makikallio TH, Ronkainen K, Kurl S (2013) Serum β-carotene and the risk of sudden cardiac death in men: a population-based follow-up study. Atherosclerosis 226:172–177
199. Kim OY, Yoe HY, Kim HJ, Park JY, Kim JY, Lee SH, Lee JH, Lee KP, Jang Y, Lee JH (2010) Independent inverse relationship between serum lycopene concentration and arterial stiffness. Atherosclerosis 208:581–586
200. Kim JY, Paik JK, Kim OY, Park HW, Lee JH, Jang Y, Lee JH (2011) Effects of lycopene supplementation on oxidative stress and markers of endothelial function in healthy men. Atherosclerosis 215:189–195
201. Karppi J, Laukkanen JA, Sivenius J, Ronkainen K, Kurl S (2012) Serum lycopene decreases the risk of stroke in men: a population based follow-up study. Neurology 79:1540–1547

202. Biddle M, Moser D, Song EK, Heo S, Payne-Emerson H, Dunbar SB, Pressler S, Lennie T (2013) Higher dietary lycopene intake is associated with longer cardiac event-free survival in patients with heart failure. Eur J Cardiovasc Nurs 12:377-384
203. Palozza P, Catalano A, Simone RE, Mele MC, Cittadini A (2012) Effect of lycopene and tomato products on cholesterol metabolism. Ann Nutr Metab 61:126–134
204. Karppi J, Kurl S, Laukkanen JA, Rissanen TH, Kauhanen J (2011) Plasma carotenoids are related to intima-media thickness of the carotid artery wall in men from eastern Finland. J Intern Med 270:478–485
205. Xu XR, Zou ZY, Huang YM, Xiao X, Ma L, Lin XM (2012) Serum carotenoids in relation to risk factors for development of artherosclerosis. Clin Biochem 45:1357–1361
206. Ceriello A, Motz E (2004) Is oxidative stress the pathogenic mechanism underlying insulin resistance, diabetes, and cardiovascular disease? The common soil hypothesis revisited. Arterioscler Thromb Vasc Biol 24:816–823
207. Ford ES, Will JC, Bowman BA, Venkat Narayan KM (1999) Diabetes mellitus and serum carotenoids: findings from the Third National Health and Nutrition Examination Survey. Am J Epidemiol 149:168–176
208. Suzuki K, Ito Y, Nakamura S, Ochiai J, Aoki K (2002) Relationship between serum carotenoids and hyperglycemia: a population-based cross-sectional study. J Epidemiol 12:357–366
209. Coyne T, Ibiebele TI, Baade PD, Dobson A, McClintock C, Dunn S, Leonard D, Shaw J (2005) Diabetes mellitus and serum carotenoids: findings of a population-based study in Queensland, Australia. Am J Clin Nutr 82:685–693
210. Montonen J, Knekt P, Järvinen R, Reunanen A (2004) Dietary antioxidant intake and risk of type 2 diabetes. Diabetes Care 27:362–366
211. Liu S, Ajani U, Chae C, Hennekens C, Buring JE, Manson JE (1999) Long-term beta-carotene supplementation and risk of type 2 diabetes mellitus: a randomized controlled trial. JAMA 282:1073–1075
212. Wang L, Liu S, Pradhan AD, Manson JE, Buring JE, Gaziano JM, Sesso HD (2006) Plasma lycopene, other carotenoids, and the risk of type 2 diabetes in women. Am J Epidemiol 164:576–585
213. Mathews-Roth MM (1993) Carotenoids in erythropoietic protoporphyria and other photosensitivity diseases. Annals NY Acad Sci 691:127–138
214. Lietz G, Henry CJK, Mulokozi G, Mugyabuso JKL, Ballart A, Ndossi GD, Lorri W, Tomkins A (2001) Comparison of the effects of supplemental red palm oil and sunflower oil on maternal vitamin A status. Am J Clin Nutr 74:501–509
215. Vuong LT, Dueker SR, Murphy SP (2002) Plasma β-carotene and retinol concentrations of children increase after a 30-d supplementation with the fruit *Momordica cochinchinensis (gac)*. Am J Clin Nutr 75: 872–879
216. Albanes D, Heinonen OP, Taylor PR, Virtamo J, Edwards BK, Rautalahti M, Hartman AM, Palmgren J, Freedman LS, Haapakoski J, Barrett MJ, Pietinen P, Malila N, Tala E, Liippo K, Salomaa E-R, Tangrea JA, Teppo L, Askin FB, Taskinen E, Erozan Y, Greenwald P, Huttunen JK (1996) α-Tocopherol and β-carotene supplements and lung cancer incidence in the Alpha-Tocopherol, Beta-Carotene Cancer Prevention Study: effects of baseline characteristics and study compliance. J Natl Cancer Inst 88:1560–1570
217. Baron JA, Cole BF, Mott L, Haile R, Grau M, Church TR, Beck GJ, Greenberg ER (2003) Neoplastic and antineoplastic effects of β-carotene on colorectal adenoma recurrence: results of a randomized trial. J Natl Cancer Inst 95:717–722
218. Goodman GE, Thornquist MD, Balmes J, Cullen MR, Meyskens FL, Omenn GS, Valanis B, Williams JH Jr (2004) The beta-carotene and retinol efficacy trial: incidence of lung cancer and cardiovascular disease mortality during 6-year follow-up after stopping β-carotene and retinol supplements. J Natl Cancer Inst 96:1743–1750
219. Palozza P (1998) Prooxidant actions of carotenoids in biologic systems. Nutr Rev 56: 257–256

220. Reich P, Shwachman H, Craig JM (1960) Lycopenemia: a variant of carotenemia. New Eng J Med 262:263–269
221. Goralczyk R, Buser S, Baush J, Bee W, Zühlke U, Barker FM (1997) Occurrence of birefringent retinal inclusions in cynomolgus monkeys after high doses of canthaxanthin. Invest Opthalmol Vis Sci 38:741–752
222. Russell RM (2004) The enigma of β-carotene in carcinogenesis: what can be learned from animal studies. J Nutr 134:262S–268S
223. Back E, Frindt C, Nohr D, Frank J, Ziebach R, Stern M, Banke M, Biesalski H (2004) Antioxidant deficiency in cystic fibrosis: when is the right time to take action? Am J Clin Nutr 80:374–384

Chapter 4
Differential Transcription Factor Networks Orchestrate Flavonoid Biosynthesis

Lílian Cristina Baldon Aizza and Marcelo Carnier Dornelas

The Flavonoid Biosynthetic Pathway

Anthocyanins are a subclass of flavonoid pigments originating from a specific branch of the phenylpropanoid pathway, frequently responsible for most diverse colors in flowers, fruits, seeds, and leaves of angiosperms. They are water-soluble molecules present in the vacuoles of epidermal cells. In flowers and fruits, these pigments play an important ecological role in pollination and seed dispersal through attracted animals [1–3]. In vegetative organs, they are produced in response to abiotic and biotic stresses related to ultraviolet (UV) radiation, temperature and drought, or/and in defense against antimicrobial agents [4,5]. Anthocyanin biosynthesis is the most characterized secondary metabolic pathway in species with divergent floral morphology, pigmentation pattern, and pollination systems, such as *Petunia hybrida, Matthiola, Dianthus, Eustoma, Gerbera hybrida, Antirrhinum majus,* and *Ipomoea* [6–14]. A representation of a general phenylpropanoid pathway, leading to the biosynthesis of anthocyanins and other flavonoids, is shown in Fig. 4.1.

All flavonoids are derived from phenylalanine via the phenylpropanoid pathway, providing precursors for several branches of the pathway and producing different secondary metabolites such as isoflavones, flavonols, phlobaphenes, proanthocyanidins, and anthocyanins depending on plant species, tissues, developmental stages, and environmental signals [2, 3]. Anthocyanin biosynthesis begins with the condensation of three molecules of acetate residues from malonlyl-coenzyme A (CoA) with one molecule of 4-coumaroyl-CoA by chalcone synthase (CHS) to form tetrahydroxychalcone, which is immediately isomerized to the colorless

M. C. Dornelas (✉)
Biologia Vegetal, Universidade Estadual de Campinas (UNICAMP), R. Monteiro Lobato, 255, Campinas, SP, 13083-862, Brazil
e-mail: dornelas@unicamp.br

L. C. Baldon Aizza
Biologia Vegetal, Instituto de Biologia, Universidade Estadual de Campinas (UNICAMP), Campinas, SP, Brazil

C. Chen (ed.), *Pigments in Fruits and Vegetables,* DOI 10.1007/978-1-4939-2356-4_4

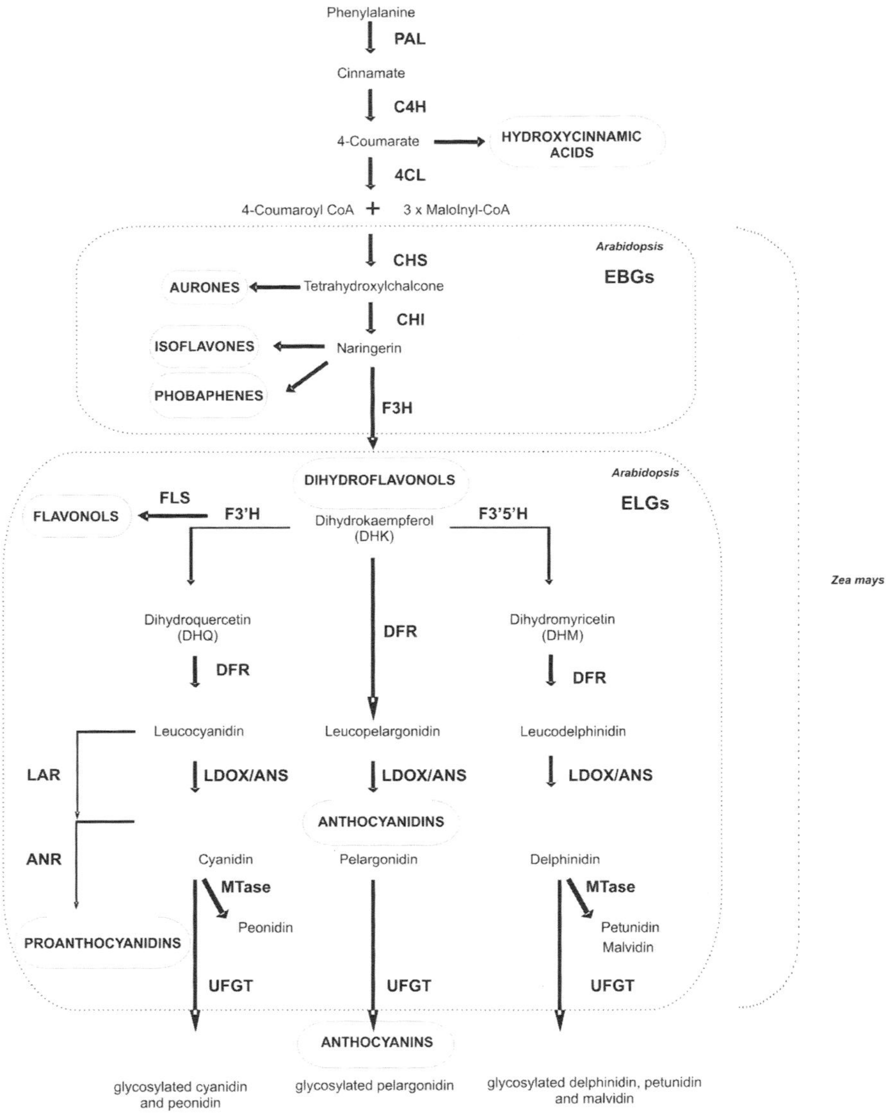

Fig. 4.1 Schematic representation of the phenylpropanoid biosynthetic pathway comprising anthocyanins and other flavonoids. Enzymes are indicated in capital letter and classes of compounds are *circled*. Anthocyanidin is further modified with glycosyl, acyl, or methyl groups, resulting in the "decorated" anthocyanin. In this case, UFGT is responsible for the glycosylation of anthocyanidins. Early and late biosynthetic genes (EBGs and ELGs) of flavonoid biosynthetic pathway in *Arabidopsis* are highlighted by *dotted rectangles*. PAL phenylalanine ammonia lyase, C4H cinnamic acid 4-hydroxylase, 4CL 4 coumarate CoA ligase, CHS chalcone syntase, CHI chalcone isomerase, F3 H flavanone 3-hydroxylase, F3′H flavanone 3′-hydroxylase, F3′5′H flavanone 3′5′-hydroxylase, FLS flavonol synthase, DFR dihydroflavonol 4-reductase, LDOX/ANS leucoanthocyanidin dioxygenase/anthocyanidin synthase, MTase methyltransferase, UFGT UDP-glucose: flavonoid 3-*O*-glucosyltransferase, LAR leucoanthocyanidin reductase, ANR anthocyanidin reductase

naringenin by chalcone isomerase (CHI). Subsequently, naringenin is converted to dihydroflavonol by flavanone 3-hydroxylase (F3H). Dihydroflavonol 4-reductase (DFR) is an important enzyme in the next step of anthocyanidin synthesis and can use any of the dihydroflavonols as substrates to produce anthocyanidin precursors. The anthocyanin biosynthetic pathway has three major branches. Each branch corresponds to the use of one, two, or three hydroxyl groups in the B-ring of the dihydrokaempferol. Flavonoid 3′hydroxylase (F3′H) and flavonoid 3′5′hydroxylase (F3′5′H) are the key enzymes responsible for these different hydroxylation patterns [15–17]. Then, these colorless molecules are converted in anthocyanidins: pelargonidin, cyanidin, or delphinidin by leucoanthocyanidin oxidase (LDOX, also known as anthocyanidin synthase, ANS). Methyltransferase (MTase) is responsible for addition of a methyl group to cyanidin, resulting in peonidin, whereas petunidin and malvidin correspond to the 3′ *O*-methyl delphinidin and 3′5 ′ *O*-dimethyl delphinidin, respectively [18]. Uridine diphosphate (UDP)-glucose: flavonoid 3-*O*-glucosyltransferases (UFGT) are essential enzymes for the glycosylation of anthocyanidins, resulting in different "decorated anthocyanins" [2].

Still, anthocyanins share the same upstream pathway with proanthocyanidins, also called condensed tannins, where anthocyanidin reductase (ANR) and leucoanthocyanidin reductase (LAR) enzymes are first committed steps to their biosynthesis (Fig. 4.1). These flavonoids accumulate mainly in seed coat and grains, providing defense against insects, herbivores, and microbial pathogens in many plants [19].

At least two clusters of genes are required for flavonoid biosynthesis in all plant species studied. The first cluster is represented by structural genes encoding enzymes for the production of flavonoid precursors, including those involved in the synthesis of particular ("decorated") anthocyanin molecules. In dicotyledonous plants, this first cluster of genes can still be divided into early biosynthetic genes (EBGs; which are involved in the production of flavones, flavonols, or phobaphenes) and late biosynthetic genes (LBGs; which are essential for anthocyanin and proanthocyanidin biosynthesis). For most species, what separates the "early" and "late" stages of the process is the action of the F3 H enzyme [12, 20], whereas in monocots such as maize, the whole process appears to be coordinately regulated as a single set, without division of EBGs and LBGs [21] as shown in Fig. 4.1.

The second cluster of genes includes R2R3-MYB and basic helix-loop-helix (bHLH) transcription factors together with WD40 proteins that directly control the transcription of the structural genes of the first cluster [2, 4, 20, 22–27]. It has been demonstrated that the combinatorial interactions among these different transcription factors, also known as the ternary MBW complex, can differentially regulate the expression of specific genes controlling multiple steps of the flavonoid/anthocyanin biosynthetic pathway among different species [22, 25].

Most of the structural genes encoding enzymes and several transcription factors regulating flavonoid biosynthetic pathway have been identified by mutational analysis, cloned and characterized in petals of *Petunia* and *Antirrhinum,* seeds of *Arabidopsis thaliana,* kernels of *Zea mays,* and in fruits of plants such as grapevine (*Vitis vinifera*) and apple (*Malus domestica*) [6, 11, 28–39].

This chapter provides an overview of recent advances in understanding the flavonoid biosynthetic pathway in plants, especially focusing on the transcription factors controlling flavonoid accumulation, with particular attention devoted to anthocyanin biosynthesis during the pigmentation of flowers and fruits in different plant species. These important findings could help us to improve desirable flower pigmentation in ornamental plants and to enhance production of anthocyanins-enriched foods.

Transcriptional Regulation of the Flavonoid Biosynthetic Pathway

The transcriptional regulation of gene expression is largely mediated by specific recognition of *cis*-acting promoter elements by *trans*-acting, sequence-specific, DNA-binding transcription factors. In general, the control of gene expression is mediated by complexes of multiple transcription factors and many of these interactions can function as either repressors or activators, increasing or decreasing transcript levels of target genes, depending on the cellular context and the promoters they bind [27, 40].

Recent progress has been made related to the transcriptional regulation of flavonoid biosynthetic pathway, especially in the identification of a remarkable network comprising MYB, bHLH, and WD40 regulatory proteins. MYB transcription factors are characterized by the MYB domain in the N-terminal with one to three imperfect repeats of almost 52 amino acids (R1, R2, and R3). In the case of plants, those MYB proteins associated to the spatial and temporal patterning of anthocyanin biosynthesis are characterized by a conserved DNA-binding domain consisting of two imperfect repeats called R2R3, which are expected to bind to promoters of the structural biosynthetic genes directly [12, 22]. Besides providing pigmentation in flowers and fruits, these MYB proteins are also involved in a myriad of physiological and biochemical processes [41–43].

Basic helix-loop-helix proteins represent a superfamily of transcription factors conserved from yeast to mammals [44]. Family members show a conserved domain with 50–60 amino acids composed of two functionally distinct regions, a basic domain located at the N-terminal end, and the hydrophobic helix-loop-helix (HLH) domain at the C-terminal end. These domains are involved in DNA binding and protein dimerization, respectively. In plants, the bHLH proteins are another group of regulatory elements linked to the regulation of flavonoid metabolism, hormonal response, cell proliferation, and, finally, cell differentiation processes such as those involved in the production of trichomes, root hair, and stomata [45–48].

MYB and bHLH proteins represent two of the largest groups of transcription factors found throughout eukaryotic organisms. In plants, proteins belonging to these two families generally interact with each other, and can be considered part of one of the best described examples of elaborate gene regulatory networks in higher plants [49, 50]. A multiple protein sequence alignment of the R2R3 domains of known

MYB proteins involved in the anthocyanin biosynthesis among different eudicot species revealed four conserved amino acids (ANDV) and conserved residues (arginine, valine, and alanine) in the R3 repeat [51, 52]. These motifs are important for the interaction between MYB and bHLH proteins in *Arabidopsis,* and another motif has recently been identified in the C-terminal region of the MYB proteins regulating anthocyanin production that is also important for protein–protein interactions [51, 52].

Likewise, WD40 proteins are highly conserved and can be found in organisms that do not biosynthesize anthocyanin pigments as algae, fungi, and animals [28, 29]. In spite of the absence of a DNA-binding domain, these proteins have conserved regions constituted of nearly 40–60 amino acids and characterized by the presence of repetitive segments in tandem of a glycine–histidine (GH) dipeptide and a tryptophan–aspartate (WD) dipeptide [25]. In plants, functional data have indicated that these proteins are involved in a wide array of developmental and biochemical processes, including facilitating protein–protein interactions, particularly between MYB and bHLH transcription factors [23, 25, 28, 29, 53].

Differential Molecular Mechanisms of the Flavonoid Biosynthetic Pathway in Monocot and Dicot Plants

Despite the reported conserved role for the MYB, bHLH, and WD40 proteins as regulators of the flavonoids' biosynthesis in several species of monocot and dicot plants, these transcription factors contribute differentially in activating different sets of target genes, resulting in distinct regulatory mechanisms in the anthocyanin pathway (Fig. 4.2).

In maize (*Z. mays*), it is known that flavonoid biosynthesis is transcriptionally regulated by an interaction between *colorless1* (*ZmC1*) or *purple leaf1* (*ZmPL1*), which encode MYB proteins, and *red1* (*ZmR1*) or *booster1* (*ZmB1*), which encode bHLH proteins [32, 33, 54–57]. Since these members of each family are functionally redundant, typically only one of the genes of each family is necessary and sufficient to activate anthocyanin/proanthocyanin production in a given tissue. For instance, the expression of the *ZmB1* gene and *ZmPL1* is sufficient to activate pigmentation [57].

Interestingly, these multiple protein–protein combinations between MYB and bHLH are possible due to direct physical interactions described to occur in the C-terminal domain of the R2R3-MYB protein ZmC1 and in the N-terminal domain of the bHLH protein ZmR1. Specific amino acids in the R3 repeat of the ZmC1 protein have also been shown to be functionally essential for the specificity of the interaction with ZmR1 protein in the control of anthocyanin accumulation in the aleurone layer of the kernel [58, 59].

In addition, *pale aleurone color1* (ZmPAC1), a WD40 protein also participates in the activation of multiple anthocyanin biosynthetic complex, responsible for the accumulation of pigments in seeds, since *pac1* mutants have a reduced level of

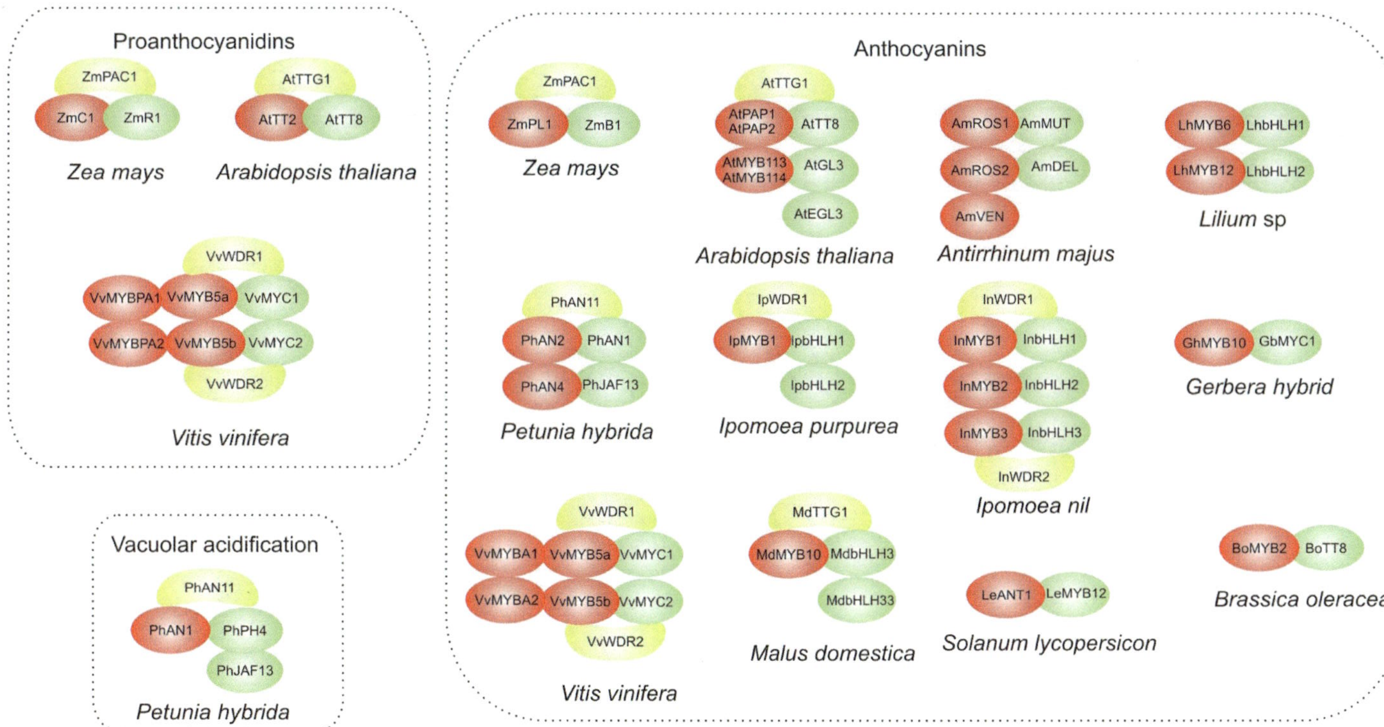

Fig. 4.2 Regulatory networks formed by MYB–bHLH–WD or MYB–bHLH complexes involved in proanthocyanidin/anthocyanin biosynthesis and vacuolar acidification identified in diverse plant species. MYB proteins are indicated in *red,* bHLH in *green,* and WD40 in *yellow*

anthocyanin production in the kernel pericarp [21, 60]. Thus, although ZmPL1 may activate the expression of the genes of the phlobaphene pathway in kernels without bHLH partners [58], the ternary MBW complex regulates the entire set of flavonoid biosynthesis genes in tissues such as leaves, seedlings, anthers, and seeds [59].

On the other hand, numerous studies have revealed that the flavonoid pathway in *Arabidopsis* requires complex combinatorial interactions between distinct MYB–bHLH–WD40 regulatory genes [61–63]. The induction of the expression of several flavonoid structural genes present in early steps of the pathway, such as *CHS, CHI,* and *F3 H* is performed by AtMYB11, AtMYB12, and AtMYB111, which are R2R3-MYB transcription factors. These seem to act redundantly and independently of bHLH proteins, since triple mutants showed downregulation of early flavonoid biosynthetic genes, affecting the formation of flavonols, but no effects were observed in anthocyanin production and accumulation [61, 62].

The production of anthocyanins in vegetative tissues and proanthocyanidins in seed coat is controlled by the interaction between MYB–bHLH–WD40 proteins during the "late" steps of the anthocyanin pathway [63–65]. Dihydroflavonols serve as common precursors for flavonol production, by the activity of flavonol synthase (FLS) enzyme, or for the formation of anthocyanidin and/or proanthocyanidin by the action of the DFR enzyme [62–65], as represented in Fig. 4.1.

Three regulators, *transparent testA2* (AtTT2), a R2R3-MYB protein,*transparent testA8* (AtTT8), a bHLH factor, and *transparent testA glabra1* (AtTTG1), a WD40 protein, are specifically responsible for the production of proanthocyanidins in developing seeds by activating the expression of the *banylus* gene, which encodes an anthocyanidin reductase enzyme [65, 66]. Additionally, these proteins activate the transcription of some "late" biosynthetic genes of the flavonoid biosynthetic pathway [66, 67].

Similarly, AtTTG1 forms another ternary complex with the bHLH proteins AtTT8, *glabra3* (AtGL3), and*enhancer of glabra3* (AtEGL3), modulating the synthesis of anthocyanin in seedlings [68]. Furthermore, this complex was shown to participate in an additional regulatory network involved in the determination of epidermal cell fates such as the formation of root hair, trichomes, and seed coat mucilage [66, 67].

Production of anthocyanin pigment1 (AtPAP1) and AtPAP2, AtMYB113 and AtMYB114 are redundant R2R3-MYB transcription factors that also control anthocyanin biosynthesis in *Arabidopsis* seedlings by upregulating the expression of LBGs such as *DFR*, *LDOX*, and *UFGT* [61, 63]. Constitutive overexpression of all these genes in *Arabidopsis* and in tobacco showed elevated pigmentation levels in vegetative organs and petals, respectively [63]. Moreover, it has been proposed that orthologs for these four genes are missing in monocots, indicating a recent divergence of these MYBs [52]. Vanderauwera et al. demonstrated that both *AtPAP1* and *AtPAP2* genes regulate anthocyanin biosynthesis in response to high-light stress [69]. Also, yeast two-hybrid assay indicated that the action of these four MYB proteins is dependent on the presence of the AtTTG1 WD40, as well as the AtTT8, AtGL3, and AtEGL3 bHLH transcription factors, confirming the formation of multiple MBW complexes [63, 68].

How do MBW Complexes Modulate the Transcriptional Regulation of the Anthocyanin Biosynthetic Pathway in Flowers?

The different color patterns observed in many ornamental plants, especially in their petals, are one of the adaptive traits associated to the attraction of pollinators [24, 26, 70, 71]. These color patterns result from the combination of two factors: one is the identity(ies) of the particular molecule(s) that is(are) synthesized; the other one is their relative proportion present in the vacuole of the epidermal cells. In addition, an interesting property of anthocyanins is to present different colors (due to their absorption spectrum) which depend largely on acidity or alkalinity (pH value) of the milieu, which results in the structural changes of anthocyanin molecules and consequently the final flower color [20, 72]. Nevertheless, the production and accumulation of these pigments, as already mentioned, require a complex array of transcription factors that participate differently in each set of the anthocyanin biosynthetic pathway among species [2, 20–25].

In *Petunia,* WD40 PhAN11 proteins interact with bHLH PhAN1 and R2R3-MYB PhAN2 or PhAN4 proteins for activation of anthocyanin production in the limbs of the petals and in the anthers, respectively [6, 30, 50]. The PhJAF13 protein, an additional bHLH regulator in *Petunia*, also interacts with PhAN2 for the activation of anthocyanin biosynthetic genes. However, by reciprocal complementation experiments, it was shown that PhJAF13 is not functionally similar to PhAN1 [6, 50]. In addition, PhAN11 is also expressed in nonpigmented tissues, indicating it has broader functions besides the control of anthocyanin biosynthesis. PhAN1 and PhAN11 are required for the acidification of the vacuolar pH in petal cells, interacting with a second R2R3-MYB transcription factor, PhPH4. Molecular analysis of mutations in the MBW complex revealed an increase of vacuolar pH and consequently, a more "bluish" *Petunia* flower [73].

The corolla of *Antirrhinum majus* flowers exhibits a venation pattern, consisting of colored stripes associated to veins that affect color intensity and perception by particular pollinators, providing specific signals within the flower and functioning as a visual guide [31]. Like in other species, the biosynthesis of anthocyanins in *A. majus* is under spatial and temporal control, coordinated by transcriptional regulation of several genes encoding the enzymes involved in the anthocyanin biosynthetic pathway. Three genes encoding R2R3-MYBs transcription factors, *rosea1* (*AmROS1*), *rosea2* (*AmROS2*), and *venosa* (*AmVEN*) in combination with two bHLH proteins, *mutabilis* (AmMUT) and *delila* (AmDEL) [11, 31] are responsible for the floral pigmentation pattern. *delila* is required for corolla tube pigmentation, while both *delila* and *mutabilis* contribute to the pigmentation of the corolla lobes. Both *delila* and *mutabilis* act redundantly in the activation of the expression of a subset of LBGs, among them *F3 H*, *DFR*, *LDOX,* and *UFGT* [12]. *rosea*1/2 regulates the pattern and intensity of pigmentation in both corolla lobes and tubes. On the other hand, *venosa* is associated with the venation patterning of the corolla, however only when *rosea1* is inactive [11]. The expression of *F3 H*, *FLS*, *F3'H*, *DFR*,

LDOX, and *UFGT* genes is highly dependent of *rosea1*. On the other hand, *rosea2* controls only *CHI* and *F3'H* genes, while *venosa* induces *CHI, F3 H, FLS, F3'H, LDOX,* and *UFGT* genes, but not *DFR*. In fact, each R2R3-MYB regulatory protein activates the transcription of distinct target promoters of both early and late genes of the pathway, controlling differentially the pattern and the intensity of pigmentation in flowers of the genus. Based on these observations, it has been proposed that this specificity conferred by MYB transcription factors can be responsible for natural variation in anthocyanin pigmentation observed among *Antirrhinum* and *Petunia* flowers [11, 30]. Furthermore, the venation pigmentation present in the *Antirrhinum* corolla is likely to be attributable to partial overlapping of the spatial regulation of these MYB and bHLH proteins [11].

In morning glory (*Ipomoea purpurea*), IpMYB1, IpbHLH1, IpbHLH2, and IpWDR1 proteins form a MBW complex which regulates not only the anthocyanin pathway in flowers but also proanthocyanidin accumulation and trichome differentiation in seed coats [13]. The lack of IpMYB1, IpbHLH2, and IpWDR1 expression leads to a partial reduction in the expression of all anthocyanin biosynthesis genes in flowers, complete loss of the expression of LBGs for proanthocyanidin production in seed coats, whereas the transcription of EBGs was unaffected. One probable explanation is that IpMYB1 activates the expression of all structural genes, while IpbHLH2 has partial control of flavonoid pathway and EBGs genes are controlled by additional regulatory genes independently of MYB transcription factors and WD40 proteins. Like *TTG1* from *Arabidopsis* and *AN11* from *Petunia, IpWDR1* is also essential in this complex for the biosynthesis of anthocyanin and proanthocyanidin, as well as for seed pigmentation and for trichome formation [13].

In *Ipomoea nil,* another species of morning glory, three R2R3-MYB (InMYB1, InMYB2, InMYB3) transcriptional regulators in combination with three bHLH proteins (InbHLH1, InbHLH2, InbHLH3) and two WD40 proteins (InWDR1 and InWDR2) represent the MYB regulatory complex involved in the regulation of anthocyanin biosynthesis in flowers [74]. Different from all other eudicots, in the two *Ipomoea* species studied, the whole flavonoid pathway appears to be controlled coordinately as a single set, as it was observed for maize [74].

G. hybrida is a common ornamental flower that also accumulates anthocyanin pigments mainly by coordinated transcriptional activation of the structural genes by R2R3-MYB (GhMYB10) and bHLH (GhMYC1) transcription factors [75, 76].

Lilium spp. flowers (Asiatic hybrid lily) are recognized because of their ornamental qualities, exhibiting a large variation in the color of their sepals, generally presenting dark red spots on their adaxial surface. Anthocyanins are responsible for that color variability, in which biosynthesis is transcriptionally regulated by a combination of two R2R3-MYB transcriptions factors, LhMYB6 and LhMYB12 (homologs of PhAN2) [77] and two bHLH proteins, LhbHLH1 and LhbHLH2 [78]. For example, LhMYB6, a light-induced anthocyanin-biosynthesis protein, regulates the presence of the dark red spots on the adaxial surface of sepals and pigmentation in leaves of juvenile shoots, while LhMYB12 controls anthocyanin production in tepals, stamens, and styles. In addition, *LhbHLH1* and *LhbHLH2* are the *Lilium* homologs of the *Petunia JAF13* and *AN1*, respectively [78].

Flavonoids and Anthocyanin Biosynthesis in Fruits and Vegetables

The transcriptional regulation of flavonoid biosynthesis has also been demonstrated to occur during the development and maturation of fleshy fruits, especially in important food crop plants such as grapevine and apple. The composition and concentration of flavonols, anthocyanins, or proanthocyanidins are important factors that influence organoleptic properties, flesh colors, and consequently their marketability and consumption. Particularly in the case of the anthocyanins, recent findings pointed out these pigments as potentially beneficial to human health due to their high antioxidant activities and therapeutic properties. Foods rich in anthocyanins may reduce risk of cancers, cardiovascular diseases, and other age-related diseases [79, 80].

Red grape (*V. vinifera*) represents an important source of several flavonoid compounds that are ultimately responsible for the astringency and final colors of red wines. Among them, proanthocyanidins are synthesized predominantly in seeds at the green berry stage [81], whereas anthocyanins are accumulated in skins during berry ripening [82]. The induction of structural genes, required for the biosynthesis of these two flavonoid compounds in developing seeds and skins of grape berries, is also triggered by a combinatorial control of the activity of MYB, bHLH, and WD40 proteins [36–38].

Initially, two R2R3-MYB genes, *VvMYBA1* and *VvMYBA2* have been identified controlling the expression of the *UFGT* gene that encodes an enzyme required for the conversion of anthocyanidins into anthocyanins, corresponding to the last step of the anthocyanin pathway [83]. Ectopic constitutive expression of both genes was sufficient to cause the synthesis and accumulation of pigments in all vegetative organs of transgenic plants, demonstrating that they play a crucial role in the induction of anthocyanin pathway [82–83]. Nevertheless, the absence of anthocyanins in a white grape cultivar has been related to a retrotransposon insertion in the promoter region of the *VvMYBA1* gene, as well as to two mutations in the coding region of the *VvMYBA2* gene [84–85].

On the other hand, *VvMYBPAP1* and *VvMYBPAP2* encode R2R3-MYB transcription factors specifically controlling proanthocyanidin synthesis in seeds by directly regulating the expression of genes encoding for the ANR and the LAR enzymes [86, 87].

Two additional R2R3-MYB genes: *VvMYB5a* and *VvMYB5b* were also identified as regulators of the production of anthocyanins and other flavonoid compounds in skin, mesocarp, and seeds, during the early phase of berry development and berry ripening, respectively [34, 35]. It has been suggested that *VvMYB5a* is specifically involved in the regulation of proanthocyanidin production in skin and seed tissues, whereas *VvMYB5b* promotes both proanthocyanidin and anthocyanin synthesis during berry development [34, 35]. Differential effects of the grapevine *VvYB5a* and *VvMYB5b* overexpression were observed in tobacco (*Nicotiana tabacum*) and tomato (*Solanum lycopersicum*), indicating that they may control different branches of the phenylpraponoid pathway in plants [34, 35]. In tobacco, the high expression levels of all flavonoid structural genes were tightly correlated with the increase of flavonoid metabolism and anthocyanin accumulation in flowers [34]. In tomato, the overexpression of flavonoid structural genes affected vegetative and reproductive traits, and the synthesis of terpenoids was upregulated, interfering in organoleptic characteristics of the fruits [88].

Recently, two bHLH transcription factors, VvMYC1 and VvMYCA1 [26], together with two WD40 proteins, VvWDR1 and VvWDR2 were identified to be necessary for the formation of the MBW regulatory complex controlling proanthocyanidin and anthocyanin biosynthesis in red grape [36, 89]. Using transient promoter and yeast two-hybrid experiments, it was demonstrated that VvMYC1 is able to interact and to form dimers with all grape MYB partners (VvMYBA1/2, VvMYBPA1, and VvMYB5a/5b), promoting the transcription of proanthocyanidin and/or anthocyanin biosynthetic genes, a function shared by its *Arabidopsis* orthologs *AtTT8*, *AtGL3,* and *AtEGL*3 [89]. In addition, *VvMYC1* showed to be involved in a positive feedback mechanism that regulates its own expression, as it was described for its *Arabidopsis* ortholog, *AtTT8* [89].

Both *V. vinifera* WD40 proteins described so far, showed high sequence similarity to AN11 and TTG1 from *Petunia* and *Arabidopsis,* respectively. It has been demonstrated that *VvWDR1* is a coactivator of anthocyanin synthesis when ectopically expressed in *Arabidopsis* plants; meanwhile no evidence has been presented to demonstrate a role for *VvWDR2* gene in the flavonoid pathway [36]. Based on studies of gene expression, at least two distinct regulatory sets participate in the early and late steps of grape berry development, and, accordingly, different R2R3-MYBs regulate separately the synthesis of flavonols, anthocyanidins, and proanthocyanidins [34, 35, 86, 87].

In apple (*M. domestica*) fruits, some transcription factors regulating the production of anthocyanin were described and apparently regulate both early and late genes of the anthocyanin pathway. Previous studies have identified three R2R3-MYB transcription factors (MdMYB10, MdMYB1, and MdMYBA) and two bHLH proteins (MdbHLH3 and MdbHLH33) as responsible for red coloration in apple fruit skin [36–39]. *MdMYB1* is an ortholog of *AtPAP1* from *Arabidopsis* and it is able to regulate anthocyanin biosynthesis in response to sunlight exposure whereby its expression was markedly enhanced [37]. Ban et al. demonstrated that UV-B irradiation and low temperature also stimulate the expression of *MdMYBA* gene, leading to an increased anthocyanin accumulation in apple fruit skin [39]. Despite the fact that MdMYB10 and MdMYB1 proteins share high homology, differing in only three amino acids in their sequences, they showed differential expression patterns in fruit tissues, and were described to control the pigmentation of the cortex (MdMYB10) or of the skin (MdMYB1 and MdMYBA) [38]. Once again, differences in the activity of MYB transcription factors indicate that these proteins may give the spatial specificity of the action of MBW complexes responsible for anthocyanin accumulation, as previously described for *Antirrhinum* and *Petunia* flowers [11, 30], as well as in the fruits of grape [34, 35] and in fruits of some Solanaceae species such as tomato, pepper, and eggplant [90].

Regulatory mechanisms controlling anthocyanin accumulation associated with the ripening phase of fruit development have been recently described [52, 91–94]. In order to further characterize tissue-specific anthocyanin accumulation with their regulation, as well to gain new knowledge of the evolution of the *MYB* gene family, orthologs of the apple *MdMYB10* gene have been analyzed in many other commercially important species from Rosaceae family such as plum, pear, peach, raspberry, cherry, and strawberry [52]. Moreover, GmMYB10, PhMYB10, and MrMYB1 are also MYBs transcriptional regulators that influence fruit pigmentation in mangosteen, pear, and *Myrica rubra* (Chinese bayberry), respectively [92–94]. In all species analyzed, it was observed that levels of expression of these R2R3-MYBs were spatially and temporally linked to anthocyanin accumulation during fruit ripening, providing strong evidence of the importance of R2R3-MYBs in the control of anthocyanin biosynthesis in these crops [52, 91–94].

VmTDR4, a *squamosa* A-class *mads*-box transcription factor was also identified and characterized in bilberry (*Vaccinium myrtillus*) and related to the control of anthocyanin accumulation during fruit ripening [91]. *VmTDR4* is orthologous to *TDR4* from tomato and to the *fruitfull* gene from *Arabidopsis*. The inhibition of *VmTDR4* expression by virus induced gene silencing (VIGS) caused a significant reduction of anthocyanin accumulation during the ripening of the fruits due to the suppression of several flavonoid biosynthesis genes, such as *CHS*, *DFR*, and *LDOX*, as well as of the expression of *VmMYB2*, a R2R3-MYB orthologous to the *AtPAP1* from *Arabidopsis*. In contrast, the expression of *VmMYB1* and *ANR* genes were significantly enhanced in early developmental stages of VmTDR4-silencied fruits [91]. The first gene shares high homology with *FaMYB1*, a known suppressor of anthocyanin biosynthesis in strawberry fruits [95], while the second one is essential for the production of proanthocyanidin, which is abundantly found in green fruits before the accumulation of anthocyanins. Taken together, these observations suggest a central role for *VmTDR4* in the regulation of anthocyanin in bilberry fruits, by the direct or indirect mediation of the expression of a R2R3-MYB transcription factor [91].

Nevertheless, several members of the MYB gene family have been reported to act as repressors of the flavonoid pathway, specifically of anthocyanins, in different plant species [95–101]. Two proteins with a single MYB domain (R3), *AtMYBL2* and *caprice* (*CPC*) have been described to be transcriptional repressors that negatively regulate the production and accumulation of anthocyanin in *Arabidopsis* by interacting with *AtTT8* [96–98]. The *FaMYB1* gene downregulates both anthocyanin and flavonol biosynthesis in strawberry (*Fragaria x ananassa*) and in tobacco [95]. Besides, it has been shown by ectopic expression that the *FaMYB1* gene can act also as a repressor of proanthocyanidin production in the leaves of *Lotus corniculatus* [99]. Recently, the *PhMYB27* and *MdMYB17* genes were suggested as putative repressors of anthocyanin synthesis in *Petunia* and apple fruits, respectively [100, 101]. A possible explanation for the inactivation of the MBW complex in the anthocyanin pathway is due to the competition for binding sites on common target promoters or between the single R3-MYB protein and other R2R3-MYB proteins for interaction with bHLH transcription factors [95–98].

Molecular mechanisms underlying pigmentation in red cabbage and purple cauliflower have also been reported [102, 103]. The coloration exhibited by red cabbage (*Brassica oleracea* L. var. *capitata*) is due to high levels of anthocyanins, cyanidin 3-diglucoside, 5-glucose [104, 105]. In *B. oleracea,* the concerted action of BoMYB2, a R2R3-MYB protein and BoTT8, a bHLH transcription factor mediates anthocyanin biosynthesis [102]. In cauliflower (*B. oleracea* var. *botrytis*), a mutation in the *purple* gene, encoding a R2R3-MYB transcription factor, resulted in a notable anthocyanin accumulation, conferring a phenotype of intense purple pigmentation in curds [103].

In purple sweet potato, anthocyanins are accumulated in several aerial tissues such as stem, leaves, and flowers, but predominantly in their tuberous roots. Two sweet potato R2R3-MYBs showed differential expression patterns: The *IbMYB1* gene is highly expressed in the tuber flesh, indicating to play an important role in pigmentation, whereas *IbMYB2* gene showed only weak expression [106]. Instead of the typical MBW complex, *IbMADS10,* a *mads*-box gene that belongs to the *squa* subfamily is also involved in the accumulation of anthocyanin in tuberous roots [106] as previously described in bilberry fruits [91].

Two related R2R3-MYB transcription regulators were characterized controlling the biosynthesis of anthocyanins and other flavonoids in the skin of tomato fruits (*S. lycopersicum*) [107, 108]. Activation tagging experiments identified a transcriptional regulator of anthocyanin synthesis, *LeANT1* (a R2R3-MYB gene) showing high homology with *PeAN2* from *Petunia* [107]. Vegetative tissues of *leant1* mutant exhibited intense purple coloration, whereas fruits showed purple spots on the epidermis and subepidermal cells. Moreover, when it was overexpressed, it resulted in the upregulation of structural genes involved in both early and late steps of the anthocyanin pathway, as well as of a gene that encodes a glutathione *S*-transferase (GST) enzyme, responsible for anthocyanin transport [107]. On the other hand, *LeMYB12* plays a central role in regulating the production of flavonol in tomato fruit, since it provides the activation of the early structural genes of the pathway, such as *CHS*, *CHI*, *F3 H*, *F3'H,* and *FLS* [108]. Furthermore, the tomato pink fruit phenotype is likely caused by a mutation affecting the *LeMYB12* gene [108].

Cultivated tomatoes do not contain anthocyanins in their fruits. The final red color displayed in the skin of tomato fruits is due mainly to the accumulation of carotenoids and naringenin chalcone, a yellow flavonol, during fruit ripening [108]. However, the introgression of genes such as dominant allele *anthocyanin fruit* (*Aft*) [109] and *aubergine* (*Abg*) [110] from the wild species *Solanum chilense* and *Solanum lycopersicoides* into cultivated tomato resulted in a purple skin pigmentation in tomato fruits. Moreover, both genes require high-light intensity to activate anthocyanin biosynthesis [111]. Other recessive gene, *atroviolaceum* (*Atv*) from wild *Solanum cheesmaniae,* also causes purple color in vegetative tissues and fruits [112]. Different combinations of these alleles produced fruits with phenotypic variation of anthocyanin pigments and, in general, the association of *Atv* activity with either *Aft* or *Abg* increased anthocyanin production in the epidermis of tomato fruits [111].

Similarly to *Aft* and *Abg* in tomato, other genes are also dependent of different environmental conditions to activate or to enhance flavonoid biosynthetic pathway.

Among them, light is one of the abiotic factors that stimulate anthocyanin production by upregulating the expression of some genes of the flavonoid pathway, as reported in apple, pear, and Chinese bayberry [37, 93, 94].

Recently, two R2R3-MYB members from *Petunia, deep purple* (*DPL*) and *purple haze* (*PHZ*) were identified and characterized as additional genes in combination with *PhAN2* and *PhAN11* to form a MBW complex involved in the regulation of anthocyanin metabolic pathway [100]. According to Albert et al., this new MBW complex requires high light induction of anthocyanin synthesis in vegetative tissues, indicating an increase of transcription levels for *PHZ* and *AN1* genes. These findings indicate that the R2R3-MYB subfamily is the primary determinant responsible for spatial localization of anthocyanins both in flowers and in fruits, resulting in different pigmentation patterns as previously reported for *Antirrhinum* and *Petunia* [11, 30]. Moreover, R2R3-MYB transcription factors are key regulators of the anthocyanin accumulation in response to light stimulus [37, 93, 94].

Metabolic Engineering of Flavonoids in Flowers and Fruits

Flavonoids comprise a wide range of roles, from flower and fruit pigmentation to human dietary nutrition. Understanding the molecular basis of the regulatory networks comprised by different families of transcription factors, undoubtedly offers a powerful gene technology for modulating flavonoid and anthocyanin content as well as the temporal and spatial accumulation of these secondary metabolites in plants. For this purpose, metabolic engineering via transgenic plants has been used as a research and breeding tool to alter flavonoid biosynthetic pathway via the manipulation of expression patterns of regulatory genes in heterologous systems and/or downregulation of endogenous genes in species of agronomic interest [24, 34–39, 102, 103, 107, 108].

For many years, anthocyanin research has been focused on flower color modification. The best known example is the generation of blue flowers, such as roses and carnations [113, 114]. In general, natural blue coloration exhibited by some flowers is due to delphinidin accumulation associated to the number of hydroxyl groups on the anthocyanin B-ring and/or linked to aromatic acyl groups present in the anthocyanin structure [115]. Both roses and carnations do not have blue varieties due to the absence of a cytochrome P450 gene that encodes F3′5′H, a key enzyme to synthesize delphinidin-class of anthocyanins [16, 115, 116]. In the case of many rose cultivars, the downregulation of endogenous *DFR* gene and the overexpression of a viola *F3′5 ′H* gene were sufficient to modify the anthocyanin structure by additional hydroxyl groups on the B-ring and, consequently, the alteration of the anthocyanin pathway [113].

The wide diversity of flower pigmentation found among angiosperms may have two main explanations. First, the inactivation of one or more enzymes involved in the anthocyanin biosynthesis, with more or less hydroxyl groups linked to anthocyanin molecules and, consequently, a shift from red to blue as mentioned above, or the production of red flowers in plant that does not produced pelargonidin-type anthocyanin (orange–red pigments) due to an inefficiency of the DFR enzyme due to the reduction of the dihydrokaempferol (DHK, Fig. 4.1) [117]. For example, the suppression of two endogenous *F3'H* and *FLS* genes by RNAi technology together with the expression of a *Gerbera DFR* gene has successfully generated red tobacco flowers [117]. Secondly, mutations ("loss of function") in the coding sequence of either structural genes or MBW transcription factor complexes involved in the anthocyanin biosynthesis may change flower color or, most notably, the transition between pigmented and white flowers [11, 14, 22–24, 30, 31]. One example is GtMYB3, a R2R3-MYB transcription factor identified in gentian flowers with remarkable homology to *PhAN2*, *AtPAP1*, *AtTT2,* and *ZmC1*, genes known to be recruited for anthocyanin and proanthocyanidin biosynthesis in other species [118]. Several mutations in the *GtMYB3* gene resulted in white gentian flowers, strengthening the central role of this gene in the regulation and accumulation of gentiodelphin, a specific polyacylated anthocyanin found in the gentian flowers [118]. Likewise, mutants by T-DNA insertion in *InMYB*1 (*c*), *bHLH* (*Ipivs*), and *InWDR1* (*ca*) displayed white morning glory flowers [13, 74].

Besides flowers, anthocyanins are also found in leaves, tuberous roots, bulbs, legumes, cereal, other flower parts, and fruits. Moreover, other flavonoids like flavonol can be found in leaves, while proanthocyanidin is present in seeds and grains [19, 62, 119]. Several studies using similar approaches and molecular tools as mentioned previously for flowers have also been performed aiming the manipulation of flavonoid biosynthesis to improve the quality of fruits and vegetables, enhancing their nutritional value.

It has been previously reported, the presence of a transcriptional complex in apple fruits, where three R2R3-MYBs together with bHLH proteins, was responsible for the coordinated regulation of fruit red skin color in fruits [36–39]. However, it has been recently demonstrated that apple has only one R2R3-MYB transcription factor, MYB10, while MYBA and MYB1 represent alleles of this MYB10 locus [120]. Moreover, apple skin variegation may be explained by distinct anthocyanin levels and differential expression of the *MYB10* gene, which apparently is associated with methylation in the promoter region of this gene [120]. A better understanding of this regulatory mechanism could result in new tools for improving the visual traits of fruits and vegetables [121].

Overexpression of two transcription factors from *Antirrhinum, delila* (a bHLH protein) *and roseal* (a R2R3-MYB protein) under the control of a tomato fruit-specific promoter resulted in a significant enhancement of anthocyanin (delphinidin) and flavonols accumulation in the flesh and skin of tomato fruits, leading to dark purple fruits, similar to blueberries and blackberries, containing high anthocyanin concentration [122].

The stringency showed by persimmon (*Diospyros kaki*) is due to high proanthocyanidin accumulation in their fleshy fruits [123]. *Pollination constant and nonastringent* (*pcna*) is a mutation that causes a reduction of proanthocyanidin accumulation in cells at early fruit developmental stage as a result of the downregulation of *ANR* and *F3'5'H* expression. Both are structural genes responsible for the synthesis of different flavonoids that comprise the proanthocyanidin profile [123, 124]. DkMYB4 is a R2R3-MYB transcription factor identified as a key determinant for the downregulation of these genes in persimmon fruits [124], while its ectopic expression in kiwifruit (*Actinidia deliciosa*) resulted in an enhanced proanthocyanidin accumulation, it did not induce anthocyanin biosynthesis [124].

Common rice (*Oryza sativa*) has white pericarp, but other varieties can show grains with colored pericarp, as is the case of the cultivated "red" rice (*Oryza rufipogon*), which have dark red pericarp due to proanthocyanidin accumulation [119]. Rc, a bHLH protein, was identified as the main regulator of proanthocyanidin biosynthesis for red pericarp in rice. Furthermore, it was demonstrated that the dominant red allele Rc contains a small deletion located in the 60th exon when compared to the recessive white allele, which can be responsible by the disruption of the production of the native bHLH transcription factor [119].

The first steps to unravel the genes controlling natural variation pigmentation in seed coat and flowers in soybean (*Glycine max* L.) have been performed [125], as well investigations in the correlation of flavonoid content and antioxidant properties were performed in Brazilian and Peruvian bean (*Phaseolus vulgaris* L.) cultivars [126], but in both cases, no genes have yet been isolated.

Concluding Remarks

Nowadays, due to the great interest related to human health through a balanced daily diet including natural products widely found in fruits and vegetables, flavonoids continue to attract the attention of researchers because of their high-antioxidant properties, their potential of disease prevention, and therapeutic effects. In this context, there has been an increasing demand in unraveling the exact molecular mechanisms underlying the biosynthesis of flavonoids, particularly of anthocyanins, to ensure an increase of the content of these compounds in economically important crops. Understanding the different enzymatic steps of the flavonoid pathway and their regulatory elements is of utmost importance in order to generate foods with enhanced antioxidants and nutrients by means of genetic approaches. Based on recent studies, a common regulatory network comprising MYB–bHLH–WD40 proteins has been described as a potential transcriptional regulator of the flavonoid biosynthetic pathway. Among them, R2R3-MYB transcription factors were shown to activate different branches of the anthocyanin biosynthetic pathway, indicating their usefulness as a potential biotechnological tool for modulating the expression of their target genes.

Acknowledgments The authors acknowledge the following Brazilian institutions for financial support: Conselho Nacional de Pesquisa e Desenvolvimento Cientifíco e Tecnológico (CNPq), Fundação de Amparo à Pesquisa do Estado de São Paulo (FAPESP), and Coordenação de Aperfeiçoamento de Pessoal de Nível Superior (CAPES).

References

1. Harborne JB (1994) The flavonoids: advances in research since 1986, 1st edn. Chapman & Hall/CRC, New York
2. Grotewold E (2006) The genetics and biochemistry of floral pigments. Ann Rev Plant Biol 57:761–780
3. Ferrer JL, Austin MB, Stewart Jr C, Noel JP (2008) Structure and function of enzymes involved in the biosynthesis of phenylpropanoids. Plant Physiol Biochem 46:356–370
4. Castellarin SD, Pfeiffer A, Sivilotti P, Degan M, Peterlunger E, Di Gaspero G (2007) Transcriptional regulation of anthocyanin biosynthesis in ripening fruits of grapevine under seasonal water deficit. Plant Cell Environ 30:1381–1399
5. Gould KS (2004) Nature's Swiss army knife: the diverse protective roles of anthocyanins in leaves. J Biomed Biotechnol 5:314–320
6. Spelt C, Quattrocchio F, Mol JNM, Koes R (2000) *Anthocyanin1* of Petunia encodes a basic helix-loop-helix protein that directly activates transcription of structural anthocyanin genes. Plant Cell 12(9):1619–1631
7. Heller W, Forkmann G, Britsch L, Grisebach H (1985) Enzymatic reduction of (+)-dihydroflavonols to flavan-3,4- cis-diols with flower extracts from *Matthiola incana* and its role in anthocyanin biosynthesis. Planta 165(2):284–287
8. Stich K, Eidenberger T, Wurst F, Forkmann G (1992). Enzymatic conversion of dihydroflavonols to flavan-3,4-diols using flower extracts of *Dianthus caryophyllus* L. (carnation). Planta 187(1):103–108
9. Davies KM, Bradley JM, Schwinn KE, Markham KR, Podivinsky E (1993) Flavonoid biosynthesis in flower petals of five lines of lisianthus (*Eustoma grandiflorum* Grise). Plant Sci 95(1):67–77
10. Helariutta Y, Elomaa P, Kotilainen M, Seppänen P, Teeri TH (1993) Cloning of cDNA coding for dihydroflavonol-4- reductase (DFR) and characterization of dfr expression in the corollas of *Gerbera hybrida* var. Regina (Compositae). Plant Mol Biol 22(2):183–193
11. Schwinn K, Venail J, Shang Y, Mackay S, Alm V, Butelli E, Oyama R, Vailey P, Davies K, Martin C (2006) A small family of MYB regulatory genes controls floral pigmentation intensity and patterning in the genus *Antirrhinum*. Plant Cell 18(4):831–851
12. Martin C, Prescott A, Mackay S, Bartlett J, Vrijlandt E (1999) Control of anthocyanin biosynthesis in flowers of *Antirrhinum majus*. Plant J 1:37–49
13. Park KI, Ishikawa N, Morita Y, Choi JD, Hoshino A, Iida S (2007) A bHLH regulatory gene in the common morning glory, *Ipomoea purpurea*, controls anthocyanin biosynthesis in flowers, proanthocyanidin and phytomelanin pigmentation in seeds, and seed trichome formation. Plant J 49(4):641–654
14. Streisfeld MA, Rausher MD (2009) Genetic changes contributing to the parallel evolution of red floral pigmentation among *Ipomoea* species. New Phytologist 183:751–763
15. Halbwirth H (2010) The creation and physiological relevance of divergent hydroxylation patterns in the flavonoid pathway. Int J Mol Sci 11:595–621
16. Holton TA, Tanaka Y (1994) Blue roses- a pigment of our imagination? Trends Biotechnol 12:40–42
17. Tanaka Y, Sasaki N, Ohmiy A (2008) Biosynthesis of plant pigments: anthocyanins, betalains and carotenoids. Plant J 54:733–749
18. Jonsson LMV, Aarsman MEG, Poulton JE, Schram AW (1984) Properties and genetic control of four methyltransferases involved in methylation of anthocyanins in flowers of *Petunia hybrida*. Planta 160:174-179
19. Lepiniec L, Debeaujon I, Routaboul J-M, Baudry A, Pourcel L, Nesi N, Caboche M (2006) Genetics and biochemistry of seed flavonoids. Ann Rev Plant Biol 57:405–430
20. Mol J, Grofewold E, Koes R (1998) How genes paint flowers and seeds. Trends Plant Sci 3:212–217

21. Selinger DA, Chandler VL (1999) A mutation in the pale aleurone color gene identifies a novel regulator of the maize anthocyanin pathway. Plant Cell 11:5–14
22. Holton TA Cornish EC (1995) Genetics and biochemistry of anthocyanin biosynthesis. Plant Cell 7:1071–1083
23. Koes K, Verweij W, Quattrocchio F (2005) Flavonoids: a colorful model for the regulation and evolution of biochemical pathways. Trends Plant Sci 10:236–242
24. Davies K (2004) Plant pigments and their manipulation. Ann Plant Rev 14:368
25. Ramsay NA, Glover BJ (2005) MYB-bHLH-WD40 protein complex and the evolution of cellular diversity. Trends Plant Sci 10:63–70
26. Winkel-Shirley B (2001) Flavonoid biosynthesis. A colorful model for genetics, biochemistry, cell biology, and biotechnology. Plant Physiol 126:485–493
27. Petroni K, Tonelli C (2011) Recent advances on the regulation of anthocyanin synthesis in reproductive organs. Plant Sci 181:219–229
28. de Vetten N, Quattrocchio F, Mol J, Koes R (1997) The *an11* locus controlling flower pigmentation in petunia encodes a novel WD-repeat protein conserved in yeast, plants and animals. Genes Dev 11:1422–1434
29. Brueggemann B, Weisshaar B, Sagasser M (2010) A WD40-repeat gene from Malus x domestica is a functional homologue of *Arabidopsis thaliana* TRANSPARENT TESTA GLABRA1. Plant Cell Rep 29:285–294
30. Quattrocchio F, Wing J, van der Woude K, Souer E, de Vetten N, Mol, J, Koes R (1999) Molecular analysis of the anthocyanin2 gene of Petunia and its role in the evolution of flower color. Plant Cell 11:1433–1444
31. Shang Y, Venail J, Mackay S, Bailey PC, Schwinn KE, Jameson PE, Martin CR, Davie KM (2011) The molecular basis for venation patterning of pigmentation and its effect on pollinator attraction in flowers of *Antirrhinum*. New Phytologist 189:602–615
32. Hernandez JM, Heine GF, Irani NG, Feller A, Kim, M-G, Matulnik T, Chandler VL, Grotewold E (2004) Different mechanisms participate in the R-dependent activity of the R2R3 MYB transcription factor C1. J Biol Chem 279(46):48205–48213
33. Grotewold E, Drummond BJ, Bowen B, Peterson T (1994) The myb-homologous *P* gene controls phlobaphene pigmentation in maize floral organs by directly activating a flavonoid biosynthetic gene subset. Cell 76(3):543–555
34. Deluc L, Barrieu F, Marchive C, Lauvergeat V, Decendit A, Richard T, Carde JP, Merillon JM, Hamdi S (2006) Characterization of a grapevine R2R3-MYB transcription factor that regulates the phenylpropanoid pathway. Plant Physiol 140:499–511
35. Deluc L, Bogs J, Walker AR, Ferrier T, Decendit A, Merillon JM, Robinson SP, Barrieu F (2008) The transcription factor VvMYB5b contributes to the regulation of anthocyanin and proanthocyanidin biosynthesis in developing grape berries. Plant Physiol 147:2041–2053
36. Matus JT, Poupin MJ, Cañón P, Bordeu E, Alcade JA, Arce-Johnson P (2010) Isolation of WDR and bHLH genes related to flavonoid synthesis in grapevine (*Vitis vinifera* L). Plant Mol Biol 72:607–620
37. Takos AM, Jaffe FW, Jacob SR, Bogs J, Robinson SP, Walker AR (2006) Light induced expression of a MYB gene regulates anthocyanin biosynthesis in red apples. Plant Physiol 142:1216–1232
38. Espley RV, Hellens RP, Putterill J, Stevenson DE, Kutty-Amma S, Allan AC (2007) Red colouration in apple fruit is due to the activity of the MYB transcription factor, MdMYB10. Plant J 49:414–427
39. Ban Y, Honda C, Hatsuyama Y, Igarashi M, Bessho H, Moriguchi T (2007) Isolation and functional analysis of a MYB transcription factor gene that is a key regulator for the development of red coloration in apple skin. Plant Cell Physiol 48:958–970
40. Wolberger C (1999) Multiprotein-DNA complexes in transcriptional regulation. Ann Rev Biophys Biomol Struct 28:29–56
41. Jin H, Martin C (1999) Multifunctionality and diversity within the plant MYB-gene family. Plant Mol Biol 41:577–585

42. Martin C, Paz-Ares J (1997) MYB transcription factors in plants. Trends Genet 13:677–673
43. Stracke R, Werber M, Weisshaar B (2001) The R2R3-MYB gene family in *Arabidopsis thaliana*. Curr Opin Plant Biol 4:447–456
44. Pires N, Dolan L (2010) Origin and diversification of basic-helix-loop-helix proteins in plants. Mol Biol Evol 27:862–874
45. Heim MA, Jakoby M, Werber M, Martin C, Weisshaar B, Bailey PC (2003) The basic helix-loop-helix transcription factor family in plants: a genome-wide study of protein structure and functional diversity. Mol Biol Evol 20:735–747
46. Toledo-Ortiz G, Hug E, Quail PH (2003) The *Arabidopsis* basic helix-loop-helix transcription factor family. Plant Cell 1749–1770
47. Serna L (2007) bHLH proteins know when to make a stomata. Trends Plant Sci 12:483–485
48. Zhang F, Gonzalez A, Zhao M, Payne CT, Lloy A (2003) A network of redundant bHLH proteins functions in all TTG1-dependent pathways of *Arabidopsis*. Development 130:4859–4869
49. Feller A, Machemer K, Braun E, Grotewold E (2011) Evolutionary and comparative analysis of MYB and bHLH plant transcription factors. Plant J 66:94–116
50. Quattrocchio F, Wing JF, van der Woude K, Mol JN, Koes R (1998) Analysis of bHLH and MYB domain proteins: species-specific regulatory differences are caused by divergent evolution of target anthocyanin genes. Plant J 13:465–488
51. Zimmerman IM, Heim MA, Weisshaart B, Uhrig JF (2004) Comprehensive identification of *Arabidopsis thaliana* MYB transcription factors interacting with R/B-like bHLH proteins. Plant J 40:22–34
52. Lin-Wang K, Bolitho K, Grafton K, Kortstee A, Karunairetnam S, McGhie TK, Espley RV, Hellens RP, Allan AC (2010) An R2R3 MYB transcription factor associated with regulation of the anthocyanin biosynthetic pathway in Rosaceae. BMC Plant Biol 10:50
53. Smith TF, Gaitatzes C, Saxena K, Neer EJ (1999) The WD repeat: a common architecture for diverse functions. Trends Biochem Sci 24:181–185
54. Cone KC, Burr FA, Burr B (1986) Molecular analysis of the maize anthocyanin regulatory locus C1. Proc Natl Acad Sci USA; 83:9631–9635
55. Paz-Ares J, Ghosal D, Wienand U, Peterson PA, Saedler H (1987) The regulatory c1 locus of *Zea mays* encodes a protein with homology to myb proto-oncogene products and with structural similarities to transcriptional activators. EMBO J 6:3553–3558
56. Ludwig SR, Habera LF, Dellaporta SL, Wessler SR (1989) Lc, a member of the maize R gene family responsible for tissue-specific anthocyanin production, encodes a protein similar to transcriptional activators and contains the myc-homology regions. Proc Natl Acad Sci USA 86:7092–7096
57. Ludwig SR, Wessler, SR (1990) Maize R gene family: tissue-specific helix–loop–helix proteins. Cell 62:849–851
58. Goff SA, Klein TM, Roth BA, Fromm ME, Cone KC, Radicella JP, Chandler VL (1990) Transactivation of anthocyanin biosynthetic genes following transfer of B regulatory genes into maize tissues. EMBO J 9:2517–2522
59. Grotewold E, Sainz MB, Tagliani L, Hernandez JM, Bowen B, Chandler VL (2000) Identification of teh residues in the Myb domain of maize C1 that specify the interaction with the bHLH cofactor R. Proc Natl Acad Sci USA 97:13579–13584
60. Carey CC, Strahle JT, Selinger DA, Chandler VL (2004) Mutations in the pale aleurone color1 regulatory gene of the *Zea mays* anthocyanin pathway have distinct phenotypes relative to the functionally similar *TRANSPARENT TESTA GLABRA1* gene in *Arabidopsis thaliana*. Plant Cell 16:450–464
61. Stracke R, Ishihara H, Huep G, Barsch A, Mehrtens F, Niehaus K, Weisshaar B (2007) Differential regulation of closely related R2R3-MYB transcription factors controls flavonol accumulation in different parts of the *Arabidopsis thaliana* seedling. Plant J 50:660–677
62. Mehrten F, Kranz H, Bednarek P, Weisshaar B (2005) The Arabidopsis transcription factor MYB12 is a flavonol-specific regulator of phenylpropanoid biosynthesis. Plant Physiol 138:1083–1096

63. Gonzalez A, Zhao M, Leavitt JM, Lloyd AM (2008) Regulation of the anthocyanin biosynthetic pathway by the TTG1/bHLH/Myb transcriptional complex in *Arabidopsis* seedlings. Plant J 53:814–827
64. Borevitz JO, Xia YJ, Blount J, Dixon RA, Lamb C (2000) Activation tagging identifies a conserved MYB regulator of phenylpropanoid biosynthesis. Plant Cell 12:2383–2393
65. Baudry A, Heim MA, Dubreucq B, Caboche M, Weisshaar B, Lepiniec L (2004) TT2, TT8, and TTG1 synergistically specify the expression of *BANYULS* and proanthocyanidin biosynthesis in *Arabidopsis thaliana*. Plant J 39:366–380
66. Nesi N, Jond C, Debeaujon I, Caboche M, Lepiniec L (2001) The *Arabidopsis TT2* gene encodes an R2R3 MYB domain protein that acts as a key determinant for proanthocyanidin accumulation in developing seed. Plant Cell 13:2099–2114
67. Nesi N, Debeaujon I, Jond C, Pelletier G, Caboche M, Lepiniec L (2000) The *TT8* gene encodes a basic helix-loop-helix domain protein required for expression of *DFR* and *BAN* genes in *Arabidopsis* siliques. Plant Cell 12:1863–1878
68. Zhang F, Gonzalez A, Zhao M, Payne CT, Lloyd A (2003) A network of redundant bHLH proteins functions in all TTG1-dependent pathways of *Arabidopsis*. Development 130:4859–4869
69. Vanderauwera S, Zimmermann P, Rombauts S, Vandenabeele S, Langebartels C, Gruissem W, Inzé D, Breusegem FV (2005) Genome-wide analysis of hydrogen peroxide-regulated gene expression in *Arabidopsis* reveals a high light-induced transcriptional cluster involved in anthocyanin biosynthesis. Plant Physiol 139:806–821
70. Fenster CB, Armbruster WS, Wilson P, Dudash MR, Thomson JD (2004) Pollination syndromes and floral specialization. Ann Rev Ecol Evol Syst 35:375–403
71. Thomson JD, Wilson P (2008) Explaining evolutionary shifts between bee and humminigbird pollination: convergence, divergence, and directionality. Int J Plant Sci 169:23–38
72. Yoshida K, Kondo T, Okazaki Y, Katou K (1995) Cause of blue petal colour. Nature 373:291
73. Quattrocchio F, Verweij W, Kroon A, Spelt C, Mol J, Koes R (2006) Ph4 of petunia is anR2R3MYB protein that activates vacuolar acidification through interactions with basic-helix-loop-helix transcription factors of the anthocyanin pathway. Plant Cell 18: 1274–1291
74. Morita Y, Saitoh M, Hoshino A, Nitasaka E, Iida S (2006) Isolation of cDNAs for R2R3-myb, bHLH and WDR transcriptional regulators and identification of c and Ca mutations conferring white flowers in the Japanese morning glory. Plant Cell Physiol 47:457–470
75. Elomma P, Mehto M, Kotilainen M, Helariutta Y, Nevalainem L, Teeri TH (1988) A bLHL transcription factor mediates organ region and flower type specific signals on dihydroflavonol-4-reductase (dfr) gene expression in the inflorescence of *Gerbera hybrid* (Asteraceae). Plant J 16:93–99
76. Elomaa P, Uimari A, Mehto M, Albert VA, Laitinem RAE, Teeri TH (2003) Activation of anthocyanin biosynthesis in *Gerbera hybrida* (Asteraceae) suggests conserved protein-protein and protein-promoter interactions between the anciently diverged monocots and eudicots. Plant Physiol 133:1831–1842
77. Yamagishi M, Shimoyamada Y, Nakatsuka T, Masuda K (2010) Two *R2R3-MYB* genes, homologs of Petunia *AN2*, regulate anthocyanin biosynthesis in flower tepals, tepal spots and leaves of Asiatic hybrid lily. Plant Cell Physiol 51:463–474
78. Nakatsuka A, Yamagishi M, Nakano M, Tasakia K, Kobayashi N (2009) Light-induced expression of basic helix-loop-helix genes involved in anthocyanin biosynthesis in flowers and leaves of Asiatic hybrid lily. Horticultural Sci 121:84–91
79. Hou DX (2003) Potential mechanisms of cancer chemoprevention by anthocyanins. Curr Mol Med 3:149–159
80. Lila MA (2004) Anthocyanins and human health: an *in vitro* investigative approach. J Biomed Biotechnol 2004:306–313
81. Kennedy JA, Hayasaka Y, Vidal S, Waters EJ, Jones GP (2001) Composition of grape skin proanthocyanidins at different stages of berry development. J Agric Food Chem 49:5348–5355

82. Boss PK, Davies C, Robinson SP (1996) Analysis of the expression of anthocyanin pathway genes in developing *Vitis vinifera* L. cv Shiraz grape berries and the implications for pathway regulation. Plant Physiol 111:1059–1066
83. Cutanda-Perez MC, Ageorges A, Gomez C, Vialet S, Terrier N, Romieu C, Torregrosa L (2009) Ectopic expression of VlmybA1 in grapevine activates a narrow set of genes involved in anthocyanin synthesis and transport. Plant Mol Biol 69:633–648
84. Kobayashi S, Goto-Yamamoto N, Hirochika H (2004) Retrotransposon-induced mutations in grape skin color. Science 304:982
85. Walker A, Lee E, Bogs J, McDavid D, Thomas M, Robinson S (2007) White grapes arose through the mutation of two similar and adjacent regulatory genes. Plant J 49:772–785
86. Bogs J, Jaffe FW, Takos AM, Walker AR, Robinson SP (2007) The grapevine transcription factor VvMYBPA1 regulates proanthocyanidin synthesis during fruit development. Plant Physiol 143:1347–1361
87. Terrier N, Torregrosa L, Ageorges A, Vialet S, Verries C, Cheynier V, Romieu C (2009) Ectopic expression of VvMybPA2 promoters proanthocyanidin biosynthesis in grapevine and suggests additional targets in the pathway. Plant Physiol 149:1028–1041
88. Manjoub A, Hernould M, Joubes J, Decendit A, Mars M, Barrieu F, Hamdi S, Delrot S (2009) Overexpression of a grapevine R2R3-MYB factor in tomato affects vegetative development, flower morphology and flavonoid and terpenoid metabolism. Plant Physiol Biochem 47: 551–561
89. Hichri I, Heppel SC, Pillet J, Leon C, Czemmel S, Delrot S, Lauvergeat V, Bogs J (2010) The basic helix–loop–helix transcription factor MYC1 is involved in the regulation of the flavonoid biosynthesis pathway in grapevine. Mol Plant 3:509–523
90. de Jong WS, Eannetta NT, De Jong DM, Bodis M (2004) Candidate gene analysis of anthocyanin pigmentation loci in the Solanaceae. Theor Appl Genet 108:423–432
91. Jaakola L, Poole M, Jones MO, Kamarainen-Karppinen T, Koskimaki JJ, Hohtola A, Haggman H, Fraser PD, Manning K, King GJ, Thomson H, Seymour GB (2010) A SQUAMOSA MADS box gene involved in the regulation of anthocyanin accumulation in bilberry fruits. Plant Physiol 153:1619–1629
92. Palapol Y, Ketsa S, Lin-Wang K, Ferguson IB, Allan AC (2009) A MYB transcription factor regulates anthocyanin biosynthesis in mangosteen (*Garcinia mangostana* L.) fruit during ripening. Planta 229:1323–1334
93. Feng S, Wang Y, Yang S, Xu Y, Chen X (2010) Anthocyanin biosynthesis in pears is refulated by a R2R3-MYB transcription factor PyMYB10. Planta 232:245–255
94. Niu SS, Xu CJ, Zhang WS, Zhang B, Li X, Ling-Wang K, Ferguson IB, Allan AC, Chen KS (2010) Coordinated regulation of anthocyanin biosynthesis in Chinese bayberry (*Myrica rubra*) fruit by a R2R3MYB transcription factor. Planta 231:887–899
95. Aharoni A, De Vos CH, Wein M, Sun Z, Greco R, Kroon A, Mol JN, O'Connell AP (2001) The strawberry FaMYB1 transcription factor suppresses anthocyanin and flavonol accumulation in transgenic tobacco. Plant J 28:319–332
96. Dubos C, Le Gourrierec J, Baudry A, Huep G, Lanet E, Debeaujon I, Routaboul JM, Alboresi A, Weisshaar B, Lepiniec L (2008) MYBL2 is a new regulator of flavonoid biosynthesis in *Arabidopsis thaliana*. Plant J 55:940–953
97. Matsui K, Umemura Y, Ohme-Takagi M (2008) AtMYBL2, a protein with a single MYB domain, acts as a negative regulator of anthocyanin biosynthesis in *Arabidopsis*. Plant J 55:954–967
98. Zhu HF, Fitzsimmons K, Khandelwal A, Kranz RG (2009) CPC, a single-repeat R3 MYB, is a negative regulator of anthocyanin biosynthesis in *Arabidopsis*. Mol Plant 2: 790–802
99. Paolocci F, Robbins MP, Passeri V, Hauck B, Morris P, Rubini A, Arcioni S, Damiani F (2011) The strawberry transcription factor FaMYB1 inhibits the biosynthesis of proanthocyanidins in *Lotus corniculatus* leaves. J Exp Bot Jan;62(3):1189–1200

100. Albert NW, Lewis DH, Zhang H, Schwinn KE, Jameson PE, Davies KM (2011) Members of an R2R3-MYB transcription factor family in *Petunia* are developmentally and environmentally regulated to control complex floral and vegetative pigmentation patterning. Plant J 65:771–784
101. Lin-Wang K, Micheletti D, Palmer J, Volz R, Lozano L, Espley R, Hellens RP, Chagne D, Rowan DD, Troggio M et al (2011) High temperature reduces apple fruit colour via modulation of the anthocyanin regulatory complex. Plant Cell Environ 34:1176–1190
102. Yuan Y, Chiu LW, Li L (2009) Transcriptional regulation of anthocyanin biosynthesis in red Babbage. Planta 23:1141–1153
103. Chiu LW, Zhou X, Burke S, Wu X, Prior RL, Li L (2010) The purple cauliflower arises from activation of a MYB transcription factor. Plant Physiol 154:1470–1480
104. Giusti MM, Rodriguez-Saona LE, Griffin D, Wrolstad RE (1999) Electrospray and tandem mass spectroscopy as tools for anthocyanin characterization. J Agric Food Chem 47:4657–4664
105. Wu X, Prior RL (2005) Systematic identification and characterization of anthocyanins by HPLC-ESI-MS/MS in common foods in the United States: fruits and berries. J Agric Food Chem 53:2589–2599
106. Lalusin AG, Nishita K, Kim SH, Ohta M, Fujimura T (2006) A new MADS-box gene (IbMADS10) from sweet potato (*Ipomoea batatas* (L.) Lam) is involved in the accumulation of anthocyanin. Mol Genet Genomics 275:44–54
107. Mathews H, Clendennen SK, Caldwell CG, Liu XL, Connors K, Matheis N, Schuster DK, Menasco DJ, Wagoner W, Lightner J, Wagner DR (2003) Activation tagging in tomato identifies a transcriptional regulator of anthocyanin biosynthesis modification and transport. Plant Cell 15:1689–1703
108. Ballester AR, Molthoff J, de Vos R, Hekkert BL, Orzaez D, Fernandez-Moreno, JP, Tripodi P, Grandillo S, Martin C, Heldens J, Ykema M, Granell A, Bovy A (2010) Biochemical and molecular analysis of pink tomatoes: deregulated expression of the gene encoding transcription factor SlMYB12 leads to pink tomato fruit color. Plant Physiol 152:71–84
109. Jones CM, Mes P, Myers JR (2003) Characterization and identification of the anthocyanin fruit (Aft) tomato. J Hered 94:449–456
110. Rick CM, Cisneros P, Chetelat RT, deVerna JD (1994) Abg– A gene on chromosome 10 for purple fruit derived from *S. lycopersicoides*. Rep Tomato Genet Coop 44:29–30
111. Mes PJ, Boches P, Myers JR, Durst R (2008) Characterization of tomatoes expressing anthocyanin in the fruit. J Am Soc Horticultural Sci 133:262–269
112. Rick CM, Reeves AF, Zobel RW (1968) Inheritance and linkage relations of four new mutants. Rep Tomato Genet Coop 18:34–35
113. Katsumoto Y, Fukuchi-Mizutani M, Fukui Y, Brugliera F, Holton TA, Karan M, Nakamura N, Yonekura-Sakakibara K, Togami J, Pigeaire A et al (2007) Engineering of the rose flavonoid biosynthetic pathway successfully generated blue-hued flowers accumulating delphinidin. Plant Cell Phys 48:1589–1600
114. Fukui Y, Tanaka Y, Kusumi T, Iwashita T, Nomoto K (2003) A rationale for the shift in colour towards blue in transgenic carnation flowers expressing the flavonoid 3′5′hydroxylase gene. Phytochemistry 63:15–23
115. Honda T, Saito N (2002) Recent progress in the chemistry of polyacylated anthocyanins as flower color pigment. Heterocycles 56:633–692
116. Holton TA, Brugliera F, Lester DR, Tanaka Y, Hyland CD, Menting JGT, Lu CY, Farcy E, Stevenson TW, Cornish EC (1993a) Cloning and expression of cytochrome P450 genes controlling flower color. Nature 366:276–279
117. Nakatsuka T, Abe Y, Kakizaki Y, Yamamura S, Nishihara M (2007) Production of red-flowered plants by genetics engineering of multiple flavonoid biosynthetic genes. Plant Cell Rep 26:1951–1959
118. Nakatsuka T, Haruta KS, Pitaksutheepong C, Abe Y, Kakizaki Y, Yamamoto K, Shimada N, Yamamura S, Nishihara M (2008) Identification and characterization of R2R3-MYB and bHLH transcription factors regulating anthocyanin biosynthesis in gentian flowers. Plant Cell Physiol 49:1818–1829

119. Sweeney MT, Thomson MJ, Pfeil BE, McCouch S (2006) Caught red-handed: rc encodes a basic helix–loop–helix protein conditioning red pericarp in rice. Plant Cell 18:283–294
120. Telias A, Lin-Wang K, Stevenson D, Cooney J, Hellens R, Allan A, Hoover E, Bradeen J (2011) Apple skin patterning is associated with differential expression of MYB10. BMC Plant Biol 11:93–107
121. Dixon RA, Liu C, Jun JH (2012) Metabolic engineering of anthocyanins and condensed tennins in plants. Curr Opin Biotechnol 24:1–7
122. Butelli E, Titta L, Giorgio M, Mock HP, Matros A, Peterek S, Schijlen EGWM, Hall RD, Bovy AG, Luo J, Martin C (2008) Enrichment of tomato fruit with health-promoting anthocyanins by expression of select transcription factors. Nature Biotechnol 26:1301–1308
123. Ikegami A, Eguchi S, Kitajima A, Inoue K, Yonemori K (2007) Identification of genes involved in proanthocyanidin biosynthesis of persimmon (*Diospyros kaki*) fruit. Plant Sci 172:1037–1047
124. Akagi T, Ikegami A, Tsujimoto T, Kobayashi S, Sato A, Kono A, Yonemori K (2009) Dk-Myb4 is a Myb transcription factor involved in proanthocyanidin biosynthesis in persimmon fruit. Plant Physiol 151:2028–2045
125. Yang K, Jeong N, Moon JK, Lee YH, Lee SH, Kim HM, Hwang CH, Back K, Palmer RG, Jeong SC (2010) Genetic analysis of genes controlling natural variation of seed coat and flower colors in soybean. J Heredity 101:757–768
126. Ranilla LG, Genovese MI, Lajolo FM (2007) Polyphenols and antioxidant capacity of seed coat and cotyledon from Brazilian and Peruvian bean cultivars (*Phaseolus vulgaris* L.). J Agric Food Chem 55:90–98

Chapter 5
Flavonoid Dietetics: Mechanisms and Emerging Roles of Plant Nutraceuticals

Arti Parihar, Erich Grotewold, and Andrea I. Doseff

Introduction

During the past few decades, the life expectancy increased dramatically throughout the world as a result of improved nutrition, sanitation, and health care. However, this extended life expectancy is accompanied with an increase in age-related pathologies that include cardiovascular and neurological diseases, obesity, and cancer, conditions that are inflicting an immense pressure on health care costs and quality of life. Thus, there has been an increased interest in recognizing and understanding the mechanisms of action of active nutritional compounds with health benefits, or nutraceuticals, for the prevention and treatment of various diseases [1]. Nutraceuticals are found in dietary supplements, beverages, and constitute main components of our diet. The beneficial effects of dietary nutraceuticals have been recognized since the beginning of human civilization, but our understanding of their modes of action remains incomplete. A better knowledge on their molecular mechanisms of action, metabolism, absorption, and bioavailability will help define their biological activities and full potential as preventive and therapeutic alternative strategies.

Flavonoids are naturally occurring polyphenolic compounds broadly found in fruits and vegetables, and constitute the largest class of nutraceuticals in our diet. To date, more than 8000 different flavonoids have been identified [2]. Among them,

A. I. Doseff (✉)
Department of Physiology and Cell Biology, Department of Molecular Genetics, Heart and Lung Research Institute,
The Ohio State University, 473 W12th Avenue, Columbus, OH, USA
e-mail: Doseff.1@osu.edu

E. Grotewold
Center for Applied Plant Sciences and Deptartment of Molecular Genetics,
The Ohio State University, Columbus, OH, USA
e-mail: Grotewold.1@osu.edu

A. Parihar
Department of Biological Sciences, Government Post Graduate College,
Center of Excellence, Ujjain, MP, India

C. Chen (ed.), *Pigments in Fruits and Vegetables,* DOI 10.1007/978-1-4939-2356-4_5

anthocyanins are responsible for providing colors to fruits and vegetables, and have dietary value as color additives with potential health benefits. So far, around 500 anthocyanins have been described and worldwide annual consumption of anthocyanins from black grape alone is estimated to be 10,000 t [3]. Clinical and epidemiological studies suggest potential beneficial effects of flavonoid dietary intervention in the prevention of several cancers [4, 5] and cardiovascular diseases [6]. These dietary flavonoids have also been shown remarkable promise in preventing obesity and signs of metabolic syndrome that can lead to type 2 diabetes (T2D) and increased risk of cardiovascular diseases by lowering lipid levels [7, 8]. Recent advances in epidemiological studies using dietary interventions, the scientific characterization of biological activities using animal models, and the acquisition of better analytical profiling methods have provided a number of new opportunities for the use of these nutraceuticals, while also posing some new future challenges. In this chapter, we discuss the recent advances on the mechanisms of action and dietary effects of flavonoids and anthocyanins and their uses in the prevention and treatment of inflammatory diseases including cardiovascular diseases, obesity, and cancers.

Flavonoids and Anthocyanins

Plant polyphenolic compounds constitute a group of diverse small molecules that can be divided into several families including phenolic acids, stilbenes, coumarins, tannins, and flavonoids [9]. Plant polyphenols have diverse structures ranging from simple compounds containing an aromatic ring with one or more hydroxyl moieties to highly complex polymeric substances such as tannins [10, 11]. Flavonoids are low molecular weight polyphenolic specialized metabolites present primarily in the cell vacuoles of many plants that often provide nutraceutical value to our diet [12].

Classification, Distribution, and Biosynthesis

The basic flavonoid structure consists of a 2-phenyl-benzo-γ-pyrane nucleus comprising two benzene rings A and B linked through a heterocyclic pyran or pyrone ring (C) (Fig. 5.1; [12, 13]). Flavonoids can be found hydroxylated at positions 3, 5, 7, 2′, 3′, 4′, and/or 5′. According to the chemical substitutions, major dietary flavonoids can be grouped into various subclasses that include flavones [(2-phenylchromen-4-one), e.g., apigenin, luteolin, and tangeretin]; flavonols [(3-hydroxy-2-phenylchromen-4-one), e.g., quercetin, kaempferol, myricetin, and rhamnazin]; flavanones [(2, 3-dihydro-2-phenylchromen-4-one), e.g., hesperetin, naringenin, and eriodictyol]; flavanol [2-phenyl-3,4-dihydro-2H-chromen (flavan)-3-ol, flavan-4-ol, flavan-3,4-diol, e.g., catechins and epicatechins; flavanonol [(3-hydroxy-2, 3-dihydro-2-phenylchromen-4-one), e.g., silibinin, taxifolin, and dihydrokaempferol]; isoflavones [(3-phenylchromen-4-one), e.g., genistein, daidzein, and glycitein];

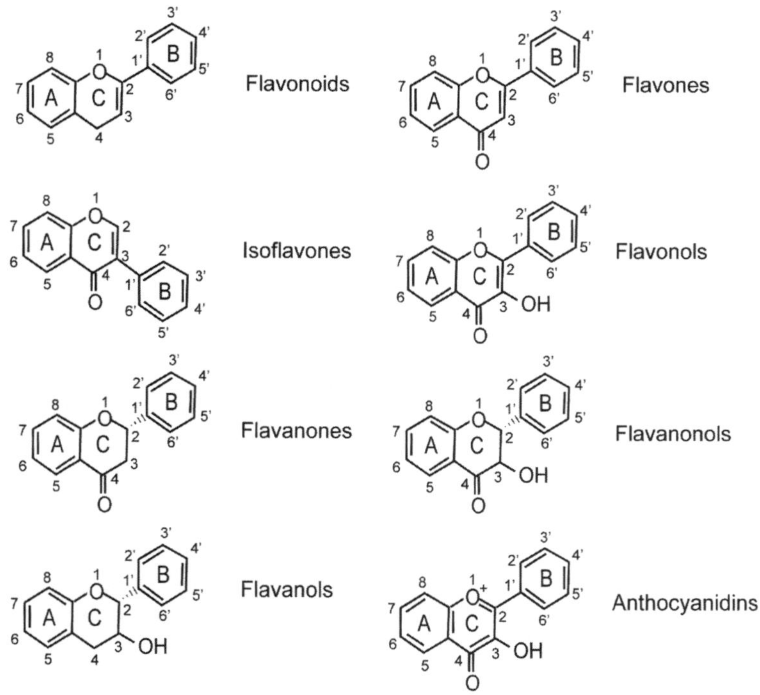

Fig. 5.1 Chemical structures of flavonoid subclasses. Basic flavonoid structure consists of two benzene rings A and B linked through a pyrone ring C. Flavonoids are classified based on the chemical substitutions occurring in the basic structure in different subclasses including the flavones, isoflavones, flavonols, flavanones, flavanonols, flavanols, and anthocyanidins

and anthocyanidins [(2-phenylchromenylium), e.g., cyanidin, delphinidin, and malvidin and red purple anthocyanin [9, 12].

Flavonoids are widely distributed in fruits and vegetables. The most common flavonoids and their distribution are listed in Table 5.1. In many fruits like grapes, apples, pears, cherries, and berries, levels range from 200 to 300 mg per 100 g fresh weight [14]. Probably, the most commonly consumed plant flavonoids with evidence of having health-modulating effects are catechins, quercetin, apigenin, and genistein found in vegetables, fruits, and beverages such as tea, wine, etc. Isoflavones such as genistein are richly found in soybean highly consumed by the Asian population [12]. High intake of tea in Holland has been associated with an increased flavan-3-ol consumption [15]. Environmental factors such as temperature, rainfall, and altitude affect polyphenolic contents of plants. In addition, the degree of ripeness considerably affects the concentrations and proportions of various polyphenols [16]. Flavonoid contents in foods are also affected by storage and cooking. It has been estimated that onions and tomatoes lose between 75 and 80 % of their initial quercetin content after boiling for 15 min, 65 % after cooking in a microwave oven, and 30 % after frying [17]. The consumption of flavonoids in the food also varies

Table 5.1 Flavonoid occurrences in foods (mg/100 g or 200 mL[a])

Flavonoids	Occurrence in foods	Flavonoids (mg)	References
Flavonols (Examples: kaemferol, myricetin, quercetin, rutin)	Onions	60.2	[276]
	Red wine	10	[277]
	Apples	4.42	[278]
	Broccoli, raw	3.21	[278]
	Tea green	10–16.7	[276]
Flavones (Examples: Apigenin, chrysin, luteolin)	Celery hearts, green	22.60	[17]
	Parsley	18.6–29 or	[279]
	Parsley dried	13,525.7	[280]
	Thyme	45.25	[281]
	Melons	7.9–9.3	[279]
	Red wines	13.8	[279]
	Citrus	21.9–24.4	[279]
Isoflavones (Examples: daidzein, genistein, glycitein)	Soy beans, legumes	94.2–97.5	[279]
	Peanuts	1.3	[279]
Flavanols (Examples: catechin, epigallocatechin)	Tea green	12–160	[16]
	Tea black	10–100	[16]
	Apple	8.14	[278]
	Red wine	10–20	[278]
	Cocoa, dark chocolate	20.13	[278]
Flavanones (Examples: eriodictyol, hesperitin, naringenin)	Blood orange juice	28	[282]
	Grapefruit juice	14.17	[278]
	Citrus Juices	96.1–99.3	[279]
Flavanonols (Examples: taxifolin)	Limon	350–7000	[283]
Anthocyanidins	Bilberry	300–698	[283]
	Blueberries	82	[278]
	Cherries	41.4–45	[279]
	Red wine	140	[277]
	Dark chocolate	165	[284]
	Apple	147	[284]
	Cranberry juice	42	[285]
	Red wine	77–103	[277]
	White wine	2–3	[277]

[a] Per 100 g (fresh weight) or 200 mL (liquids); 100 g is comparable to about 3.5 ounces; 100 mL is comparable to about 3.5 fluid ounces

in different regions of the world. The Mediterranean diet, rich in parsley and celery, has a high content of the flavone apigenin, and has been associated with lower prevalence of hypertension, cardiovascular diseases, obesity, cancers, and diabetes [18]. Thus, the health effects of flavonoids and anthocyanins may depend on the amount consumed, the distribution, and the bioavailability.

Anthocyanins typically produce bright to deep red, purple, and blue colors for many berries, black currants, purple grapes, and red wine [19, 20]. The colors may change in the full visible spectrum with pH value. These chemical properties of anthocyanin complicate their exact quantitation. Berries contain higher levels of anthocyanins than most other fruits and vegetables, and it has been suggested that anthocyanins are the major nutraceuticals in those fruits responsible for their health benefits. Comparative studies of the phenolic and flavonoid contents in blueberries, blackberries, and strawberries suggested that blueberries posses the highest total phenolic content (TPC, 9.44 mg gallic acid/g dry weight), total flavonoid content (TFC, 36.08 mg rutin/g dry weight), and total anthocyanidin content (TAC, 24.38 mg catechin/g dry weight; [21]).

The basic flavonoid biosynthesis pathway has been comprehensively studied at the biochemical, molecular, and genetic levels in several plants including maize, *Arabidopsis thaliana,* Petunia, snapdragon, soybean, and barley [22–30]. Chalcone synthase (CHS) uses malonyl-CoA and *p*-coumaroyl-CoA to generate naringenin chalcone, the precursor for all the flavonoids, including the anthocyanins. Flavanones such as naringenin provide a central branch point in the flavonoid pathway, serving as substrates for the introduction of –OH groups at the 3′ and 5′ positions of the B-ring (e.g., F3′H and F3′5′H), or for the hydroxylation of the C-ring by the soluble monooxygenase flavanone 3-hydroxylase (F3H). Dihydroflavonol 4-reductase (DFR) provides one entry step to the biosynthesis of anthocyanins. Depending on plant species, DFR can utilize different dihydroflavonols (e.g., dihydromyricetin, dihydrokaempferol or dihydroquercetin) to synthesize leucoanthocyanidins, which are then converted into the corresponding anthocyanidins [31]. In addition, naringenin can be converted by isoflavone synthase (IFS) directly to the isoflavone genistein or into flavones by a flavone synthase (FNS). The conversion of flavanones (naringenin or eriodictyol) to flavones occurs by different mechanisms, depending on whether the product is the flavone aglycone (apigenin or luteolin) or the corresponding *C*-glycosides [32].

The complex regulation of flavonoid biosynthesis continues to provide a paradigm for the combinatorial control of plant gene expression, providing interesting insights into how other plant metabolic pathways are regulated. Many of the enzymes of the flavonoid pathway have been functionally manipulated in microorganisms and in plants, creating multiple opportunities for metabolic engineering of designed flavonoids with unique nutraceutical properties [33].

Bioavailability and Metabolism

The concentration of flavonoids in circulation after consumption depends in part on the nature of the compounds, food sources, bioavailability, and metabolism. The plasma concentrations of intact flavonoids rarely exceed 1 μM [34]. Pharmacokinetic studies showed that the plasma level of the flavanol quercetin increased to 0.289 μmol/L when given for 2 weeks at 90 mg/day [35]. The plasma level of naringenin (obtained from either orange or grapefruit juice) reached 0.64 and

5.98 μmol/L when given as a single 23 or 199 mg dose, respectively [36]. Plasma concentrations of epigallocatechin, derived from green tea extracts, increased to 0.025 μmol/L when given as a single 17.5 mg dose [37]. Anthocyanins from black currant juice given as a single dose of 20 mg/kg body weight increased the plasma level to 0.155 μM/L [38]. Metabolites of flavonoids have also been shown to circulate in the blood bound to various proteins, in particular albumin [14]. Flavonoids, including anthocyanins, were found in tissue organs, especially in those in which they are metabolized (e.g., intestine and liver), but also in kidney, liver, brain, eyes, cortex, and cerebellum [39–41]. In colorectal cancer patients who received oral curcumin for 7 days (3.6 g daily) prior to surgery, levels of curcumin were around 12.7 ± 5.7 nmol/g in cancer tissue [42]. In some instances, the concentrations of flavonoids in plasma do not directly correlate with the levels present in target tissues. A clinical dietary intervention study in breast cancer patients showed high levels of equol (a metabolite of the isoflavone daidzein) and low levels of genistein in breast tissues, but an inverse correlation in plasma [43]. These results suggest that flavonoids and anthocyanins cross the blood–tissue barrier to potentially provide protection at different organ levels. These studies highlight the ability of flavonoids to reach organ locations and the need to determine their bioavailability not only in plasma but also in tissues to properly define the levels of effectiveness. However, few studies are available that assess the bioavailability of flavonoids in various tissues of a same animal.

Flavonoids are usually stable once consumed, some reaching the small intestine where aglycones and glucose-conjugated flavonoids are absorbed and metabolized. In order to be absorbed, these compounds must be hydrolyzed by intestinal enzymes such as β-glucosidases and lactase-phlorizin hydrolases, or by the colonic microflora to then undergo extensive modifications [44]. Catechol-*O*-methyl transferase catalyzes the transfer of a methyl group from *S*-adenosyl-L-methionine to polyphenols such as quercetin, luteolin, caffeic acid, catechins, and cyanidin [45]. The methylation occurs at the C3′ and C4′ positions [46, 47]. Sulfotransferases, mainly found in the liver, catalyze the transfer of a sulfate from 3′-phosphoadenosine-5′-phosphosulfate to a hydroxyl group on various polyphenols, but the position of sulfation in flavonoids has not been clearly identified yet [48, 49]. Uridine 5′-diphospho (UDP)-glucuronosyltransferases, membrane-bound enzymes located in the endoplasmic reticulum, catalyze the transfer of glucuronic acid from UDP-glucuronic acid to dietary polyphenols constituents. Glucuronidation commonly occurs at the C3 position and takes place in the intestine and liver [50, 51]. Glucuronidation is an important pathway in the metabolism of flavonoids in humans and for the detoxification of xenobiotics in general. Notably, high levels of glucuronidase have been reported in neutrophils and increased deglucuronidation of the flavone luteolin was found in lipopolysaccharide (LPS)-treated rats [52]. We found increased glucuronidase levels in human monocytes exposed to LPS (Doseff, unpublished), suggesting that the metabolism of flavonoids may be different under inflammatory conditions.

Isoflavone glucosides are hydrolyzed to the corresponding aglycones or daidzein metabolized to equol by microflora [53, 54]. Some flavonoids are absorbed at the gastric level [55, 56]. However, their rate and extent of intestinal absorption are determined by the chemical structure of polyphenols. Analytical results show that the quantities of these polyphenols found intact in urine may vary from one phenolic compound to another. Phenolics also show high individual variation probably due to microflora diversity and metabolism [57–59]. Microbial colon populations are likely to modify the anthocyanins ingested promoting deglycosylation and demethylation [60]. Humans lack esterases, thus require the enzymatic activity of colonic microflora for the metabolism of anthocyanins conjugated to rhamnose or rutinose [61], or for acylated anthocyanins [62]. Anthocyanidins generate simple phenolic acids, or are taken up from the gastrointestinal tract lumen and subsequently metabolized to glucuronides, sulfates, and methylates by phase II drug-metabolizing enzymes in the liver [63]. Anthocyanins have been found at nanomolar or micromolar levels in serum [64–66]. They can be absorbed intact as glycosides [67], but the proportion of anthocyanins absorbed and excreted in the urine as a percentage of the intake is less than 0.1 % [45]. Extensively conjugated metabolites are more likely to be eliminated in the bile, whereas small conjugates, such as monosulfates, are preferentially excreted in urine. The total amount of metabolites excreted in the urine is roughly correlated with maximum plasma concentrations [44]. Studies showed that the three conjugated metabolites of anthocyanins are likely to be excreted in urine, but alternatively a portion of them may reenter the jejunum with the bile and later either being absorbed by the colon entering the enterohepatic circulation [68]. The absorbed intact anthocyanins and anthocyanidins are largely excreted in urine whereas the nonabsorbed anthocyanins are excreted through feces.

Glucosides can be transported also into enterocytes by the sodium-dependent glucose transporter SGLT1 [69] and then hydrolyzed by a cytosolic β-glucosidase [70]. The polyphenols that are not absorbed in the small intestine reach the colon, where the glycosides are hydrolyzed into aglycones by microflora which then extensively metabolize the aglycones into various aromatic acids [63–73].

In the case of flavones, our own studies found that dietary flavones aglycones are more efficiently absorbed found in the serum at higher concentrations than their corresponding glycosides [74]. As a result of extensive chemical modifications, it is feasible that the biological active forms reaching the blood and tissues are different from those present in foods [75–77]. Thus, it will be important to better define the biological activities of flavonoid derivatives, their rate of absorption, their distribution and bioavailability to properly understand their role in health and establish nutritional recommendations for consumers.

Mechanisms of Action of Flavonoids and Anthocyanins

Antioxidant Properties

Body cells and tissues are continuously targets of the damages caused by reactive oxygen species (ROS) including free radicals, which are produced during normal oxygen metabolism or induced by exogenous damage [78, 79]. Important mechanisms by which free radicals are harmful are by lipid peroxidation increasing inflammatory mediators and contributing to a general inflammatory response and tissue damage. Cellular defense mechanisms against ROS include enzymes such as superoxide dismutases, catalase, and glutatione peroxidase, and also nonenzymatic mechanisms such as the scavenging action of free radicals by glutathione and ascorbic acid [80].

Flavonoids enhance powerful endogenous cellular antioxidant strategies, as shown in numerous cell-based studies. However, there is no clear evidence of their antioxidant properties in human studies. The antioxidant action of flavonoids is a consequence of their structure, which permits them to chelate metals, to directly bind to proteins, to act as direct radical scavengers, and thus to reduce lipid peroxidation [81–84]. The highly conjugated system and certain hydroxylation patterns, such as the 3-hydroxy group in flavonols, are considered important for the antioxidant activities. Flavonoids are powerful inducers of endogenous antioxidant defense mechanisms stimulating the activity of glutathione peroxidase, catalase, and superoxide dismutases, and inhibit the expression of enzymes such as xanthine oxidase [85]. Structural–functional relationship analyses reported that luteolin (3′,4′,5,7-tetrahydroxyflavone) is one of the most potent inhibitors of xanthine oxidase activity, which is also inhibited by quercetin and silibin [86–89]. Numerous studies have described the antioxidant efficacy of flavonoids in inhibiting the lipid peroxidation and protecting other biomolecules, such as proteins and DNA, from oxidation [90–92]. The flavones and catechins seem to be the most powerful flavonoids for protecting the body against ROS. Flavonoids display antioxidant activities that are relevant to the prevention of atherosclerosis. For example, the isoflavone genistein prevented low-density lipoprotein (LDL) oxidation and protected vascular cells against oxidized LDL particles [93]. Nobiletin, a citrus polymethoxylated flavone has extensive antioxidant activities suppressing the formation of oxidants by modulating the xanthine oxidase system, 12-*O*-tetradecanoylphorbol-13-acetate (TPA)-induced oxidative stress, and NO (nitric oxide) generation in macrophage cell lines [94, 95]. The flavonols myricetin and quercetin prevent lipid peroxidation and may be active in the regeneration of α-tocopherol [96]. Several flavonoids, including quercetin, interfere with inducible nitric oxide synthase (iNOS) activity, resulting in reduction of ischemia–reperfusion injury [97]. Silibin, a naturally occurring flavanone obtained from *Silybum marianum* (milk thistle) has been reported to inhibit NO production [98]. Flavonoids also act as metal chelators which results in the direct reduction in the rate of the Fenton reaction, thus preventing oxidation caused by highly reactive hydroxyl radicals [83, 99]. Anthocyanins are powerful antioxidants by virtue of their electron-donating properties [100]. In addition, they enhance the absorbing capacity of oxygen radical in cells, stimulating the expression of phase II detoxification enzymes, minimizing the formation of oxidative adducts in DNA and decreasing lipid peroxidation [101]. In general, the anthocyanidins (aglycons) have superior radical scavenging activity than the respective anthocyanins (glycosides), which decreases as the number of sugar moieties increase. However, at high concentrations, flavonoids and other polyphenols may also be cytotoxic, causing an increase in mitochondrial permeability, cytochrome c release, the activation of caspases, an increase in levels of p53 and p21, the suppression of Bcl-2, apoptosis induction, and necrotic cell death [102–104].

Role as Signaling Molecules

The role of flavonoids as signaling molecules is not yet well characterized in plants. In plants, flavonoids modulate the transport of the plant hormone auxin and have also been shown to modulate the jasmonic acid pathway [105, 106]. Jasmoates in plants and prostaglandins in animals correspond to fatty acid oxidation products and participate in similar wound responses in both kingdoms. Conversely, expression of early genes of the flavonoid biosynthetic pathway induced by jasmonic acid and activating mitogen-activated protein kinases (MAPK) were observed after wounding, suggesting their role in defense signaling pathways [107, 108]. In addition, flavonoids in root exudates were found to modulate expression of nodule-gene expression [109–111]. In mammalian systems, flavonoid-sensitive signaling networks have been described extensively [112] regulating numerous cell processes, including growth, proliferation, and death. Flavonoids can interact selectively within MAPK signaling cascades [113, 114]. We have shown that the flavone apigenin, but not naringenin, induced the activation of protein kinase Cδ (PKCδ) leading to apoptosis of leukemia cells and differential phosphorylation of Hsp27, an inhibitor of the caspase cascade [102, 115]. Flavonoids such as apigenin and luteolin, among others, inhibit the NF-κB signaling pathway in response to inflammation in vivo [116, 117]. Epicatechin and kaempferol attenuate the activation of extracellular signal-related kinases (ERK1/2) and c-Jun N-terminal kinase (JNK) and protect against neuronal cell death [118]. Multiple flavonoids and anthocyanins modulate cell proliferation by affecting various kinases. Resveratrol causes the activation of glycogen synthase kinase 3β (GSK-3β) regarded as an important component of cell signaling, protein synthesis, cell proliferation, cell differentiation, cell adhesion, and apoptosis [119]. In addition, resveratrol induced the mitochondrial adenosine triphosphate (ATP)-sensitive K+channel (mK_{ATP}) and large conductance Ca^{2+}-activated K^+channel (BK_{Ca}) channels [120], enhanced autophagy via the mammalian target of rapamycin (mTOR) pathway [121], activated the transcription factor Nrf2 [122], the expression of heme oxygenase 1 [123], and upregulated endothelial nitric oxide synthase (eNOS) [124]. Anthocyanins, such as delphinidin and cyanidin, halt the activation of p38 MAPK and JNK, thereby preventing the expression of vascular endothelial growth factor (VEGF) stimulated by platelet-derived growth factor in vascular smooth muscle cells [125]. Using mice models, resveratrol was shown to counteract the increased levels of pro-inflammatory biochemical markers, such as tumor necrosis factor (TNF) α, interleukin (IL)-1β, and IL-6. In addition, resveratrol downregulates the cytokines MCP-1, cyclooxygenase (COX)-2, and iNOS when stimulated by the pro-inflammatory treatment [126]. *Trans*-resveratrol was demonstrated to prevent copper-catalyzed oxidation in human LDL cholesterol by chelation [127]. In human lung epithelial cells, resveratrol inhibits IL-8, granulocyte-macrophage colony-stimulating factor (GM-CSF), iNOS, and COX-2 [128]. In human alveolar macrophages, resveratrol decreases IL-8 and GM-CSF, further underscoring its anti-inflammatory potential [129].

Role in Gene Expression and Epigenetic Regulation

Genome-wide changes in gene expression and epigenetic modifications have been linked to the etiology of cancers, cardiovascular diseases, and obesity. Numerous studies in cell cultures have shown that nutraceuticals influence the expression of genes involved in energy metabolism, cell growth, and cell differentiation [130]. Epicatechin, the main flavonoid found in cocoa, induced gene expression changes in Caco-2 cells (human colon carcinoma cells), and downregulated genes included several transcription factors; oncogenes, tumor suppressors, and modulators of the stress response; cell receptors and cell surface antigens; metabolism; apoptosis-related proteins; DNA synthesis, recombination, and repair [130]. Epicatechin induced antioxidant response element (ARE)-mediated gene expression. ARE promotes the expression of genes including those required for glutathione synthesis, suggesting that flavonoids may be cytoprotective by increasing antioxidant gene expression [131]. Quercetin affected the expression of multidrug resistance gene-1 (MDR1) in the human hepatocarcinoma cell line HepG2 [132]. Limited studies have shown changes in gene expression related to flavonoid consumption. Exposure of mice to silymarin, which is a mixture of bioactive flavonoids isolated from the seeds and fruits of milk thistle, was found to modulate the differentiation and cell selection in the thymus via increased expression of c-myc [133]. Red wine polyphenolic compounds have been shown to be associated with an increase of endothelium-dependent relaxation and a modest induction of gene expression of iNOS and COX-2 within the arterial wall [134]. COX-2 is the enzyme that catalyses the biosynthesis of prostaglandins at inflammatory sites. COX-2 is a very interesting drug target because of its role in the development and the resolution of the inflammatory response. Thus, the modulation of COX-2 expression may constitute a novel therapeutic strategy in inflammatory diseases. In intestinal epithelial cells, the effectiveness of flavonoids in COX-2 activity was found to depend on the cellular status. All flavonols and flavones were found to induce COX-2 in the basal state, whereas most flavonoids, including kaempferol, quercetin, and luteolin, which exhibited the greatest induction effect in quiescent cells, did not increase COX-2 levels [135]. Endothelial cells treated with low doses of flavonoids (at levels found in plasma; 0.1–1 μM) showed differences in gene expression. Resveratrol and quercetin increased eNOS and VEGF expression and reduced H_2O_2 induction of endothelin-1 (ET-1) [136]. Other flavonoids tested at these low concentrations showed no effect suggesting interesting structural relationships between flavonoids and specific gene expression changes.

Daily supplementation of quercetin (150 mg/day) induced extensive changes in human monocyte gene expression profiles in healthy individuals as shown by microarray analysis [137]. However, in a double-blind crossover study in subjects exhibiting a cardiovascular risk phenotype receiving quercetin or placebo daily for 6 weeks showed no differences in the 24 verification genes tested [137]. Intervention studies with healthy subjects consuming two varieties of polyphenol-rich fruit juices in an 8-week trial showed that juice intervention significantly reduced oxidative DNA damage in leukocytes modulating the expression of glutathione S-transferase (GST) [138].

In recent years, numerous reports have shown that nutrients can influence gene expression and epigenetic changes [139]. Changes in epigenetic patterns in fetus were influenced by maternal nutrition [140]. Isoflavones, flavonols, and catechins have received much attention due to their ability to influence activity of chromatin-modifying enzymes. Epigallocatechin-3-gallate, for example, was shown to inhibit activity of histone acetyltransferase and DNA methyltransferase (DNMT) [141]. Several bioactive food components, including tea polyphenols (catechin, epicatechin, and epigallocatechin-3-*O*-gallate, EGCG) and bioflavonoids (quercetin, fisetin, and myricetin), inhibit DNMT-1-mediated DNA methylation in a dose-dependent manner [142]. Our studies have shown that apigenin-induced changes in histone phosphorylation leading to DNA damage and cell cycle in leukemia cells [143]. Several other dietary compounds alter DNA methylation in vitro, in animal models, or in humans [144]. Mixtures of bioactive compounds naturally present in foods may act synergistically and might be more active than the individual compounds isolated from food [47]. Some dietary compounds, for example butyrate, were found to affect posttranslational modifications of histones through its action as an inhibitor of class 1 histone deacetylases [145–147]. In addition, several other dietary components have been identified, which modulate the acetylation state of histones or affect the activities of HDACs and/or histone acetyl transferases [148]. Among them, there is strong evidence for effects of organosulfur compounds from garlic (diallyl disulfide, allyl mercaptan, and *S*-allylmercaptocysteine) and of isothiocyanates (sulforaphane and 6-methylsulfinylhexyl isothiocyanate) from cruciferous vegetables [148]. The role of resveratrol in the activation of class III HDACs (sirtuins) has been much studied in recent years because of their potential role in lifespan and in delaying, age-related diseases including cancers [149]. Resveratrol activates PGC1α in a SIRT1-dependent manner [150] and improves health and survival of mice fed a high-fat diet [151].

It is expected that in the next few years the full potential of nutrigenomics will permit us to gain a better understanding of the changes in gene expression triggered by nutraceuticals in dietary intervention studies using animal models or human studies.

Preventive Dietetics in Disease Control

Cardiovascular Diseases

Flavonoids exhibit potent cardioprotective activities in many cardiovascular diseases. These effects have been ascribed to their antioxidant, antithrombotic, anti-apoptotic, anti-arhythmic, anti-ischemic, antihypertensive, and anti-inflammatory activities shown in several model systems [152–163]. Thus, flavonoids have been suggested as promising therapeutic and preventive compounds in cardiovascular diseases.

Cardioprotective effects of flavonoids have been shown in several epidemiological studies. The intake of vegetables, legumes, and fruits with high flavonoid content was significantly associated with reduced risks of CVD mortality [164]. Inverse association between dietary flavonoids intake and incidence of coronary heart disease (CHD) has been observed [165, 166]. The incidence of myocardial infarction was estimated to decrease by 11 % with an increase in tea consumption of three cups per day [167]. In cross-sectional and prospective population studies, the intake of tea and intake of flavonoids found in tea have been associated with a reduced risk of cardiovascular disease. Isolated flavonoids found in tea have also been consistently shown to inhibit the development of atherosclerosis in animal models [168]. The major catechin in green tea, EGCG, decreases vascular inflammation by increasing the synthesis of NO, which blocks endothelial exocytosis, the initial step in leukocyte trafficking and vascular inflammation [169]. Green tea catechins (mainly EGCG) selectively activate peroxisome proliferator-activated receptor β/δ (PPAR-β/δ) [170]. The chronic administration during the 3rd and 5th month of decaffeinated green tea extracts (catechins tablet; 455 mg/day comparable to four cups of green tea) in hemodialysis patients decreased atherosclerotic factors such as TNF-α, soluble intercellular adhesion molecule 1 (sICAM-1), monocyte chemoattractant protein 1 (MCP-1), and C-reactive protein (CRP) with respect to the control group receiving placebo [171]. Green tea EGCG (50 and 100 mg/L) inhibits cardiomyocyte apoptosis, a critical factor during the transition from compensatory hypertrophy to heart failure from oxidative stress in vitro, and the molecular mechanisms implicated might be related to the inhibitory effects of EGCG on p53 induction and Bcl-2 decrease [171, 172].

The ingestion of flavanol-rich cocoa (821 mg of flavanols/day, quantitated as (−)-epicatechin, (+)-catechin, and related procyanidin oligomers) by healthy human subjects induces vasodilatation via activation of the NO system, providing a plausible mechanism for the protection against coronary events [173]. The intake of dietary flavonols (quercetin+kaempferol+myricetin) was inversely related with fatal myocardial infarction [174]. In an extensive study involving several countries for 25 years, the average flavonol and flavone intake was inversely correlated with CHD mortality [175]. In a study of elderly subjects, dietary flavonol and flavone intake (12–42 mg/day) was found to reduce CHD mortality risk by 50 % [174]. Similarly, in a 26-year follow-up study in Finland, the dietary intake of flavonoids was associated with reduced risk of CHD [176–178]. These multigroup studies involving several countries strongly support the preventive effects of flavonoids in cardiovascular diseases.

Randomized controlled trials on flavonoids and flavonoid-rich foods provided evidence that some subclasses of flavonoids are associated with a significant reduction in blood pressure. Short-term interventions with cocoa flavan-3-ols have shown reduced systolic blood pressure (SBP) and diastolic blood pressure significantly [179]. However, only few studies have examined the potential effects of flavonoids on blood pressure systematically and there are no population-based studies that examined the relative effect of a habitual or usual intake of different subclasses of flavonoids on incident hypertension.

Several studies in humans have investigated the effects of flavonoid-rich foods and beverages on platelet function. The data for cocoa and red grape supplements generally suggest reduced platelet activation [180]. The flavonoids present in Ginkgo extract inhibit excessive production of *platelet-activating factor* (PAF) [155, 161, 181]. In addition, the Ginkgo extracts also exhibit anti-atherosclerotic properties by regulating the transformation of cholesterol to plaque formation. Effects of cardiovascular protective effects of flavonoids were also demonstrated in several in vitro studies. Polyphenolic constituents of red wine have long been recognized to have cardioprotective properties [153]. Flavonoid content derived from grape seeds tested against ischemia/reperfusion (IR) injury showed that proanthocyanidins directly scavenge peroxyl and hydroxyl radicals and reduce oxidative stress developed during IR in perfused hearts [182]. The most studied flavonoid, resveratrol, may trigger a broad spectrum of cardioprotective pathways, including different signaling pathways (see the section, "Role as Signaling Molecules"). Red wine was shown to increase vasorelaxation and blood pressure reduction by enhancing NOS activity and NO production [183–185]. Moderate wine consumption (15.5–31 g alcohol/day) has been shown to improve both vascular and nonvascular diseases [186]. In a 12-week intervention human trial, the consumption of red wine (30 g of ethanol/day (~160 mL)) exhibited anti-inflammatory effect and the potential to reduce myocardial fibrosis, prevent aortic thickening, attenuate the increase of aortic reactivity to norepinephrine, and prevent the decrease in acetylcholine-induced endothelium-dependent relaxation; and in rats, it led to the relaxation of resistant arteries and reduce blood pressure during NO deficiency [187–189].

Mitochondrial dysfunction and excessive production of ROS contribute to pathogenesis in many cardiovascular diseases. Resveratrol, quercetin, procyanidins, catechins, hydroxytyrosol, and pyrroloquinoline quinine (PQQ) were found to enhance mitochondrial function [190]. Furthermore, resveratrol and quercetin (doses of ~500–1000 mg/day) improved the levels of anti-inflammatory mediators. We recently showed that apigenin restores normal mitochondrial complex I activity in endothelial cells under inflammatory stress conditions [191]. Recently, using a novel genome-wide approach, phage display coupled with second-generation sequencing (PD-Seq), we identified the direct targets of apigenin [192]. The effect of apigenin in metabolic function may be mediated by the ability of apigenin to interact with isocitrate dehydrogenase 3 (IDH3) and other proteins involved in the regulation of main metabolism [192].

Studies have shown that regular ingestion of anthocyanins increased cardiac glutathione concentrations in rats [193], suggesting increasing reducing power of cells. Hydroxytyrosol, a flavonoid constituent of olive oil, stimulates mitochondrial function, increasing production of the mitochondriogenesis protein PGC-1α, perhaps constituting a new mechanism by which olive oil lowers the risk of various diseases [194–197].

Regular ingestion of anthocyanins increased cardiac glutathione concentrations in rats [193] suggesting increasing reducing power of cells. A single anthocyanin cyanindin-3-*O*-glucose chloride has been shown to has a greater efficacy in protecting against oxidant degradation of LDLs than vitamins C, E, and β-carotene combined [198]. In addition, anthocyanin extracts prepared from chokeberry, bilberry, or elderberry have direct vasoactive and vasoprotective properties [199]. Using various animal models, it was demonstrated that proanthocyanidins have a dose-dependent vasorelaxant effect (0.1–100 μg/mL of proanthocyanidin-rich fraction from *Croton* sp.), involving the activation of eNOS. Consumption of more than one serving of blueberries and strawberries per week, both considered rich sources of anthocyanins, was associated with a 10% reduction in hypertension [200]. The blueberries, blackcurrants, or blood orange juice can contain very high amounts of anthocyanins (>500 mg anthocyanins per serving; [201]). Collectively, these studies highlight the value of epidemiological studies helping define levels of commonly consumed fruits rich in flavonoids, such as blueberries, cranberries, and strawberries, which can be readily incorporated into the diet.

In summary, the use of flavonoids as dietary supplements may be beneficial in fighting cardiovascular diseases by virtue of improving mitochondrial functions and protecting damaging oxidative reactions in cells.

Emerging Role of Nutraceuticals in Obesity

Obesity, the major risk factor for cardiovascular disease, insulin resistance, and T2D, is reaching epidemic levels [202]. The lack of effectiveness of current treatments to reduce obesity is prompting an increasing interest in nutraceuticals. Consumption of more than five servings of blueberries, apples, or pears was associated with a lower risk of T2D by 23% [203]. Grape skin or whole grapes, cherries, and berries were found to enhance insulin secretion as well as selectively inhibiting COX-2, which might be useful for the prevention of T2D [204]. Blueberries have been found to increase adiponectin production, a key hormone made in white adipose tissue that prevents liver from developing insulin resistance [203, 205]. Obese mice supplemented with cranberry extract enriched in flavonoids showed an amelioration of insulin resistance and plasma lipid profile, and a reduction of visceral fat mass [206]. These effects might be mediated by the activation of the adiponectin-AMP-activated protein kinase (AMPK) pathway.

Flavonoids have been reported to inhibit nutrient absorption including saponins from different sources such as ginseng and tea [207, 208]. Similar activities were reported for licochalcone A from roots of *Glycyrrhiza uralensis*, platycodin D from fresh roots of *Platycodon grandiflorum*, dioscin from *Dioscorea nipponica*, seeds of *Achyranthes aspera*, and phenolic constituents from the leaves of *Nelumbo nucifera* [209–213]. Administration of grapefruit or orange juice to hypercholesterolemic casein-fed rabbits reduced LDL cholesterol (LDL-C) and hepatic lipid accumulation suggesting components of citrus may have lipid-lowering properties [214].

In vitro studies addressing the potential antiobesity action of quercetin using cells from adipose tissue and the 3T3-L1 cell line showed a reduction in triacylglycerol levels [206]. In addition, quercetin decreased the expression of genes involved in adipocyte differentiation, including the sterol regulatory element-binding proteins (SREBP)-1 and fatty acid synthase (FAS), and by increasing acetyl-CoA carboxylase (ACC) phosphorylation [206]. Various dietary bioactive flavonoids like genistein, epigallocatechin gallate, quercetin, resveratrol, and ajoene affect adipocytes during specific stages of development, resulting in either inhibition of adipogenesis or induction of apoptosis.

It has been demonstrated that anthocyanins enhances adipocytokine (adiponectin and leptin) secretion in isolated rat adipocytes [215]. Human adipocytes treated with anthocyanins showed significant changes in adipocytokine gene expression (upregulation of adiponectin and downregulation of plasminogen activator inhibitor-1 and IL-6 [216, 217]). Cyanidin-3-glucoside (C3G), a typical anthocyanin, significantly suppressed the development of obesity in C57BL/6 mice induced by a high-fat diet [218]. Thus, these studies suggest that anthocyanins modulate gene expression of adipocytokines and may have future therapeutic advantage for the regulation of adipocyte function [3]. Future studies on the mechanisms of action will help define the value of flavonoids in obesity.

Dietary Nutraceuticals and Cancer

The antiproliferative, anti-metastatic, and anti-angiogenic effects of flavonoids have been extensively illustrated in cell and animal models. Nutraceuticals found in red wine such as catechin, epicatechin, quercetin, and resveratrol reduced the proliferation of human breast and prostate cancer cells [219, 220]. Quercetin and apigenin inhibit melanoma growth and influence the invasive and metastatic potential in mice [221]. Genistein controls cell cycle progression in human breast and prostate cancer cell lines [222, 223]. Furthermore, genistein inhibited EGFR tyrosine kinase activity in human epidermoid carcinoma A431 cells [224]. Luteolin and quercetin also effectively suppress EGFR tyrosine kinase actively [46, 225]. We reported that apigenin, with an IC50 of 20 μM, induced caspase-3-dependent and ROS-independent apoptosis in leukemia cells, yet had no effect in nonmalignant cells [102]. The flavonol quercetin, probably one of the best studied, induced apoptosis in HL-60 leukemia cell line with an IC50 ~ 80 μM. Similar to our findings, quercetin induced apoptosis in a caspase-3-dependent fashion, but independent of its antioxidant activity [226]. While similarities between modes of action seem to exist, the human cellular targets of these flavonoids remain unknown. Quercetin acted synergistically with cisplatin, possibly due to inhibition of PKC [227]. Recently, we showed that apigenin induces apoptosis in a PKCδ-dependent manner [102, 115, 145]. Quercetin also exerted antiproliferative effects in other malignant tumor cell lines [228–230]. In addition, using a novel genome-wide approach for the target identification of small molecules, we identified ~ 160 targets of apigenin [192]. Notably, among these targets we identified the RNA-binding protein hnRNPA2 (heterogeneous nuclear ribonucleoprotein A2). HnRNPA2 expression is elevated in several cancers [231–234] and has been shown to play fundamental roles in the progression of tumorigenesis by regulating splicing [235]. These findings are paradigm shifting as they identified new mechanisms by which dietary components can regulate tumorigenesis and suggest that nutraceuticals exert the biological effects by impacting multiple cellular targets [192]. Anthocyanins and anthocyanin-rich extracts from fruits and vegetable have shown antiproliferative effect in multiple cancer cell types [236, 237]. Interestingly, anthocyanins selectively inhibited growth of cancer cells with relatively slight or no effect on the growth of normal cells [238, 239]. However, the selective effect of the anthocyanins on the growth of cancer cells versus normal cells is yet to be explored. Ethanol extract of black raspberries exhibited selective inhibition of the growth and stimulated apoptosis of a highly tumorigenic rat esophageal epithelial cell line (RE-149-DHD), when compared to its low tumorigenic precursor line, RE-149 [240]. The antiproliferative activity of anthocyanin was correlated with the inhibition of various cell cycles regulating proteins, such as p53, p21, p27, and cyclins with concomitant upregulation of the tumor suppressor genes WAF1/p21, KIP1/p27, INK4a/p16, and INK4c/p18. Moreover, anthocyanins inactivates the survival kinases AKT, the inhibitors of apoptosis (IAPs) and MAPK [241, 242]. Anthocyanins with an ortho-dihydroxyphenyl structure on the B-ring suppressed TPA-induced cell transformation and activator protein-1 transactivation, suggesting that the ortho-dihydroxyphenyl structure may contribute to their inhibitory action. Delphinidin and peonidin are anthocyanidins, conferring blue and purplish red hues to the flowers and fruits, respectively. Delphinidin but not peonidin has been shown to block the phosphorylation of the ERK pathway at early times and JNK signaling pathway at later times without affecting p38 MAPK activity [241].

Flavonoids are able to induce cell cycle arrest, disrupt mitochondrial function, and inhibit NF-κB [143, 243, 244]. Inflammatory mediators that include NF-κB and COX-2 are abnormally upregulated in various cancers and their inhibition using anthocyanin exhibited significant chemopreventive potential [245, 246]. We showed that apigenin and food diets rich in apigenin inhibit NF-κB [74, 116]. Anthocyanins also inhibit NF-κB and COX-2, and various ILs and causes anti-inflammatory effects in multiple cell types in vitro [237]. Mouse epidermal cells (JB6 CI41), when treated with an anthocyanin-rich extract from black raspberries, downregulated benzoapyrene diol-epoxide (BaPDE)-induced expression of NF-κB [247]. However, experimental studies have shown that the amounts of anthocyanins needed to elicit effects in vitro far exceed the amounts observed in human plasma in vivo. Future studies aimed at enhancing the absorption of anthocyanins and/or their metabolites may be necessary for their optimal use in the chemoprevention of human cancers [101]. In vivo studies recently showed that, when mice were fed with food formulations enriched in apigenin aglycones, the absorption of aglycone flavones increases significantly, suggesting the importance of deglycosylation in preventing inflammatory diseases [74].

Despite the great amount of information obtained using pure flavonoids, studies with diets rich in flavonoids remain scarce. Several studies addressed the association of flavonoids intake from beverages (such as tea and wine), fruits, vegetables, and herbals with anticarcinogenic effects [98, 219, 248, 249]. There are inverse correlations between dietary flavonoid intake and subsequent lung cancer occurrence [248]. Human epidemiological data come primarily from studies in which decreased cancer rates are associated with consumption of foods rich in isoflavones or flavones. In a study involving 9959 Finish men and women, quercetin intake from onions and apples was inversely associated with the risk of having lung cancer [250]. Epidemiological and experimental studies have suggested that a high intake of flavonoids may be linked to a reduced risk of cancer [251]. In addition, diets rich in flavones have been reported to reduce ovarian cancer risk or enhance the efficacy of cancer therapy in a large population-based case–control study, which correlated with the presence of apigenin [252]. Studies have shown the involvement of flavonoids with the modulation of P-glycoproteins, the proteins associated with multidrug resistance, inhibition of several enzymes of the CYP450 family and modulation of phase II enzymes. Because both classes of enzymes are involved in drug metabolism and in the process of chemical carcinogenesis, interaction of flavonoids with these enzyme systems hold immense therapeutic potential for detoxification, chemoprevention, and the suppression of drug resistance [253].

Human Clinical Studies

Epidemiological studies have provided important understanding on the possible etiologic factors in several disease states. Population and case–control studies suggest that many of the common diseases can be significantly modify by the diet [254]. The increasing interest in identifying alternative preventive and therapeutic approaches to disease control has intensified research on flavonoids including anthocyanins. However, despite the robust evidence provided by in vitro and animal models, the effectiveness of flavonoids in human health remains controversial. Insufficient evidence mostly resides in the use of whole foods where the presence of combination of flavonoids makes difficult to define individual compound contribution.

Probably, one of the most studied areas focused in assessing the role of dietary nutraceuticals is in cardiovascular diseases. Several studies in humans and animals have shown that a diet rich in soy, that has high levels of isoflavones, reduces plasma total cholesterol and LDL cholesterol. Daily soy intake showed a reduction of total cholesterol and LDL-C as well as of ischemic and cerebrovascular events in different populations [255, 256]. Studies conducted in human subjects with or without diabetes showed that soy consumption is beneficial in obesity and diabetes by lowering hyperglycemia and reducing body weight, hyperlipidemia, and hyperinsulinemia [257].

Ongoing human stage I and II clinical trials are evaluating the effect of cocoa uptake (ranging from 6.5 to 50 g/day for 18 weeks) in arterial hypertension, a major contributor of cardiovascular disease. Levels of inflammatory mediators including LDL, TNF, and platelet and monocyte function will be some of the outcomes studied (NCT01276951). Similar studies are testing the effects of dark chocolate and green tea (NCT0055963). Studies in coronary artery disease (CAD) investigated the effect of almond flavonoids or grape seed extract as well as vitamin C and grape extract (NCT0103862 and NCT00554242, respectively). A phase I study investigated the effect of oligonol (lychee fruit extract) on endothelial function (NCT1162174). Risk of incidence of CVD was associated with food sources in an extensive study involving more than 39,000 subjects (NCT00006504) finished in 2005, but no results have been reported yet. Related to obesity, the FLAVO trial (NCT00677599) evaluated the effect of chocolate and soybean enriched with epicatechin versus non-enriched in type 2 postmenopausal women is ongoing. A phase IV trial study tested the effect of oligomeric proanthocyanidins derived from red wine, citrus, among others (OPC-3 by NutraMetrix) in fasting state and after a high-fat meal in lipid and inflammatory markers, but results have not been released. A recently funded study (NCT01622309) investigated the effect in cardiovascular and obesity markers in relationship with the consumption of *Solanum melongena*, known as eggplant fruit in Brazil in a 120-days diet eggplant–flour intervention study including 180 subjects. The effects of tea consumption were studied in a group of 1000 individuals between ages 35 and 60 years where a strong inverse relationship between tea drinking and stroke was reported [258, 259]. A study of tea consumption in Japan on a group of 203 patients who had undergone coronary angioplasty and the other 94 without coronary stenosis showed an inverse relationship between tea intake and incidence of CAD [260]. EGCG has received particular attention with regard to its anti-sclerotic effects. Catechin reduces the CVD risk through antioxidant, anti-inflammatory, and endothelial function-enhancing properties [260]. In an another randomized crossover trial in Japan, the initial and final values of some blood parameters were compared in a group of 22 CAD patients who drank either 100 mL/day mixture of green and black tea or water for 1 month. Their results showed a significantly increased plasma adiponectin level in the tea intake group compared to the subjects who drank water [261]. The effects of dietary supplementation of flavonoid-rich food on SBP in an 8-year trial of 2200 individuals concluded that the population who received flavonoid-rich foods had a lower risk of CVD [262]. Human studies using 150–300 mg of quercetin-4′-*O*-β-D-glucoside supplement found to inhibit platelet signaling and aggregation (thrombus formation) [263]. An extensive review of the therapeutic potential of probably the four most widely investigated flavonoids: flavopiridol (a novel semisynthetic flavone analogue of rohitukine), catechins, genistein, and quercetin have been conducted [264]. In crossover trial studies, consumption of double-strength cranberry juice (54 % juice, 835 mg total polyphenols, and 94 mg anthocyanins) for 4 weeks with a 2-week rest period, followed by a matched placebo beverage (480 mL/day) showed a reduced carotid femoral pulse wave velocity, a clinically relevant measure of arterial stiffness in patients with CAD [265]. Some flavonoids including isoflavones present at high concentrations in soy are considered phytoestrogens, behaving as estrogen mimics [266]. Epidemiologic data in humans and animals suggest that dietary phytoestrogens have protective effects against menopausal symptoms and a variety of disorders, including cardiovascular disease, cancer, hyperlipidemia, osteoporosis, and various forms of chronic renal disease [267]. A multicenter, randomized, double-blind, placebo-controlled 24-month trial showed that the daily supplementation with 120 mg soy hypocotyl isoflavones reduced whole-body bone loss [268, 269]. Flavopiridol is one of the most successful examples of the use of flavonoids for health. Flavopiridol's anticancer activities are based on the inhibition of cyclin-dependent kinases (CDKs) and displays unique anticancer properties in human clinical trials conducted by Aventis Pharma (formerly Hoechst Marion Roussel) and the National Cancer Institute (NCI). By July 1999, the compound had entered phase II trials for gastric cancer and leukemia (CLL), and phase I/II trials for esophageal tumor and nonsmall cell lung cancer (NSCLC) [270]. Recently, in a randomized controlled double-blind crossover trial, the effectiveness of two dietary flavonoids, quercetin, and genistein in prostate cancer was conducted (NCT01538316). In addition, flavonoid-treated patients showed no colon cancer recurrence against 20% recurrence found in controls [271].

As the results of these trails become available, a clearer picture of the effects of flavonoids in health is expected to emerge.

Next Generation Nutraceuticals: Functional Foods as a Novel Preventive and Therapeutic Approach for Disease Control

In traditional medicine, whether part of the cultural folklore or historical knowledge, consumption of plant-derived foods has been associated with health benefits. In the past decade, prolonged life expectancy accompanied with the increased costs of health insurance has triggered great interest in the use of plant nutraceuticals as therapeutic or preventive alternatives. In addition, adverse effects and lack of efficacy related to current therapies together with growing interest in the use of more natural interventions helped sustain significantly increased consumption of plant supplements and nutritional foods. However, a paramount concern is whether the levels of flavonoids found in diets, metabolized, and ultimately absorbed reach the required concentration needed for disease control. Moreover, and most likely, the beneficial effect of nutraceuticals will be a result of a combination of specific flavonoids (and other phytochemicals) at optimal concentrations.

In this regard, the formulation of "functional foods" or "smart diets" with defined flavonoid content may be a part of future strategies in flavonoid health intervention. Functional foods can be tailored to increase flavonoids absorption and availability. Few examples of functional foods effectiveness are available so far. Soy isoflavone bioavailability was increased in a bread-based diet versus juice diets, suggesting a relevant role in the mode of administration to increase flavonoid concentrations in vivo [272]. In a randomized crossover design study, different food matrix showed significantly increased soy isoflavone metabolism in hypercholesterolemia individuals [273]. Daily consumption of soy bread for a month was shown to favorably influence prostate-specific antigen (PSA) levels in men diagnosed with prostate cancer [274]. Isoflavone-rich functional foods such as breakfast cereals, energy and snack bars, frozen desserts to soy-germ-fortified tomato juice have also been developed [275]. We showed that food formulations designed to increased the aglycone apigenin level increased bioavailability in vivo and reduced inflammatory mediators, consistent with a higher absorption of aglycones compared with glycosides [74]. Future efforts in this area are expected as novel "functional foods" may provide new opportunities to reach the demand and the localization required for maximum effectiveness of nutraceutical flavonoids.

Concluding Remarks

The value of flavonoids and anthocyanins in health is well established. However, the limited information from human studies, including the identification of hundreds of compounds that are likely to provide the greatest protection in the context of preventive nutrition in a diet, furnishes one of the main future challenges. In addition, understanding the molecular mechanisms of action, the metabolism, and the distribution of flavonoids are the key to defining ideal flavonoid-containing diets. Technological advances in the selection and bioengineering of plants with highest nutraceutical levels combined with improvements in flavonoid content through the design of better food formulations are likely to have a major impact. All these factors will open the door to future epidemiologic studies targeted to better define the relationships between the intake of nutraceuticals and the role in the treatment and prevention of disease.

Acknowledgments Research in Dr. Doseff's laboratory is supported by grants NIH (R01 HL075040-01) and by state funds appropriated to the Ohio Plant Biotechnology Consortium through The Ohio State University, Ohio Agricultural Research and Development Center. Dr. Grotewold's research on anthocyanins is supported by grant from the USDA (2010-65115-20408). Research in the regulatory networks of phenolics biosynthesis in Drs. Doseff and Grotewold's laboratories is supported by NSF (IOS-1125620).

References

1. DeFelice SL Nutraceuticals: opportunities in an emerging market. Scrip Mag. Sept 1992
2. Bravo L (1998) Polyphenols: chemistry, dietary sources, metabolism, and nutritional significance. Nutr Rev 56:317–333
3. Ghosh D, Konishi T (2007) Anthocyanins and anthocyanin-rich extracts: role in diabetes and eye function. Asia Pac J Clin Nutr 16:200–208
4. Luo H, Jiang B, Li B, Li Z, Jiang BH, Chen YC (2012) Kaempferol nanoparticles achieve strong and selective inhibition of ovarian cancer cell viability. Int J Nanomedicine 7:3951–3959
5. Trudel D, Labbe DP, Bairati I, Fradet V, Bazinet L, Tetu B (2012) Green tea for ovarian cancer prevention and treatment: a systematic review of the in vitro, in vivo and epidemiological studies. Gynecol Oncol 126:491–488
6. Garcia-Lafuente A, Guillamon E, Villares A, Rostagno MA, Martinez JA (2009) Flavonoids as anti-inflammatory agents: implications in cancer and cardiovascular disease. Inflamm Res 58:537–552

7. Jung M, Park M, Lee HC, Kang YH, Kang ES, Kim SK (2006) Antidiabetic agents from medicinal plants. Curr Med Chem 13:1203–1218
8. Gorinstein S, Caspi A, Libman I, Katrich E, Lerner HT, Trakhtenberg S (2004) Fresh israeli jaffa sweetie juice consumption improves lipid metabolism and increases antioxidant capacity in hypercholesterolemic patients suffering from coronary artery disease: studies in vitro and in humans and positive changes in albumin and fibrinogen fractions. J Agric Food Chem 52:5215–5222
9. Harborne JB, William CA (1975) Flavone and flavonol glycoside. In: Harborne JB, Mabry TJ., Mabry H (eds) The flavonoids. Chapman and Hall, London, pp 376–441
10. Dragsted LO, Strube M, Larsen JC (1993) Cancer-protective factors in fruits and vegetables: biochemical and biological background. Pharmacol Toxicol 72(Suppl 1):116–135
11. Harborne JB (1986) Nature, distribution and function of plant flavonoids. Prog Clin Biol Res 213:15–24
12. Beecher GR (2003) Overview of dietary flavonoids: nomenclature, occurrence and intake. J Nutr 133:3248S–3254S
13. Lotito SB, Frei B (2006) Consumption of flavonoid-rich foods and increased plasma antioxidant capacity in humans: cause, consequence, or epiphenomenon? Free Radic Biol Med 41:1727–1746
14. Pandey KB, Rizvi SI (2009) Plant polyphenols as dietary antioxidants in human health and disease. Oxid Med Cell Longev 2:270–278
15. Arts IC, Hollman PC, Feskens EJ, Bueno de Mesquita HB, Kromhout D (2001) Catechin intake and associated dietary and lifestyle factors in a representative sample of Dutch men and women. Eur J Clin Nutr 55:76–81
16. Manach C, Scalbert A, Morand C, Remesy C, Jimenez L (2004) Polyphenols: food sources and bioavailability. Am J Clin Nutr 79:727–747
17. Crozier IA, Lean M, McDonald M, Black C (1997) Quantitative analysis of the flavonoid content of commercial tomatoes, onions, lettuce, and celery. J Agric Food Chem 45:590–595
18. Stafford HA (1990) Flavonoid metabolism. CRC Press, Boca Raton
19. Delgado-Vargas F, Jimenez AR, Paredes-Lopez O (2000) Natural pigments: carotenoids, anthocyanins, and betalains–characteristics, biosynthesis, processing, and stability. Crit Rev Food Sci Nutr 40:173–289
20. Winkel-Shirley B (2002) Biosynthesis of flavonoids and effects of stress. Curr Opin Plant Biol 5:218–223
21. Huang WY, Zhang HC, Liu WX, Li CY (2012) Survey of antioxidant capacity and phenolic composition of blueberry, blackberry, and strawberry in Nanjing. J Zhejiang Univ Sci B 13:94–102
22. Dixon RA, Xie DY, Sharma SB (2005) Proanthocyanidins–a final frontier in flavonoid research? New Phytol 165:9–28
23. Xie DY, Dixon RA (2005) Proanthocyanidin biosynthesis–still more questions than answers? Phytochemistry 66:2127–2144
24. Dixon RA, Steele CL (1999) Flavonoids and isoflavonoids – a gold mine for metabolic engineering. Trends Plant Sci 4:394–400
25. Koes R, Verweij W, Quattrocchio F (2005) Flavonoids: a colorful model for the regulation and evolution of biochemical pathways. Trends Plant Sci 10:236–242
26. Mol J, Grotewold E, Koes R (1998) How genes paint flowers and seeds. Trends Plant Sci 3:212–217
27. Winkel BSJ (2006) The biosynthesis of flavonoids. In: Grotewold E (ed) The science of flavonoids. Springer, New York, pp 71–95
28. Winkel-Shirley B (2001) It takes a garden. How work on diverse plant species has contributed to an understanding of flavonoid metabolism. Plant Physiol 127:1399–1404
29. Winkel-Shirley B (2001) Flavonoid biosynthesis. A colorful model for genetics, biochemistry, cell biology, and biotechnology. Plant Physiol 126:485–493
30. Grotewold E (2006) The genetics and biochemistry of floral pigments. Annu Rev Plant Biol 57:761–780

31. Nakajima J, Tanaka Y, Yamazaki M, Saito K (2001) Reaction mechanism from leucoanthocyanidin to anthocyanidin 3-glucoside, a key reaction for coloring in anthocyanin biosynthesis. J Biol Chem 276:25797–25803
32. Brazier-Hicks M, Evans KM, Gershater MC, Puschmann H, Steel PG, Edwards R (2009) The C-glycosylation of flavonoids in cereals. J Biol Chem 284:17926–17934
33. Ververidis F, Trantas E, Douglas C, Vollmer G, Kretzschmar G, Panopoulos N (2007) Biotechnology of flavonoids and other phenylpropanoid-derived natural products. Part II: reconstruction of multienzyme pathways in plants and microbes. Biotechnol J 2:1235–1249
34. van het Hof KH, Wiseman SA, Yang CS, Tijburg LB (1999) Plasma and lipoprotein levels of tea catechins following repeated tea consumption. Proc Soc Exp Biol Med 220:203–209
35. Young JF, Nielsen SE, Haraldsdóttir J, Daneshvar B, Lauridsen ST, Knuthsen P et al (1999) Effect of fruit juice intake on urinary quercetin excretion and biomarkers of antioxidative status. Am J Clin Nutr 69:87–94
36. Vitaglione P, Sforza S, Galaverna G, Ghidini C, Caporaso N, Vescovi PP et al (2005) Bioavailability of trans-resveratrol from red wine in humans. Mol Nutr Food Res 49:495–504.
37. Erlund I, Silaste ML, Alfthan G, Rantala M, Kesaniemi YA, Aro A (2002) Plasma concentrations of the flavonoids hesperetin, naringenin and quercetin in human subjects following their habitual diets, and diets high or low in fruit and vegetables. Eur J Clin Nutr 56:891–898
38. Spencer JP, Kuhnle GG, Williams RJ, Rice-Evans C (2003) Intracellular metabolism and bioactivity of quercetin and its in vivo metabolites. Biochem J 372:173–181
39. Felgines C, Talavera S, Texier O, Gil-Izquierdo A, Lamaison JL, Remesy C (2005) Blackberry anthocyanins are mainly recovered from urine as methylated and glucuronidated conjugates in humans. J Agric Food Chem 53:7721–7727
40. Talavera S, Felgines C, Texier O, Besson C, Gil-Izquierdo A, Lamaison JL et al (2005) Anthocyanin metabolism in rats and their distribution to digestive area, kidney, and brain. J Agric Food Chem 53:3902–3908
41. Kalt W, Blumberg JB, McDonald JE Vinqvist-Tymchuk MR, Fillmore SA, Graf BA et al (2008) Identification of anthocyanins in the liver, eye, and brain of blueberry-fed pigs. J Agric Food Chem 56:705–712
42. Garcea G, Berry DP, Jones DJ, Singh R, Dennison AR, Farmer PB et al (2005) Consumption of the putative chemopreventive agent curcumin by cancer patients: assessment of curcumin levels in the colorectum and their pharmacodynamic consequences. Cancer Epidemiol Biomarkers Prev 14:120–125
43. Maubach J, Bracke ME, Heyerick A, Depypere HT, Serreyn RF, Mareel MM et al (2003) Quantitation of soy-derived phytoestrogens in human breast tissue and biological fluids by high-performance liquid chromatography. J Chromatogr B Analyt Technol Biomed Life Sci 784:137–144
44. D'Archivio M, Filesi C, Di Benedetto R, Gargiulo R, Giovannini C, Masella R (2007) Polyphenols, dietary sources and bioavailability. Ann Ist Super Sanita 43:348–361
45. Wu X, Cao G, Prior RL (2002) Absorption and metabolism of anthocyanins in elderly women after consumption of elderberry or blueberry. J Nutr 132:1865–1871
46. Lee MJ, Maliakal P, Chen L, Meng X, Bondoc FY, Prabhu S et al (2002) Pharmacokinetics of tea catechins after ingestion of green tea and (−)-epigallocatechin-3-gallate by humans: formation of different metabolites and individual variability. Cancer Epidemiol Biomarkers Prev 11:1025–1032
47. Meng X, Lee MJ, Li C, Sheng S, Zhu N, Sang S et al (2001) Formation and identification of 4′-O-methyl-(−)-epigallocatechin in humans. Drug Metab Dispos 29:789–793
48. Piskula MK, Terao J (1998) Accumulation of (−)-epicatechin metabolites in rat plasma after oral administration and distribution of conjugation enzymes in rat tissues. J Nutr 128:1172–1178
49. Falany CN (1997) Enzymology of human cytosolic sulfotransferases. FASEB J 11:206–216
50. Spencer JP, Chowrimootoo G, Choudhury R, Debnam ES, Srai SK, Rice-Evans C (1998) The small intestine can both absorb and glucuronidate luminal flavonoids. FEBS Lett 458: 224–230

51. Boersma MG, van der WH, Bogaards J Boeren S, Vervoort J, Cnubben NH et al (2002) Regioselectivity of phase II metabolism of luteolin and quercetin by UDP-glucuronosyl transferases. Chem Res Toxicol 15:662–670
52. Shimoi K, Saka N, Nozawa R, Sato M, Amano I, Nakayama T et al (2001) Deglucuronidation of a flavonoid, luteolin monoglucuronide, during inflammation. Drug Metab Dispos 29:1521–1524
53. Atkinson C, Frankenfeld CL, Lampe JW (2005) Gut bacterial metabolism of the soy isoflavone daidzein: exploring the relevance to human health. Exp Biol Med (Maywood) 230:155–170
54. Franke AA, Custer LJ, Hundahl SA (2004) Determinants for urinary and plasma isoflavones in humans after soy intake. Nutr Cancer 50:141–154
55. Crespy V, Morand C, Besson C, Manach C, Demigne C, Remesy C (2002) Quercetin, but not its glycosides, is absorbed from the rat stomach. J Agric Food Chem 50:618–621
56. Piskula MK, Yamakoshi J, Iwai Y (1999) Daidzein and genistein but not their glucosides are absorbed from the rat stomach. FEBS Lett 447:287–291
57. Tian Q, Giusti MM, Stoner GD, Schwartz SJ (2006) Urinary excretion of black raspberry (*Rubus occidentalis*) anthocyanins and their metabolites. J Agric Food Chem 54:1467–1472
58. Fito M, Guxens M, Corella D, Saez G, Estruch R, de la Torre R et al (2007) Effect of a traditional Mediterranean diet on lipoprotein oxidation: a randomized controlled trial. Arch Intern Med 167:1195–1203
59. Marrugat J, Covas MI, Fito M, Schröder H, Miro-Casas E, Gimeno E et al (2004) Effects of differing phenolic content in dietary olive oils on lipids and LDL oxidation–a randomized controlled trial. Eur J Nutr 43:140–147
60. Keppler K, Humpf HU (2005) Metabolism of anthocyanins and their phenolic degradation products by the intestinal microflora. Bioorg Med Chem 13:5195–5205
61. Scalbert A, Deprez S, Mila I, Albrecht AM, Huneau JF, Rabot S (2002) Proanthocyanidins and human health: systemic effects and local effects in the gut. Biofactors 13:115–120
62. Plumb GW, Pascual-Teresa S, Santos-Buelga C, Cheynier V, Williamson G (1998) Antioxidant properties of catechins and proanthocyanidins: effect of polymerisation, galloylation and glycosylation. Free Radic Res 29:351–358
63. He J, Giusti MM (2010) Anthocyanins: natural colorants with health-promoting properties. Annu Rev Food Sci Technol 1:163–187
64. Tulipani S, Romandini S, Busco F, Bompadre S, Mezzetti B, Battino M (2009) Ascorbate, not urate, modulates the plasma antioxidant capacity after strawberry intake. Food Chemistry 117:181–188
65. Mazza G, Kay CD, Cottrell T, Holub BJ (2002) Absorption of anthocyanins from blueberries and serum antioxidant status in human subjects. J Agric Food Chem 50:7731–7737
66. Miller MG, Shukitt-Hale B (2012) Berry fruit enhances beneficial signaling in the brain. J Agric Food Chem. [Epub ahead of print]
67. Bub A, Watzl B, Heeb D, Rechkemmer G, Briviba K (2001) Malvidin-3-glucoside bioavailability in humans after ingestion of red wine, dealcoholized red wine and red grape juice. Eur J Nutr 40:113–120
68. Ichiyanagi T, Shida Y, Rahman MM et al (2005) Metabolic pathway of cyanidin 3-O-beta-D-glucopyranoside in rats. J Agric Food Chem 53:145–150
69. Wright EM, Loo DD, Hirayama AB (2011) Biology of human sodium glucose transporters. Physiol Rev 91:733–794
70. Day AJ, DuPont MS, Ridley S, Rhodes M, Rhodes MJ, Morgan MR et al (1998) Deglycosylation of flavonoid and isoflavonoid glycosides by human small intestine and liver beta-glucosidase activity. FEBS Lett 436:71–75
71. Booth AN, Deeds F, Jones FT, Murray CW (1956) The metabolic fate of rutin and quercetin in the animal body. J Biol Chem 223:251–257
72. Baba S, Furuta T, Horie M, Nakagawa H (1981) Studies on drug metabolism by use of isotopes XXVI: determination of urinary metabolites of rutin in humans. J Pharm Sci 70: 780–782

73. Kuhnau J (1976) The flavonoids. A class of semi-essential food components: their role in human nutrition. World Rev Nutr Diet 24:117–191
74. Hostetler G, Riedl K, Cardenas H, Diosa-Toro M, Arango D, Schwartz S et al (2012) Flavone deglycosylation increases their anti-inflammatory activity and absorption. Mol Nutr Food Res 56:558–569
75. Day AJ, Williamson G (2001) Biomarkers for exposure to dietary flavonoids: a review of the current evidence for identification of quercetin glycosides in plasma. Br J Nutr 86 (Suppl 1):S105–S110
76. Natsume M, Osakabe N, Oyama M, Sasaki M, Baba S, Nakamura Y et al (2003) Structures of (−)-epicatechin glucuronide identified from plasma and urine after oral ingestion of (−)-epicatechin: differences between human and rat. Free Radic Biol Med 34:840–849
77. Zhang Y, Hendrich S, Murphy PA (2003) Glucuronides are the main isoflavone metabolites in women. J Nutr 133:399–404.
78. de Groot H (1994) Reactive oxygen species in tissue injury. Hepatogastroenterology 41: 328–332
79. Grace PA (1994) Ischaemia-reperfusion injury. Br J Surg 81:637–647
80. Halliwell B (1995) How to characterize an antioxidant: an update. Biochem Soc Symp 61:73–101
81. Kong JM, Chia LS, Goh NK, Chia TF, Brouillard R (2003) Analysis and biological activities of anthocyanins. Phytochemistry 64:923–933
82. Rice-Evans CA, Miller NJ, Paganga G (1996) Structure-antioxidant activity relationships of flavonoids and phenolic acids. Free Radic Biol Med 20:933–956
83. Pietta PG (2000) Flavonoids as antioxidants. J Nat Prod 63:1035–1042
84. Guo JJ, Hsieh HY, Hu CH (2009) Chain-breaking activity of carotenes in lipid peroxidation: a theoretical study. J Phys Chem B 113:15699–15708
85. Du Y, Guo H, Lou H (2007) Grape seed polyphenols protect cardiac cells from apoptosis via induction of endogenous antioxidant enzymes. J Agric Food Chem 55:1695–1701
86. Cos P, Ying L, Calomme M, Hu JP, Cimanga K, Van Poel B et al (1998) Structure-activity relationship and classification of flavonoids as inhibitors of xanthine oxidase and superoxide scavengers. J Nat Prod 61:71–6
87. Shoskes DA (1998) Effect of bioflavonoids quercetin and curcumin on ischemic renal injury: a new class of renoprotective agents. Transplantation 66:147–152
88. Chang WS, Lee YJ, Lu FJ, Chiang HC (1993) Inhibitory effects of flavonoids on xanthine oxidase. Anticancer Res 13:2165–2170
89. Iio M, Ono Y, Kai S, Fukumoto M (1986) Effects of flavonoids on xanthine oxidation as well as on cytochrome c reduction by milk xanthine oxidase. J Nutr Sci.Vitaminol (Tokyo) 32:635–642
90. Castelluccio C, Paganga G, Melikian N, Bolwell GP, Pridham J, Sampson J et al (1995) Antioxidant potential of intermediates in phenylpropanoid metabolism in higher plants. FEBS Lett 368:188–192
91. Salah N, Miller NJ, Paganga G, Tijburg L, Bolwell GP, Rice-Evans C (1995) Polyphenolic flavanols as scavengers of aqueous phase radicals and as chain-breaking antioxidants. Arch Biochem Biophys 322:339–346
92. Boyle SP, Dobson VL, Duthie SJ, Kyle JA, Collins AR (2000) Absorption and DNA protective effects of flavonoid glycosides from an onion meal. Eur J Nutr 39:213–223
93. Kapiotis S, Hermann M, Held I, Seelos C, Ehringer H, Gmeiner BM (1997) Genistein, the dietary-derived angiogenesis inhibitor, prevents LDL oxidation and protects endothelial cells from damage by atherogenic LDL. Arterioscler Thromb Vasc Biol 17:2868–2874
94. Murakami A, Nakamura Y, Ohto Y, Yano M, Koshiba T, Koshimizu K et al (2000) Suppressive effects of citrus fruits on free radical generation and nobiletin, an anti-inflammatory polymethoxyflavonoid. Biofactors 12:187–192
95. Murakami A, Nakamura Y, Torikai K, Tanaka T, Koshiba T, Koshimizu K et al (2000) Inhibitory effect of citrus nobiletin on phorbol ester-induced skin inflammation, oxidative stress, and tumor promotion in mice. Cancer Res 60:5059–5066

96. Gordon MH, Roedig-Penman A (1998) Antioxidant activity of quercetin and myricetin in liposomes. Chem Phys Lipids 97:79–85
97. Shoskes DA (1998) Effect of bioflavonoids quercetin and curcumin on ischemic renal injury: a new class of renoprotective agents. Transplantation 66:147–152
98. Nijveldt RJ, van Nood E, van Hoorn DE, Boelens PG, van Norren K, van Leeuwen PA (2001) Flavonoids: a review of probable mechanisms of action and potential applications. Am J Clin Nutr 74:418–425
99. Perron NR, Brumaghim JL (2009) A review of the antioxidant mechanisms of polyphenol compounds related to iron binding. Cell Biochem Biophys 53:75–100
100. Wang SY, Jiao H (2000) Scavenging capacity of berry crops on superoxide radicals, hydrogen peroxide, hydroxyl radicals, and singlet oxygen. J Agric Food Chem 48:5677–5684
101. Wang LS, Stoner GD (2008) Anthocyanins and their role in cancer prevention. Cancer Lett 269:281–290
102. Vargo MA, Voss OH, Poustka F, Cardounel AJ, Grotewold E, Doseff AI (2006) Apigenin-induced-apoptosis is mediated by the activation of PKCdelta and caspases in leukemia cells. Biochem Pharmacol 72:681–692
103. Salvi M, Brunati AM, Clari G, Toninello A (2002) Interaction of genistein with the mitochondrial electron transport chain results in opening of the membrane transition pore. Biochim Biophys Acta 1556:187–196
104. Shen SC, Ko CH, Tseng SW, Tsai SH, Chen YC (2004) Structurally related antitumor effects of flavanones in vitro and in vivo: involvement of caspase 3 activation, p21 gene expression, and reactive oxygen species production. Toxicol Appl Pharmacol 197:84–95
105. Murphy A, Peer WA, Taiz L (2000) Regulation of auxin transport by aminopeptidases and endogenous flavonoids. Planta 211:315–324
106. Pourcel L, Irani NG, Koo AJK, Bohorquez-Restrepo A, Howe GA, Grotewold E (2013) A chemical complementation approach reveals genes and interactions of flavonoids with other pathways. Plant J 74:383–397
107. Wingender R, Rohrig H, Horicke C, Wing D, Schell J (1989) Differential regulation of soybean chalcone synthase genes in plant defence, symbiosis and upon environmental stimuli. Mol Gen Genet 218:315–322
108. Seo S, Sano H, Ohashi Y (1999) Jasmonate-based wound signal transduction requires activation of WIPK, a tobacco mitogen-activated protein kinase. Plant Cell 11:289–298
109. Hungria M, Joseph CM, Phillips DA (1991) Anthocyanidins and flavonols, major nod gene inducers from seeds of a black-seeded common bean (*Phaseolus vulgaris* L.). Plant Physiol 97:751–758
110. Novak K, Chovanec P, Skrdleta V, Kropacova M, Lisa L, Nemcova M (2002) Effect of exogenous flavonoids on nodulation of pea (*Pisum sativum* L.). J Exp Bot 53:1735–1745
111. Yeh KC, Peck MC, Long SR (2002) Luteolin and GroESL modulate in vitro activity of NodD. J Bacteriol 184:525–530
112. Spencer JP, Rice-Evans C, Williams RJ (2003) Modulation of pro-survival Akt/protein kinase B and ERK1/2 signaling cascades by quercetin and its in vivo metabolites underlie their action on neuronal viability. J Biol Chem 278:34783–34793
113. Kobuchi H, Roy S, Sen CK, Nguyen HG, Packer L (1999) Quercetin inhibits inducible ICAM-1 expression in human endothelial cells through the JNK pathway. Am J Physiol 277:C403–C411
114. Kong AN, Yu R, Chen C, Mandlekar S, Primiano T (2000) Signal transduction events elicited by natural products: role of MAPK and caspase pathways in homeostatic response and induction of apoptosis. Arch Pharm Res 23:1–16
115. Gonzalez-Mejia ME, Voss OH, Murnan EJ, Doseff AI (2010) Apigenin-induced apoptosis of leukemia cells is mediated by a bimodal and differentially regulated residue-specific phosphorylation of heat-shock protein-27. Cell Death Dis 1:e64
116. Nicholas C, Batra S, Vargo MA, Voss OH, Gavrilin MA, Wewers MD et al (2007) Apigenin blocks lipopolysaccharide-induced lethality in vivo and proinflammatory cytokines

expression by inactivating NF-kappaB through the suppression of p65 phosphorylation. J Immunol 179:7121–7127

117. Kim JS, Jobin C (2005) The flavonoid luteolin prevents lipopolysaccharide-induced NF-kappaB signalling and gene expression by blocking IkappaB kinase activity in intestinal epithelial cells and bone-marrow derived dendritic cells. Immunology 115:375–387
118. Schroeter H, Spencer JP, Rice-Evans C, Williams RJ (2001) Flavonoids protect neurons from oxidized low-density-lipoprotein-induced apoptosis involving c-Jun N-terminal kinase (JNK), c-Jun and caspase-3. Biochem J 358:547–557
119. Xi J, Wang H, Mueller RA, Norfleet EA, Xu Z (2009) Mechanism for resveratrol-induced cardioprotection against reperfusion injury involves glycogen synthase kinase 3beta and mitochondrial permeability transition pore. Eur J Pharmacol 604:111–116
120. Goh SS, Woodman OL, Pepe S, Cao AH, Qin C, Ritchie RH (2007) The red wine antioxidant resveratrol prevents cardiomyocyte injury following ischemia-reperfusion via multiple sites and mechanisms. Antioxid Redox Signal 9:101–113
121. Gurusamy N, Lekli I, Mukherjee S, Ray D, Ahsan MK, Gherghiceanu M et al (2010) Cardioprotection by resveratrol: a novel mechanism via autophagy involving the mTORC2 pathway. Cardiovasc Res 86:103–12
122. Ungvari Z, Bagi Z, Feher A, Recchia FA, Sonntag WE, Pearson K et al (2010) Resveratrol confers endothelial protection via activation of the antioxidant transcription factor Nrf2. Am J Physiol Heart Circ Physiol 299:H18–H24
123. Das S, Fraga CG, Das DK (2006) Cardioprotective effect of resveratrol via HO-1 expression involves p38 map kinase and PI-3-kinase signaling, but does not involve NFkappaB. Free Radic Res 40:1066–1075
124. Hung LM, Su MJ, Chen JK (2004) Resveratrol protects myocardial ischemia-reperfusion injury through both NO-dependent and NO-independent mechanisms. Free Radic Biol Med 36:774–781
125. Oak MH, Bedoui JE, Madeira SV, Chalupsky K, Schini-Kerth VB (2006) Delphinidin and cyanidin inhibit PDGF(AB)-induced VEGF release in vascular smooth muscle cells by preventing activation of p38 MAPK and JNK. Br J Pharmacol 149:283–290
126. Vang O, Ahmad N, Baile CA, Baur JA, Brown K, Csiszar A et al (2011) What is new for an old molecule? Systematic review and recommendations on the use of resveratrol. PLoS ONE 6:e19881
127. Fremont L (2000) Biological effects of resveratrol. Life Sci 66:663–673
128. Donnelly LE, Newton R, Kennedy GE, Fenwick PS, Leung RH, Ito K et al (2004) Anti-inflammatory effects of resveratrol in lung epithelial cells: molecular mechanisms. Am J Physiol Lung Cell Mol Physiol 287:L774–L783
129. Culpitt SV, Rogers DF, Fenwick PS, Shah P, De Matos C, Russell RE et al (2003) Inhibition by red wine extract, resveratrol, of cytokine release by alveolar macrophages in COPD. Thorax 58:942–946
130. Noe V, Penuelas S, Lamuela-Raventos RM, Permanyer J, Ciudad CJ, Izquierdo-Pulido M (2004) Epicatechin and a cocoa polyphenolic extract modulate gene expression in human Caco-2 cells. J Nutr 134:2509–2516
131. Bahia PK, Rattray M, Williams RJ (2008) Dietary flavonoid (-)epicatechin stimulates phosphatidylinositol 3-kinase-dependent anti-oxidant response element activity and up-regulates glutathione in cortical astrocytes. J Neurochem 106:2194–2204
132. Kioka N, Hosokawa N, Komano T, Hirayoshi K, Nagata K, Ueda K (1992) Quercetin, a bioflavonoid, inhibits the increase of human multidrug resistance gene (MDR1) expression caused by arsenite. FEBS Lett 301:307–309
133. Johnson VJ, Osuchowski MF, He Q, Sharma RP (2002) Physiological responses to a natural antioxidant flavonoid mixture, silymarin, in BALB/c mice: II. Alterations in thymic differentiation correlate with changes in c-myc gene expression. Planta Med 68:961–965
134. Diebolt M, Bucher B, Andriantsitohaina R (2001) Wine polyphenols decrease blood pressure, improve NO vasodilatation, and induce gene expression. Hypertension 38:159–165

135. Lopez-Posadas R, Ballester I, Mascaraque C, Suarez MD, Zarzuelo A, Martinez-Augustin O et al (2010) Flavonoids exert distinct modulatory actions on cyclooxygenase 2 and NF-kappaB in an intestinal epithelial cell line (IEC18). Br J Pharmacol 160:1714–1726
136. Nicholson SK, Tucker GA, Brameld JM (2010) Physiological concentrations of dietary polyphenols regulate vascular endothelial cell expression of genes important in cardiovascular health. Br J Nutr 103:1398–1403
137. Boomgaarden I, Egert S, Rimbach G, Wolffram S, Muller MJ, Doring F (2010) Quercetin supplementation and its effect on human monocyte gene expression profiles in vivo. Br J Nutr 104:336–345
138. Bub A, Watzl B, Blockhaus M, Briviba K, Liegibel U, Muller H et al (2003) Fruit juice consumption modulates antioxidative status, immune status and DNA damage. J Nutr Biochem 14:90–98
139. Choi SW, Friso S (2010) Epigenetics: a new bridge between nutrition and health. Adv Nutr 1:8–16
140. Chmurzynska A (2010) Fetal programming: link between early nutrition, DNA methylation, and complex diseases. Nutr Rev 68:87–98
141. Mathers JC, Strathdee G, Relton CL (2010) Induction of epigenetic alterations by dietary and other environmental factors. Adv Genet 71:3–39
142. Lee WJ, Shim JY, Zhu BT (2005) Mechanisms for the inhibition of DNA methyltransferases by tea catechins and bioflavonoids. Mol Pharmacol 68:1018–1030
143. Arango D, Parihar A, Villamena FA, Wang L, Freitas MA, Grotewold E et al (2012) Apigenin induces DNA damage through the PKCδ-dependent activation of ATM and H2AX causing down-regulation of genes involved in cell cycle control and DNA repair. Biochem Pharmacol 84:1571–1580
144. Mathers JC, Ford D (2009) Nutrition, epigenetics and aging. In: Chio SW, Friso S (eds) Nutrients and epigenetics. CRC Press/Taylor & Francis Group, Boca Raton, pp 175–206
145. Davie JR (2003) Inhibition of histone deacetylase activity by butyrate. J Nutr 133:2485S–2493S
146. Kruh J (1982) Effects of sodium butyrate, a new pharmacological agent, on cells in culture. Mol Cell Biochem 42:65–82
147. Vidali G, Boffa LC, Mann RS, Allfrey VG (1978) Reversible effects of Na-butyrate on histone acetylation. Biochem Biophys Res Commun 82:223–227
148. Delage B, Dashwood RH (2008) Dietary manipulation of histone structure and function. Annu Rev Nutr 28:347–366
149. Baur JA (2010) Resveratrol, sirtuins, and the promise of a DR mimetic. Mech Ageing Dev 131:261–269
150. Lagouge M, Argmann C, Gerhart-Hines Z, Meziane H, Lerin C, Daussin F et al (2006) Resveratrol improves mitochondrial function and protects against metabolic disease by activating SIRT1 and PGC-1α. Cell 127:1109–1122
151. Baur JA, Pearson KJ, Price NL, Jamieson HA, Lerin C, Kalra A et al (2006) Resveratrol improves health and survival of mice on a high-calorie diet. Nature 444:337–342
152. Lopez-Lopez G, Moreno L, Cogolludo A, Galisteo M, Ibarra M, Duarte J et al (2004) Nitric oxide (NO) scavenging and NO protecting effects of quercetin and their biological significance in vascular smooth muscle. Mol Pharmacol 65:851–859
153. Bagchi D, Sen CK, Ray SD, Das DK, Bagchi M, Preuss HG et al (2003) Molecular mechanisms of cardioprotection by a novel grape seed proanthocyanidin extract. Mutat Res 523–524:87–97
154. Park C, So HS, Shin CH, Baek SH, Moon BS, Shin SH et al (2003) Quercetin protects the hydrogen peroxide-induced apoptosis via inhibition of mitochondrial dysfunction in H9c2 cardiomyoblast cells. Biochem Pharmacol 66:1287–1295
155. Pietri S, Maurelli E, Drieu K, Culcasi M (1997) Cardioprotective and anti-oxidant effects of the terpenoid constituents of *Ginkgo biloba* extract (EGb 761). J MolCell Cardiol 29: 733–742

156. Haramaki N, Aggarwal S, Kawabata T, Droy-Lefaix MT, Packer L (1994) Effects of natural antioxidant *Ginkgo biloba* extract (EGB 761) on myocardial ischemia-reperfusion injury. Free Radic Biol Med 16:789–794
157. Liu X, Wei J, Tan F, Zhou S, Wurthwein G, Rohdewald P (2004) Pycnogenol, French maritime pine bark extract, improves endothelial function of hypertensive patients. Life Sci 74:855–862
158. Galisteo M, Garcia-Saura MF, Jimenez R, Villar IC, Zarzuelo A, Vargas F et al (2004) Effects of chronic quercetin treatment on antioxidant defence system and oxidative status of deoxycorticosterone acetate-salt-hypertensive rats. Mol Cell Biochem 259:91–99
159. Read MA (1995) Flavonoids: naturally occurring anti-inflammatory agents. Am J Pathol 147:235–237
160. Rump AF, Schussler M, Acar D, Cordes A, Ratke R, Theisohn M et al (1995) Effects of different inotropes with antioxidant properties on acute regional myocardial ischemia in isolated rabbit hearts. Gen Pharmacol 26:603–611
161. Varga E (2002) The protective effect of EGb 761 in isolated isochemical/reperfused rat hearts. Acta Pharm Hung 72:265–271
162. Osman HE, Maalej N, Shanmuganayagam D, Folts JD (1998) Grape juice but not orange or grapefruit juice inhibits platelet activity in dogs and monkeys. J Nutr 128:2307–2312
163. Belinky PA, Aviram M, Mahmood S, Vaya J (1998) Structural aspects of the inhibitory effect of glabridin on LDL oxidation. Free Radic Biol Med 24:1419–1429
164. Nothlings U, Schulze MB, Weikert C, Boeing H, van der Schouw YT, Bamia C et al (2008) Intake of vegetables, legumes, and fruit, and risk for all-cause, cardiovascular, and cancer mortality in a European diabetic population. J Nutr 138:775–781
165. Maron DJ (2004) Flavonoids for reduction of atherosclerotic risk. Curr Atheroscler Rep 6:73–78
166. Huxley RR, Neil HA (2003) The relation between dietary flavonol intake and coronary heart disease mortality: a meta-analysis of prospective cohort studies. Eur J Clin Nutr 57:904–908
167. Peters U, Poole C, Arab L (2001) Does tea affect cardiovascular disease? A meta-analysis. Am J Epidemiol 154:495–503
168. Hodgson JM, Croft KD (2010) Tea flavonoids and cardiovascular health. Mol Aspects Med 31:495–502
169. Lin YL, Lin JK (1997) (−)-Epigallocatechin-3-gallate blocks the induction of nitric oxide synthase by down-regulating lipopolysaccharide-induced activity of transcription factor nuclear factor-kappaB. Mol Pharmacol 52:465–472
170. Danesi F, Di Nunzio M, Boschetti E, Bordoni A (2009) Green tea extract selectively activates peroxisome proliferator-activated receptor beta/delta in cultured cardiomyocytes. Br J Nutr 101:1736–1739
171. Pascual-Teresa S, Moreno DA, Garcia-Viguera C (2010) Flavanols and anthocyanins in cardiovascular health: a review of current evidence. Int J Mol Sci 11:1679–1703
172. Sheng R, Gu ZL, Xie ML, Zhou WX, Guo CY (2007) EGCG inhibits cardiomyocyte apoptosis in pressure overload-induced cardiac hypertrophy and protects cardiomyocytes from oxidative stress in rats. Acta Pharmacol Sin 28:191–201
173. Fisher ND, Hughes M, Gerhard-Herman M, Hollenberg NK (2003) Flavanol-rich cocoa induces nitric-oxide-dependent vasodilation in healthy humans. J Hypertens 21:2281–2286
174. Katan MB (1997) Flavonoids and heart disease. Am J Clin Nutr 65:1542–1543
175. Hollman PC, Katan MB (1999) Dietary flavonoids: intake, health effects and bioavailability. Food Chem Toxicol 37:937–942
176. Hertog MG, Hollman PC, Katan MB, Kromhout D (1993) Intake of potentially anticarcinogenic flavonoids and their determinants in adults in The Netherlands. Nutr Cancer 20:21–29
177. Knekt P, Jarvinen R, Reunanen A, Maatela J (1996) Flavonoid intake and coronary mortality in Finland: a cohort study. BMJ 312:478–481

178. Rimm EB, Katan MB, Ascherio A, Stampfer MJ, Willett WC (1996) Relation between intake of flavonoids and risk for coronary heart disease in male health professionals. Ann Intern Med 125:384–389
179. Hooper L, Kroon PA, Rimm EB et al (2008) Flavonoids, flavonoid-rich foods, and cardiovascular risk: a meta-analysis of randomized controlled trials. Am J Clin Nutr 88:38–50
180. Hodgson JM (2008) Tea flavonoids and cardiovascular disease. Asia Pac J Clin Nutr 17(Suppl 1):288–290
181. Akisu M, Kultursay N, Coker I, Huseyinov A (1998) Platelet-activating factor is an important mediator in hypoxic ischemic brain injury in the newborn rat. Flunarizine and *Ginkgo biloba* extract reduce PAF concentration in the brain. Biol Neonate 74:439–444
182. Sato M, Maulik G, Ray PS, Bagchi D, Das DK (1999) Cardioprotective effects of grape seed proanthocyanidin against ischemic reperfusion injury. J Mol Cell Cardiol 31: 1289–1297
183. Leikert JF, Rathel TR, Wohlfart P, Cheynier V, Vollmar AM, Dirsch VM (2002) Red wine polyphenols enhance endothelial nitric oxide synthase expression and subsequent nitric oxide release from endothelial cells. Circulation 106:1614–1617
184. Rathel TR, Samtleben R, Vollmar AM, Dirsch VM (2007) Activation of endothelial nitric oxide synthase by red wine polyphenols: impact of grape cultivars, growing area and the vinification process. J Hypertens 25:541–549
185. Kondrashov A, Vrankova S, Dovinova I, Sevcik R, Parohova J, Barta A et al (2012) The effects of new Alibernet red wine extract on nitric oxide and reactive oxygen species production in spontaneously hypertensive rats. Oxid Med Cell Longev 2012:806285
186. Saleem TS, Basha SD (2010) Red wine: a drink to your heart. J Cardiovasc Dis Res 1: 171–176
187. Estruch R, Sacanella E, Badia E et al (2004) Different effects of red wine and gin consumption on inflammatory biomarkers of atherosclerosis: a prospective randomized crossover trial. Effects of wine on inflammatory markers. Atherosclerosis 175:117–123
188. Pechanova O, Bernatova I, Babal P, Martinez MC., Kysela S, Stvrtina S et al (2004) Red wine polyphenols prevent cardiovascular alterations in L-NAME-induced hypertension. J Hypertens 22:1551–1559
189. Duarte J, Andriambeloson E, Diebolt M, Andriantsitohaina R (2004) Wine polyphenols stimulate superoxide anion production to promote calcium signaling and endothelial-dependent vasodilatation. Physiol Res 53:595–602
190. Shenoy SF, Chowanadisai W, Sharman E, Keen CL, Liu J, Rucker RB (2011) Biofactors in food promote health by enhancing mitochondrial function. Cal Agri 65:141–147
191. Duarte S, Arango D, Parihar A, Hamel P, Yasmeen R, Doseff AI (2013) Apigenin protects endothelial cells from lipopolysaccharide (LPS)-induced inflammation by decreasing caspase-3 activation and modulating mitochondrial function. Int J Mol Sci 14:17664–17679
192. Arango D, Morohashi K, Yilmaz A, Kuramochi K, Parihar A, Brahimaj B, Grotewold E, Doseff AI (2013) Molecular basis for the action of a dietary flavonoid revealed by the comprehensive identification of apigenin human targets. Proc Natl Acad Sci U S A 110:E2153–E2162
193. Toufektsian MC, de Lorgeril M, Nagy N, Salen P, Donati MB, Giordano L et al (2008) Chronic dietary intake of plant-derived anthocyanins protects the rat heart against ischemia-reperfusion injury. J Nutr 138:747–752
194. Bendini A, Cerretani L, Carrasco-Pancorbo A, Gomez-Caravaca AM, Segura-Carretero A, Fernandez-Gutierrez A et al (2007) Phenolic molecules in virgin olive oils: a survey of their sensory properties, health effects, antioxidant activity and analytical methods. An overview of the last decade. Molecules 12:1679–1719
195. Bertelli AA (2007) Wine, research and cardiovascular disease: instructions for use. Atherosclerosis 195:242–247
196. Fito M, de la TR, Covas MI (2007) Olive oil and oxidative stress. Mol Nutr Food Res 51:1215–1224

197. Hao J, Shen W, Yu G, Jia H, Li X, Feng Z et al (2010). Hydroxytyrosol promotes mitochondrial biogenesis and mitochondrial function in 3T3-L1 adipocytes. J Nutr Biochem 21:634–644
198. Vinson JA, Dabbagh YA (1998) Effect of green and black tea supplementation on lipids, lipid oxidation and fibrinogen in the hamster: mechanisms for the epidemiological benefits of tea drinking. FEBS Lett 433:44–46
199. Bell DR, Gochenaur K (2006) Direct vasoactive and vasoprotective properties of anthocyanin-rich extracts. J Appl Physiol 100:1164–1170
200. Cassidy A, De Vivo, I, Liu Y et al (2010) Associations between diet, lifestyle factors, and telomere length in women. Am J Clin Nutr 91:1273–1280
201. Wu X, Beecher GR, Holden JM, Haytowitz DB, Gebhardt SE, Prior RL (2006) Concentrations of anthocyanins in common foods in the United States and estimation of normal consumption. J Agric Food Chem 54:4069–4075
202. Gross LS, Li L, Ford ES, Liu S (2004) Increased consumption of refined carbohydrates and the epidemic of type 2 diabetes in the United States: an ecologic assessment. Am J Clin Nutr 79:774–779
203. Wedick NM, Pan A, Cassidy A, Rimm EB, Sampson L, Rosner B et al (2012) Dietary flavonoid intakes and risk of type 2 diabetes in US men and women. Am J Clin Nutr 95:925–933
204. Zhang Y, Jayaprakasam B, Seeram NP, Olson LK, DeWitt D, Nair MG (2004) Insulin secretion and cyclooxygenase enzyme inhibition by cabernet sauvignon grape skin compounds. J Agric Food Chem 52:228–233
205. Song Y, Manson JE, Buring JE, Sesso HD, Liu S (2005) Associations of dietary flavonoids with risk of type 2 diabetes, and markers of insulin resistance and systemic inflammation in women: a prospective study and cross-sectional analysis. J Am Coll Nutr 24:376–384
206. Shabrova EV, Tarnopolsky O, Singh AP, Plutzky J, Vorsa N, Quadro L (2011) Insights into the molecular mechanisms of the anti-atherogenic actions of flavonoids in normal and obese mice. PLoS ONE 6:e24634
207. Yamamoto M, Kumagai A, Yamamura Y (1983) Plasma lipid-lowering and lipogenesis-stimulating actions of ginseng saponins in tumor-bearing rats. Am J Chin Med 11:88–95
208. Han LK, Kimura Y, Kawashima M, Takaku T, Taniyama T, Hayashi T et al (2001) Anti-obesity effects in rodents of dietary teasaponin, a lipase inhibitor. Int J Obes Relat Metab Disord 25:1459–1464
209. Won S, Kim S, Kim Y, Lee P, Ryu J, Kim J (2007) Licochalcone A: a lipase inhibitor from the roots of *Glycyrrhiza uralensis*. Food Res Intern 40:1046–1050
210. Zhao HL, Kim YS (2004) Determination of the kinetic properties of platycodin D for the inhibition of pancreatic lipase using a 1,2-diglyceride-based colorimetric assay. Arch Pharm Res 27:1048–1052
211. Kwon CS, Sohn HY, Kim SH, Kim JH, Son KH, Lee JS et al (2003) Anti-obesity effect of *Dioscorea nipponica* Makino with lipase-inhibitory activity in rodents. Biosci Biotechnol Biochem 67:1451–1456
212. Rani N, Sharma SK, Vasudeva N (2012) Assessment of antiobesity potential of *Achyranthes aspera* Linn. Seed. Evid Based Complement Alternat Med 2012:715912
213. Ono Y, Hattori E, Fukaya Y, Imai S, Ohizumi Y (2006) Anti-obesity effect of *Nelumbo nucifera* leaves extract in mice and rats. J Ethnopharmacol 106:238–244
214. Kurowska EM, Spence JD, Jordan J, Wetmore S, Freeman DJ, Piche LA et al (2000) HDL-cholesterol-raising effect of orange juice in subjects with hypercholesterolemia. Am J Clin Nutr 72:1095–1100
215. Tsuda T, Ueno Y, Aoki H, Koda T, Horio F, Takahashi N et al (2004) Anthocyanin enhances adipocytokine secretion and adipocyte-specific gene expression in isolated rat adipocytes. Biochem Biophys Res Commun 316:149–157
216. Tsuda T, Ueno Y, Kojo H, Yoshikawa T, Osawa T (2005) Gene expression profile of isolated rat adipocytes treated with anthocyanins. Biochim Biophys Acta 1733:137–147

217. Tsuda T, Ueno Y, Yoshikawa T, Kojo H, Osawa T (2006) Microarray profiling of gene expression in human adipocytes in response to anthocyanins. Biochem Pharmacol 71: 1184–1197
218. Tsuda T, Horio F, Uchida K, Aoki H, Osawa T (2003) Dietary cyanidin 3-O-beta-D-glucoside-rich purple corn color prevents obesity and ameliorates hyperglycemia in mice. J Nutr 133:2125–2130
219. Le Marchand L (2002) Cancer preventive effects of flavonoids–a review. Biomed Pharmacother 56:296–301
220. Messina MJ, Persky V, Setchell KD, Barnes S (1994) Soy intake and cancer risk: a review of the in vitro and in vivo data. Nutr Cancer 21:113–131
221. Caltagirone S, Rossi C, Poggi A, Ranelletti FO, Natali PG, Brunetti M, Aiello FB et al (2000) Flavonoids apigenin and quercetin inhibit melanoma growth and metastatic potential. Int J Cancer 87:595–600
222. Peterson G, Barnes S (1991) Genistein inhibition of the growth of human breast cancer cells: independence from estrogen receptors and the multi-drug resistance gene. Biochem Biophys Res Commun 179:661–667
223. Kyle E, Neckers L, Takimoto C, Curt G, Bergan R (1997) Genistein-induced apoptosis of prostate cancer cells is preceded by a specific decrease in focal adhesion kinase activity. Mol Pharmacol 51:193–200
224. Merlino GT, Xu YH, Ishii S, Clark AJ, Semba K, Toyoshima K et al (1984) Amplification and enhanced expression of the epidermal growth factor receptor gene in A431 human carcinoma cells. Science 224:417–419
225. Huang YT, Hwang JJ, Lee PP, Ke FC, Huang JH, Huang CJ et al (1999) Effects of luteolin and quercetin, inhibitors of tyrosine kinase, on cell growth and metastasis-associated properties in A431 cells overexpressing epidermal growth factor receptor. Br J Pharmacol 128:999–1010
226. Dodurga Y, Avci CB, Satiroglu-Tufan L, Susluer SY, Sigva Z OD, Saydam G et al (2012) Effects of quercetin induced cell death on a novel Gene "*URG4/URGCP*" expression in leukemia cells. J Cancer Sci Ther 4:006–009
227. Hofmann J, Ueberall F, Posch L, Maly K, Herrmann DB, Grunicke H (1989) Synergistic enhancement of the antiproliferative activity of cis-diamminedichloroplatinum(II) by the ether lipid analogue BM41440, an inhibitor of protein kinase C. Lipids 24:312–317
228. Suolinna EM, Buchsbaum RN, Racker E (1975) The effect of flavonoids on aerobic glycolysis and growth of tumor cells. Cancer Res 35:1865–1872
229. Edwards JM, Raffauf RF, Le Quesne PW (1979) Antineoplastic activity and cytotoxicity of flavones, isoflavones, and flavanones. J Nat Prod 42:85–91
230. Wei YQ, Zhao X, Kariya Y, Fukata H, Teshigawara K, Uchida A (1994) Induction of apoptosis by quercetin: involvement of heat shock protein. Cancer Res 54:4952–4957
231. Zhou J, Allred DC, Avis I, Martínez A, Vos MD, Smith L, Treston AM, Mulshine JL (2001) Differential expression of the early lung cancer detection marker, heterogeneous nuclear ribonucleoprotein-A2/B1 (hnRNP-A2/B1) in normal breast and neoplastic breast cancer. Breast Cancer Res Treat 66:217–224
232. Golan-Gerstl R, Cohen M, Shilo A, Suh SS, Bakacs A, Coppola L, Karni R (2011) Splicing factor hnRNP A2/B1 regulates tumor suppressor gene splicing and is an oncogenic driver in glioblastoma. Cancer Res 71:4464–4472
233. Sueoka E, Sueoka N, Iwanaga K, Sato A, Suga K, Hayashi S, Nagasawa K, Nakachi K (2005) Detection of plasma hnRNP B1 mRNA, a new cancer biomarker, in lung cancer patients by quantitative real-time polymerase chain reaction. Lung Cancer 48:77–83
234. Wu S, Sato M, Endo C, Sakurada A, Dong B, Aikawa H, Chen Y, Okada Y, Matsumura Y, Sueoka E, Kondo T (2003) hnRNP B1 protein may be a possible prognostic factor in squamous cell carcinoma of the lung. Lung Cancer 41:179–186
235. Cooper TA, Wan L, Dreyfuss G (2009) RNA and disease. Cell 136:777–793

236. Rodrigo KA, Rawal Y, Renner RJ, Schwartz SJ, Tian Q, Larsen PE et al (2006) Suppression of the tumorigenic phenotype in human oral squamous cell carcinoma cells by an ethanol extract derived from freeze-dried black raspberries. Nutr Cancer 54:58–68
237. Seeram NP, Adams LS, Zhang Y, Lee R, Sand D, Scheuller HS et al (2006) Blackberry, black raspberry, blueberry, cranberry, red raspberry, and strawberry extracts inhibit growth and stimulate apoptosis of human cancer cells in vitro. J Agric Food Chem 54:9329–9339
238. Hakimuddin F, Paliyath G, Meckling K (2004) Selective cytotoxicity of a red grape wine flavonoid fraction against MCF-7 cells. Breast Cancer Res Treat 85:65–79
239. Galvano F, La Fauci L, Lazzarino G, Fogliano V, Ritieni A, Ciappellano S et al (2004) Cyanidins: metabolism and biological properties. J Nutr Biochem 15:2–11
240. Zikri NN, Riedl KM, Wang LS, Lechner J, Schwartz SJ, Stoner GD (2009) Black raspberry components inhibit proliferation, induce apoptosis, and modulate gene expression in rat esophageal epithelial cells. Nutr Cancer 61:816–826
241. Hou DX, Kai K, Li JJ, Lin S, Terahara N, Wakamatsu M et al (2004) Anthocyanidins inhibit activator protein 1 activity and cell transformation: structure-activity relationship and molecular mechanisms. Carcinogenesis 25:29–36
242. Shin DY, Lee WS, Lu JN, Kang MH, Ryu CH, Kim GY et al (2009) Induction of apoptosis in human colon cancer HCT-116 cells by anthocyanins through suppression of Akt and activation of p38-MAPK. Int J Oncol 35:1499–1504
243. Shukla S, Gupta S (2004). Molecular mechanisms for apigenin-induced cell-cycle arrest and apoptosis of hormone refractory human prostate carcinoma DU145 cells. Mol Carcinog 39:114–126
244. Nam NH (2006) Naturally occurring NF-kappaB inhibitors. Mini Rev Med Chem 6: 945–951
245. Martin S, Giannone G, Andriantsitohaina R, Martinez MC (2003) Delphinidin, an active compound of red wine, inhibits endothelial cell apoptosis via nitric oxide pathway and regulation of calcium homeostasis. Br J Pharmacol 139:1095–1102
246. Chang YC, Huang HP, Hsu JD, Yang SF, Wang CJ (2005) Hibiscus anthocyanins rich extract-induced apoptotic cell death in human promyelocytic leukemia cells. Toxicol Appl Pharmacol 205:201–212
247. Huang C, Huang Y, Li J, Hu W, Aziz R, Tang MS et al (2002) Inhibition of benzo(a)pyrene diol-epoxide-induced transactivation of activated protein 1 and nuclear factor kappaB by black raspberry extracts. Cancer Res 62:6857–6863
248. Christensen KY, Naidu A, Parent ME, Pintos J, Abrahamowicz M, Siemiatycki J et al (2012) The risk of lung cancer related to dietary intake of flavonoids. Nutr Cancer 64:964–974
249. Arts IC (2008) A review of the epidemiological evidence on tea, flavonoids, and lung cancer. J Nutr 138:1561S–1566S
250. Le Marchand L, Murphy SP, Hankin JH, Wilkens LR, Kolonel LN (2000) Intake of flavonoids and lung cancer. J Natl Cancer Inst 92:154–160
251. Kohno H, Tanaka T, Kawabata K, Hirose Y, Sugie S, Tsuda H et al (2002) Silymarin, a naturally occurring polyphenolic antioxidant flavonoid, inhibits azoxymethane-induced colon carcinogenesis in male F344 rats. Int J Cancer 101:461–468
252. Gates MA, Vitonis AF, Tworoger SS, Rosner B, Titus-Ernstoff L, Hankinson SE et al (2009) Flavonoid intake and ovarian cancer risk in a population-based case-control study. Int J Cancer 124:1918–1925
253. Kale A, Gawande S, Kotwal S (2008) Cancer phytotherapeutics: role for flavonoids at the cellular level. Phytother Res 22:567–577
254. Willett WC, Koplan JP, Nugent R, Dusenbury C, Puska P, Gaziano TA (2006) Prevention of chronic disease by means of diet and lifestyle changes. In: Jamison DT, Breman JG, Measham AR et al (eds) Disease control priorities in developing countries, 2nd edn. World Bank, Washington, DC, pp 833–850
255. Burslem J, Schonfeld G, Howald MA, Weidman SW, Miller JP (1978) Plasma apoprotein and lipoprotein lipid levels in vegetarians. Metabolism 27:711–719

256. Zhang X, Shu XO, Gao YT, Yang G, Li Q, Li H et al (2003) Soy food consumption is associated with lower risk of coronary heart disease in Chinese women. J Nutr 133:2874–2878
257. Bhathena SJ, Velasquez MT (2002) Beneficial role of dietary phytoestrogens in obesity and diabetes. Am J Clin Nutr 76:1191–1201
258. Chen Z, Li Y, Zhao LC, Zhou BF, Yang J, Wang ZW et al (2004) A study on the association between tea consumption and stroke. Zhonghua Liu Xing Bing Xue Za Zhi 25:666–670
259. Sano J, Inami S, Seimiya K, Ohba T, Sakai S, Takano T et al (2004) Effects of green tea intake on the development of coronary artery disease. Circ J 68:665–670
260. Frei B, Higdon JV (2003) Antioxidant activity of tea polyphenols in vivo: evidence from animal studies. J Nutr 133:3275S–3284S
261. Shimada K, Kawarabayashi T, Tanaka A, Fukuda D, Nakamura Y, Yoshiyama M et al (2004) Oolong tea increases plasma adiponectin levels and low-density lipoprotein particle size in patients with coronary artery disease. Diabetes Res Clin Pract 65:227–234
262. Mennen LI, Sapinho D, de Bree A, Arnault N, Bertrais S, Galan P et al (2004) Consumption of foods rich in flavonoids is related to a decreased cardiovascular risk in apparently healthy French women. J Nutr 134:923–936
263. Hubbard GP, Wolffram S, Lovegrove JA, Gibbins JM (2004) Ingestion of quercetin inhibits platelet aggregation and essential components of the collagen-stimulated platelet activation pathway in humans. J Thromb Haemost 2:2138–2145
264. Wang HK (2000) The therapeutic potential of flavonoids. Expert Opin Investig Drugs 9:2103–2119
265. Dohadwala MM, Holbrook M, Hamburg NM et al (2011) Effects of cranberry juice consumption on vascular function in patients with coronary artery disease. Am J Clin Nutr 93:934–940
266. Makela S, Santti R, Salo L, McLachlan JA (1995) Phytoestrogens are partial estrogen agonists in the adult male mouse. Environ Health Perspect 103(Suppl 7):123–127
267. Setchell KD (1998) Phytoestrogens: the biochemistry, physiology, and implications for human health of soy isoflavones. Am J Clin Nutr 68:1333S–1346S
268. Amato P, Young RL, Steinberg FM, Murray MJ, Lewis RD, Cramer MA et al (2013) Effect of soy isoflavone supplementation on menopausal quality of life. Menopause 20(4): 443–447
269. Wong WW, Lewis RD, Steinberg FM, Murray MJ, Cramer MA, Amato P et al (2009) Soy isoflavone supplementation and bone mineral density in menopausal women: a 2-y multicenter clinical trial. Am J Clin Nutr 90:1433–1439
270. Ren W, Qiao Z, Wang H, Zhu L, Zhang L (2003) Flavonoids: promising anticancer agents. Med Res Rev 23:519–534
271. Hoensch H, Groh B, Edler L, Kirch W (2008) Prospective cohort comparison of flavonoid treatment in patients with resected colorectal cancer to prevent recurrence. World J Gastroenterol 14:2187–2193
272. Walsh KR, Zhang YC, Vodovotz Y, Schwartz SJ, Failla ML (2003) Stability and bioaccessibility of isoflavones from soy bread during in vitro digestion. J Agric Food Chem 51:4603–4609
273. Ahn-Jarvis J, Clinton SK, Riedl KM, Vodovotz Y, Schwartz SJ (2012) Impact of food matrix on isoflavone metabolism and cardiovascular biomarkers in adults with hypercholesterolemia. Food Funct 3:1051–1058
274. Dalais FS, Meliala A, Wattanapenpaiboon N et al (2004) Effects of a diet rich in phytoestrogens on prostate-specific antigen and sex hormones in men diagnosed with prostate cancer. Urology 64:510–515
275. Tiziani S, Vodovotz Y (2005) Rheological characterization of a novel functional food: tomato juice with soy germ. J Agric Food Chem 53:7267–7273
276. Lean ME, Noroozi M, Kelly I, Burns J, Talwar D, Sattar N et al (1999) Dietary flavonols protect diabetic human lymphocytes against oxidative damage to DNA. Diabetes 48: 176–181
277. Waterhouse AL (2002) Wine phenolics. Ann N Y Acad Sci 957:21–36

278. United Stated Department of Agriculture (USDA) (2013) USDA database for the flavonoid content of selected foods, Release 3.1 (2013). https://www.ars.usda.gov/Services/docs.htm?docid=6231. Last Accessed on 31 Oct 2013
279. Maras JE, Talegawkar SA, Qiao N, Lyle B, Ferrucci L, Tucker KL (2011) Flavonoid intakes in the Baltimore longitudinal study of aging. J Food Compost Anal 24:1103–1109
280. Mattila P, Astola J, Kumpulainen J (2000) Determination of flavonoids in plant material by HPLC with diode-array and electro-array detections. J Agric Food Chem 48:5834–5841
281. USDA Database for the Flavonoid Content of Selected Foods, Release 3.0 (2011) Nutrient data laboratory. U.S. Department of Agriculture, Agricultural Research Service. http://www.ars.usda.gov/nutrientdata/flav. Accessed on 14 Jan 2013
282. Rouseff RL (1988) Liquid chromatographic determination of naringin and neohesperidin as a detector of grapefruit juice in orange juice. J Assoc Off Anal Chem 71:798–802
283. Teselkin YuO, Babenkova IV, Tjukavkina NA, Rulenko IA, Kolesnik Yu, Kolhir VK et al (1998) Influence of dihydroquercetin on the lipid peroxidation of mice during post-radiation period. Phytother Res 12:519
284. Hammerstone JF, Lazarus SA, Schmitz HH (2000) Procyanidin content and variation in some commonly consumed foods. J Nutr 130:2086S–2092S
285. Prior RL, Lazarus SA, Cao G, Muccitelli H, Hammerstone JF (2001) Identification of procyanidins and anthocyanins in blueberries and cranberries (Vaccinium spp.) using high-performance liquid chromatography/mass spectrometry. J Agric Food Chem 49:1270–1276

Chapter 6
The Betalain Secondary Metabolic Network

Gregory J. Hatlestad and Alan Lloyd

Introduction

Beets are red because of an evolutionary left turn. Virtually, all extant flowering plants produce the red/violet, phenylalanine-based, secondary metabolic anthocyanin pigments, with the exception of a handful of families in a single order, the Caryophyllales. These families produce unrelated red/violet/yellow betalain or beet pigments. While the families of almost all the major crops produce anthocyanins, the betalain families contain crop species that are widely grown and form important staples in many agricultural economic systems including beets, Swiss chard, spinach, *Amaranthus*, *Chenopodium*-quinoa, and prickly pear.

The betalain pigments are mutually exclusive with anthocyanins; no known taxa produce both pigments [1]. And, where they are produced, betalains physically and functionally replace the anthocyanins in all biological contexts. Betacyanins, the red betalains, have a visible absorption spectrum very similar to anthocyanins. Betalain biosynthesis is developmentally regulated in flowers and fruits to produce diverse colors to attract pollinators and seed dispersal animals. The betalain pathway responds in the same way to the same biotic and abiotic signals that regulate anthocyanins. These signals include light, temperature, pathogens, and other environmental signaling events. Like the phenylpropanoids such as the anthocyanins, betalains are free oxygen radical scavengers and are nutritionally beneficial in ways similar to the phenylpropanoids. Betalain pigments were originally referred to as "nitrogenous anthocyanins" in recognition that these nitrogen-containing compounds have replaced the biological functions of anthocyanins in the plants where they occur.

While the betalain pathway is considered agriculturally and nutritionally important, and fascinating from an evolutionary perspective, relatively little is known at the molecular level about the enzymatic steps in the betalain pathway, about

A. Lloyd (✉) · G. J. Hatlestad
Institute for Cell and Molecular Biology, The University of Texas at Austin,
Austin, TX 78712, USA
e-mail: Lloyd@uts.cc.utexas.edu

C. Chen (ed.), *Pigments in Fruits and Vegetables,* DOI 10.1007/978-1-4939-2356-4_6

how steps in the network are regulated, or about how this new metabolic pathway evolved. A complete set of structural and regulatory network genes has not yet been identified, our understanding of the regulation, both environmental and developmental, is at its infancy, and the evolution of the pathway is still largely a black box.

The phytochemical literature is full of papers identifying the structures of specific betalain compounds (reviewed in [2]), measuring their antioxidant properties [3], measuring their stability (reviewed in [4]), and other chemical properties. These will not be reviewed here.

Phylogenetic Restriction

As noted, within plants, betalain pigments are restricted to the Caryophyllales order and within this order to the monophyletic core families as defined by Cuenoud et al. [5]. A recent analysis overlaid the occurrence of betalains, anthocyanins, or the absence of pigments within this monophyletic group on a molecular phylogeny [6]. The analysis led to the conclusion that betalains may have arisen either one time or two times within this single restricted group, with equal probability for either scenario. Betalains and anthocyanins can occur in sister families within the larger group and over time one or both pigment pathways have been lost multiple times independently. It should be noted that the molecular phylogenies for this group were constructed with markers that had no relationship to the betalain pathway. It will be interesting, and will soon be possible, to produce phylogenies based on the actual betalain biosynthetic and regulatory genes.

Betalain pigments are also produced in basidiomycete fungi, notably the red, orange, or yellow capped *Amanita muscaria*, or fly agaric. Molecular evidence indicates that this occurrence is the result of an independent evolutionary origin unrelated to the plant pigments ([7], and discussed below).

What are the Betalains

The chemical precursors, intermediates, and products of the betalain ring structure biosynthetic pathway have been determined. Unlike the unrelated phenylalanine-based anthocyanins, betalains derive from tyrosine. The betalains are water-soluble, vacuole-localized pigments that range from red/violet betacyanins, to yellow betaxanthins. The synthesis of the betalain ring structure is proposed to require three enzyme-mediated steps (Fig. 6.1). A molecule of L-3,4-dihydroxyphenylalanine (L-DOPA) is the substrate in step 2, the first ring biosynthetic gene reported, where a ring-opening extradiol cleavage enzyme, DOPA-4,5-dioxygenase (DODA) produces 4,5-seco-DOPA which spontaneously cyclizes to betalamic acid [7]. The gene/enzyme responsible for step 3 in Fig. 6.1 has been recently identified [8] as a cytochrome P450 enzyme, CYP76AD1. This enzyme also uses L-DOPA as a substrate

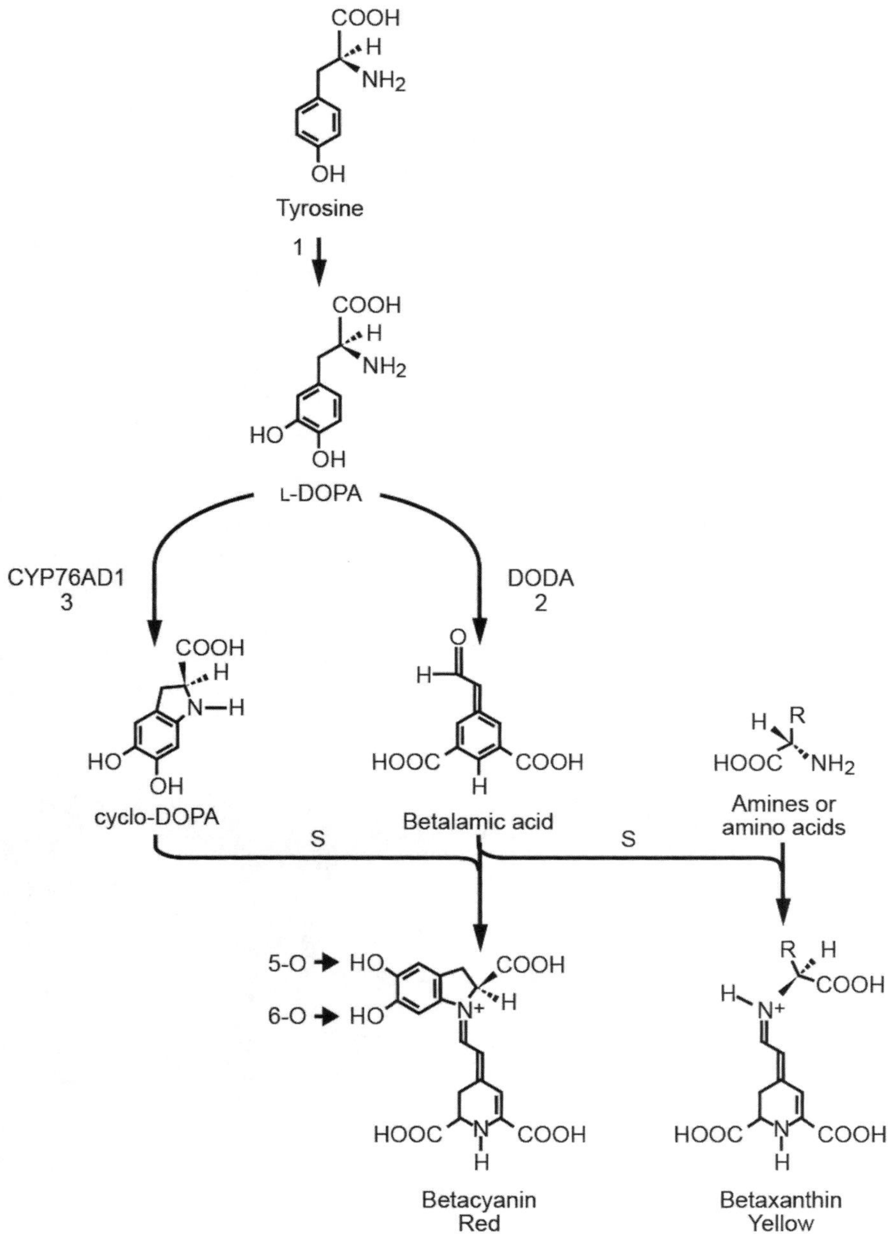

Fig. 6.1 The proposed betalain pathway. Tyrosine is converted by an unknown enzyme to L-DOPA. L-DOPA is the substrate in two enzyme-mediated steps to produce cyclo-DOPA and betalamic acid by CYP76AD1 and DODA, respectively. Betalamic acid then spontaneously condenses with either an amine to produce the yellow pigment betaxanthin or with cyclo-DOPA to produce the red pigment betacyanin

molecule to produce cyclo-DOPA via a DOPAquinone intermediate. The gene(s) or enzyme(s) responsible for step 1, the conversion of tyrosine to L-DOPA (tyrosinase), has not been positively identified.

Betalamic acid is itself a yellow and fluorescent compound that is required to produce both the yellow and red betalains. Betalamic acid condenses with the product of step 3, cyclo-DOPA, to form the red betanidin pigment, or condenses with amino acids or other amine groups to form the yellow fluorescent betaxanthin pigments [9]. The ring structure is often glycosylated or acylated on the cyclo-DOPA moiety (see 5-O and 6-O arrows in Fig. 6.1) and genes encoding UDP glycosyltransferases (UGT) with in vitro activity on betanidin are known from *Dorotheanthus* [10] and beet [11] though their specificity is not high. In vitro studies have also shown that the cyclo-DOPA structure can be glycosylated [12] and that the glycosylated-cyclo-DOPA can participate in the spontaneous condensation with betalamic acid. The order of modification and condensation does not appear to be fixed.

Steps 1 and 3 were earlier proposed to be performed by the same enzyme, a polyphenol oxidase (PPO), because a single PPO-type enzyme, tyrosinase, performs both reactions during melanogenesis in animals [13], and PPO enzymes will perform steps 1 and 3 in vitro [14]. But, it has not been demonstrated, genetically or otherwise, that these steps are performed by a PPO in the plant [15]. Arguments against a PPO enzyme performing steps 1 and 3 are: all plant PPO enzymes are reported to be plastid localized [16] while the betalain pathway is cytoplasmic, and it is relatively common to obtain mutants that only make yellow betaxanthin pigments (Fig. 6.2), i.e., they are missing step 3, but step 1 is intact. Yellow mutants are

Fig. 6.2 Betalain genetic phenotypes. *Top row*, 4 o'clocks, *left to right*: RR CC, rr CC, rr cc. *Bottom row*, table beets, RR YY, rr YY, RR yy

known from many betalain species, perhaps all that have been in cultivation for any length of time, including beet, *Celosia*, 4 o'clock, *Bougainvillea*, cactus, *Portulaca*, and others. Yellow should be a rare or nonexistent mutant type if a single enzyme were responsible for both steps.

Genetics of Betalains-Beets, 4 o'clocks, and Portulaca

Wesley Keller performed a genetic analysis of beets (*Beta vulgaris*) in the 1930s and identified two loci, R and Y, involved in the production of betalains [17]. R alleles are responsible for the red versus yellow shift (Fig. 6.2) corresponding to step 3 in Fig. 6.1. The R locus is now identified as encoding CYP76AD1 ([8]; discussed in detail below). Red table beets contain a dominant R allele while yellow beets are homozygous for the recessive r allele.

Y alleles direct whether pigment or no pigment is produced in the interior of the beetroot, regardless of whether the pigment is red or yellow (Fig. 6.2). White beets, which include virtually all sugar beets, are homozygous for the recessive y allele; however, it should be noted that "white" beets are still competent to produce betalains and that the pigment is largely restricted to the epidermal layers. The big Y allele responsible for red table beets appears to be a dominant gain-of-function allele that was discovered during cultivation.

Similarly, in 4 o'clock (*Mirabilis jalapa*), there was classic work from Correns (published in German in the early 1900s and reviewed by Rheinberger [18]) and more recent, but still old work [19–21] demonstrating two main betalain loci, also called R and Y. Again, the R alleles dictate red versus yellow (Fig. 6.2), and the yellow (rr) mutant has been shown to be complemented to red by expression of the beet R gene [8]. Dominant Y versus recessive y 4 o'clock alleles are responsible for the presence (either red or yellow) or complete absence of pigment, respectively (white flowers and absence of betalains in leaves and stems (Fig. 6.2)), not tissue placement. Thus, for both beets and 4 o'clock, the R locus is consistent with step 3 in the betalain biosynthetic scheme (Fig. 6.1). The 4 o'clock Y locus is consistent with step 2, the DODA gene, because 4 o'clock yy mutants do not make any betalain pigments; while the beet Y locus is more consistent with some type of upstream regulator because yy beets are fully capable of producing epidermally restricted red betalain pigments. The beet recessive y allele can be considered to direct the "normal" pigment state that mimics the epidermal placement of anthocyanins in the vast majority of flowering plants.

In beets, Keller [17] reported that the Y and R loci were linked at about 7 cM and this distance has more recently been verified [22]. The work of Engels et al. [21] appears to show that the 4 o'clock Y and R loci are not linked.

In *Portulaca*, three loci direct flower color, C, R, and I. Dominant C is required to make any pigment, red/violet or yellow, dominant R is required to make red/violet (rr flowers are yellow), while dominant I inhibits the formation of yellow pigments [23]. The white or cc mutant is complemented by expression of DOPA dioxygenases from either *Amanita* mushrooms [24] or *Portulaca* [7]. So, the C locus undoubtedly encodes the DODA gene, though the *Portulaca* DODA gene has not yet been shown to be genetically linked to the C locus and mutations giving rise to any c alleles have not yet been reported. The *Portulaca* R locus probably encodes a paralog of the beet and 4 o'clock R loci, encoding a cytochrome P450 protein, as the r mutation has the same yellow phenotype in all three species. However, no molecular analysis has been reported. Nothing is presently known about the identity of the *Portulaca* I gene.

External Signals are Transduced to Output of Betalain Pigment Phenotypes

The regulation of betalain biosynthesis by environmental signals and applied bioactive compounds has been investigated by several groups. A commonly used system for these studies is hairy root cultures of beets and other species. Most of these studies were conducted with an eye towards maximizing pigment production in culture for potential commercial use. Blue plus far red light [25], cytokinin, nitric oxide [26], methyl jasmonate [27], Ca2+[28], and phosphate [29] have all been shown to strongly influence betalain pigment accumulation in cultures. There are fewer studies using whole seedlings or plants, but those generally agree with the in vitro studies. For example, high white or UV-A light induced betalain biosynthesis in ice plant [30]. There is a single gene expression paper showing that a DODA gene homolog in calli of the betalain producer, *Suaeda salsa,* is upregulated by light [31].

Cloning the Betalamic Acid Biosynthetic Gene, 4,5-DOPA-dioxygenase

The first betalain ring structure biosynthetic gene to be cloned was DODA from the mushroom, *A. muscaria* [32]. To clone this gene, Zrÿd and coworkers purified the protein from mushroom caps, raised antibodies against it, and used the anti-DODA antibodies to screen an *Amanita* cDNA expression library in *Escherichia coli.* They went on to show that expression of this mushroom gene complemented white *Portulaca* tissue (cc genotype), allowing it to produce pigment, either red or yellow depending on the genotype at the R/r locus [24].

This same group was successful in cloning the first plant betalain 4,5-DOPA-dioxygenase (DODA; [7]). Unfortunately, the mushroom gene and protein were not similar enough to the plant versions to allow the use of the antibody or the mushroom gene sequence as probes. They produced subtractive libraries to isolate differentially expressed cDNAs and identified genes highly expressed in red *Portulaca* flowers. Through sequencing of clones, they identified a putative translation product with high similarity to the LigB ring cleavage domain of prokaryotic proteins with potential extradiol 4,5-dioxygenase activity, the same activity that a betalain 4,5-DOPA-dioxygenase was predicted to have. They went on to show that expression of this gene in white *Portulaca* petal tissue resulted in red or yellow pigmentation, depending on the genetic background, showing that they had the right gene. Sequence analysis of this gene indicated that it has a different phylogenetic origin than the mushroom gene. They also point out that all plants have homologs of this gene and that there are a common set of amino acid changes near the putative active site that separates the betalain and nonbetalain-producing plant DODA genes. It remains to be conclusively shown whether these amino acid changes are responsible for a change in substrate specificity or are the result of the common ancestry of these caryophyllaceous plants.

Cloning the Cyclo-DOPA Biosynthetic Gene, CYP76AD1

The enzymatic steps 1 and 3 (Fig. 6.1) were proposed to be performed by a single enzyme, a PPO. This prediction was supported by the fact that a single PPO-like enzyme, tyrosinase, performs both of these reactions during the formation of dark melanin pigments in animals, and the fact that plant PPO enzymes will also perform these reactions in vitro [14]. Arguments against a PPO are that all known plant PPO enzymes are chloroplast localized while the betalain pathway is cytoplasmic, and that it is commonplace to obtain mutants that only make yellow betaxanthin pigments (in beets, cockscomb, 4 o'clocks, *Portulaca*, cactus, and others), suggesting that these mutants have an intact step 1 but cannot supply the cyclo-DOPA ring structure required for red pigment in the proposed step 3.

Recently, a cytochrome P450 enzyme (CYP76AD1) was identified that is required for the conversion of L-DOPA to the cyclo-DOPA moiety in beet (step 3; [8]). CYP76AD1 was identified through analysis of next generation sequencing of red beet transcripts, highlighting the value of high-throughput next-generation sequencing in studies on nonmodel organisms. cDNA from red table beet seedling hypocotyl sections was sequenced using Roche 454 pyrosequencing. These stem sections produce high concentrations of betalain pigments from the outer epidermis to the inner core and eventually give rise to the swollen red beet. It was hypothesized that betalain pathway genes would be highly represented among these transcripts. This was proven to be true as the previously identified betalain biosynthesic gene, DODA, was the 14th most expressed contig in the data set of nearly 10,000 contigs.

The 454 database was queried for other highly expressed genes that could be candidates for enzymatic steps 1 and/ or 3. PPO- and laccase-encoding cDNAs were found but they were not expressed at the high levels expected for betalain biosynthetic genes. The contig database was also searched for cytochrome P450-encoding sequences because step 1 resembles the canonical cytochrome P450 reaction and it was possible that step 3 could be performed by an unusual cytochrome P450 activity [33]. A cytochrome P450 cDNA, CYP76AD1, was identified as the 33rd most highly expressed contig.

After similar transcript profiling was performed in several other betalain-producing Caryophyllales, CYP76AD1 became the founding member of a new subfamily of cytochrome P450 enzymes that are most similar to CYP76T and CYP76C. It is predicted that CYP76AD2 and CYP76AD3 are responsible for the same step in *Amaranthus* and 4 o'clocks, respectively.

CYP76AD1's role in the betalain pathway was verified through a series of genetic and biochemical experiments. Its expression is correlated with the red phenotype, CYP76AD1 is expressed at high levels in red beets but low levels in yellow and white beets. Mutant analysis using virus-induced gene silencing (VIGS) to suppress gene expression in very red beets resulted in the loss of the red pigment and appearance of yellow pigment. The change from betanin (red) to betaxanthin (yellow) pigments was verified using mass spectrometry.

As already stated there are many yellow mutants and CYP76AD1 was able to complement yellow mutants in beets, cockscomb, and 4 o'clocks. In each species, overexpression of CYP76AD1 resulted in the loss of the yellow phenotype and the production of red betalain pigments.

These data showed that CYP76AD1 is responsible for the enzymatic step 3 in the pathway, biosynthesis of cyclo-DOPA from L-DOPA. Loss of step 3 should result in the inability to synthesize red pigments but should not affect the ability to synthesize yellow pigments consistent with the function of the beet R gene. This led to the hypothesis that CYP76AD1 was the beet R gene genetically defined by Keller in 1936 (discussed above, [17]).

The CYP76AD1 alleles were sequenced in a sugar beet variety, C869 [34], which segregates red and yellow hypocotyls (segregating for *R*/*r*) [17, 22]. A 5-bp insertion was identified, 325 bp before the stop codon in the yellow (rr) segregants (TAAAT), that shows complete linkage to the R phenotype. This insertion and the resulting frameshift introduced an early stop codon, which causes the deletion of the heme-ring-binding site and results in an inactive protein. On the basis of this genetic data and the above described functional data, *CYP76AD1* has been identified as the *R* locus described more than 70 years ago [8].

Modification Enzymes

Betalain pigments, like Anthocyanins, are usually highly decorated. In different species, there is a vast array of different combinations of modifications that produce pigment end products (reviewed in [2]). These modifications are most easily observed on the stable final products; however, there is some debate about when these decorations are added. Two groups have identified enzymes from betalain-producing species that are 5-*O*-glucosyltransferases [10, 12]. One is predicted to modify the unglycosylated betacyanin, betanidin, and the other is predicted to modify cyclo-DOPA, an unstable intermediate (see Fig. 6.1). It is likely that both scenarios are occurring simultaneously.

Enzymes that modify the final product betanidin consist of two types so far, 5-*O*-glucosyltransferase (betanidin 5-GT) and 6-*O*-glucosyltransferase (betanidin 6-GT; *arrows* in Fig. 6.1 point to the 5-O and 6-O positions). Betanidin 5-GT transfers glucose from UDP-glucose to the 5-hydroxyl group of betanidin. Betanidin 6-GT performs the glucose transfer to the 6-hydroxyl group. However, both betanidin 5-GT and 6-GT are not very specific and can transfer glucose to several flavonols and anthocyanidins, and, based on activities and protein sequence, it has been hypothesized that these enzymes evolved from flavonoid GTs [10, 35].

There is one enzyme/cDNA identified in 4 o'clock and cockscomb that is able to glycosylate cyclo-DOPA in vitro, UDP-glucose:cyclo-DOPA 5-*O*-glucosyltransferase, cDOPA5GT [12], and it has been proposed to perform this same function in vivo. Again, this activity was not highly specific for cyclo-DOPA and it was speculated that these types of GTs evolved from flavonoid GTs.

The added glucose molecules can be further decorated at multiple positions with a variety of groups including acyl, malonyl, apiosyl, feruloyl, glucosyl, hydroxylcinnamoyl, and other moieties. Although some biochemistry has been performed on some of these reactions, nothing has yet been reported on the genes or genetics associated with these reactions.

Expression in Heterologous Species

Evolution has restricted betalain pigments to the order Caryophyllales and a few fungi; however, there has been work to transfer the betalain pathway to heterologous species such as *E. coli*, yeast, *Solanum tuberosum* (potato), *Antirrhinum majus* (snapdragon), and *Arabidopsis thaliana*. Most of the work to express betalains in heterologous species has centered around the DODA gene. The first attempts made with plant betalain pathway genes in *E. coli* proved unsuccessful [7]. However, Sasaki and coworkers were finally able to produce betalamic acid using an *E. coli* expressed DODA gene from 4 o'clocks (MjDODA) and its substrate L-DOPA [36]. They also reported preliminary experiments expressing DODA in yeast; however, they did not show any data from these experiments. Gandía-Herrero and García-Carmona recently expressed a *B. vulgaris* DODA gene with its codons optimized for *E. coli* and they also successfully produced betalamic acid and the yellow betaxanthin pigments [37]. The method used by this group was to express the DODA in a heterologous system and then produced the pigment in vitro using protein extracts.

Betalains were first produced in vivo in heterologous systems in 2012 by two different groups. Harris and colleagues expressed a DODA gene from *Portulaca grandiflora,* PgDODA, in cell cultures of *S. tuberosum* (potato) and petals of *A. majus* (snapdragon) using biolistic introduction of the overexpression constructs [38]. They also created stable transgenic lines of *A. thaliana* expressing PgDODA using *Agrobacterium* transformation. Upon feeding the transgenic tissues with L-DOPA, yellow, orange, and red betalain pigments were formed in vivo. It is interesting that with just the addition of one gene (DODA) and one substrate (L-DOPA), nonbetalain-producing plants can be transformed to produce both yellow and red betalain pigments. This suggests that anthocyanin producing plants contain an activity to perform step 3, the production of cyclo-DOPA from L-DOPA. At present, it is unknown how this is being accomplished in the anthocyanin plants.

Hatlestad et al. [8] also had success with heterologous expression in vivo, this time in yeast and with both the DODA and the newly discovered step 3 enzyme, CYP76AD1. They expressed a beet DODA, BvDODA1, in yeast and fed L-DOPA, which resulted in the production of yellow betalain pigments. CYP76AD1 was also expressed by itself in yeast as a check of its enzymatic function. CYP76AD1 was shown to use L-DOPA to produce cyclo-DOPA, consistent with the proposed step 3 activity. In addition they were able to recreate part of the betalain pathway in vivo in yeast by expressing the BvDODA1 and CYP76AD1 genes, simultaneously and feeding with L-DOPA, resulting in the production of betanidin, the undecorated red beet pigment [8].

Genetic and Genomic Resources for Betalain Research

Unfortunately, none of the betalain-producing species has the resources of a model genetic organism. However, there are resources available and the amount of these resources is slowly growing. The National Center for Genetic Resources Preservation (NCGRP) maintains extensive germplasm collections and seeds can be requested through Germplasm Resources Information Network (GRIN; http://www.ars-grin.gov) administered by the United States Department of Agriculture. The *B. vulgaris* collection includes table beet, sugar beet, swiss chard, fodder beet, wild or sea beet (the undomesticated wild precursor), as well as segregating populations from controlled crosses. These five distinct plant types are all the same species. They also maintain extensive collections of *Amaranthus* including cultivated varieties and many species, and collections of *Chenopodium* including *C. quinoa* (the pseudograin, quinoa) accessions as well as other *Chenopodium* species. GRIN also maintains representative collections of other betalain-producing genera like *Opuntia*, *Portulaca*, *Boehravia*, *Tetragonia*, and *Basella*. The second largest *Beta* germplasm collection is maintained at the Institute of Plant Genetics and Crop Plant Research (IPK), Gatersleben, Germany (http://eurisco.ecpgr.org/).

Beet recombinant inbred lines (RILs) have been produced from a sugar beet/table beet cross and these are available by request. The RILs were developed from a cross used to produce a molecular-based genetic map [39]. In addition, many other RIL populations are in earlier stages of completion by this same research group.

The McGrath group has also produced a sugar beet bacterial artificial chromosome (BAC) library that can be screened by PCR to identify genomic clones for genes [40].

The Gene Index Project maintains a database of beet ESTs and beet genes that can be searched by basic local alignment search tool (BLAST; http://compbio.dfci.harvard.edu/cgi-bin/tgi/gimain.pl?gudb=beet). They also maintain a database for *Mesembryanthemum crystallinum* (ice plant) a betalain producer in the Aizoaceae family.

The Genome Analysis of the Plant Biological System (GABI) is a public/private group with several goals regarding the sugar beet genome including a BAC-based physical map and a complete annotated genome (http://www.gabi.de/projekte-alle-projekte-neue-seite-144.php). The beet genome is approximately 758 Mb and

886 cM [41]. The GABI-based group has assembled 535 chromosomally anchored contigs that comprise about 600 Mb [41]. These can be accessed through (http://bvseq.molgen.mpg.de/index.shtml).

Future Needs

There is much still to be discovered about betalains. This is not limited to the minutiae of the pathway, but major elements, including biosynthetic steps and regulation are still to be discovered. Currently, there is nothing known about the gene(s) required for the first step in the synthesis of betalains, the tyrosine hydroxylase or L-DOPA synthesis step, or the genes involved in the regulation of the pathway. There is very little known about the genes involved in the modification of the pigments, and nothing known about the transport of these pigments into the vacuole, nor any possible degradation pathways.

The step 1 enzyme that catalyzes the conversion of tyrosine to L-DOPA is unknown and is a glaring hole in our knowledge of how betalains are made. There are reports of biochemical isolation of this activity from betalain plants [14], but no reports on gene or protein identity. This step has long been hypothesized to be a tyrosinase-like or PPO enzyme, as in vitro experiments and analysis of animal systems have shown that these types of enzymes can produce L-DOPA from tyrosine; however, it was also long thought that steps 1 and 3 would be performed by the same enzyme. This last theory was challenged when CYP76AD1 (the beet R locus) was discovered to be required for step 3 but not for step 1. While it has not been demonstrated that CYP76AD1 cannot do step 1 (if it does, it is redundant with other genes), the genetic and heterologous expression data are clear that it is only the R locus that is required for step 3 activity, production of cyclo-DOPA from L-DOPA. At the present time, it is possible that step 1 is performed by another cytochrome P450, or another type of enzyme or enzymes, including PPO-like enzymes. It is interesting and perhaps revealing to note that there are no reported step 1 mutants among the many betalain mutants in several cultivated species. Unless, the step 1 enzyme(s) is somehow required for life, this mutant-deficit implies that there is genetic redundancy for this activity.

There are a few studies about how betalain pigments are produced in response to different stimuli (discussed briefly above) but there is nothing known about the genes responsible for this environmental regulation or the developmental regulation needed to place betalains in specific tissues. Identifying the regulatory genes will provide insight into the evolution of the pathway and understanding of developmental placement and environmental regulation. It will be important for possible improvements to betalain-producing fruits and decorative plants by genetic or molecular means. Because betalains are expressed in the same patterns (temporally/spatially) and respond similarly to the same stresses as anthocyanins, it has been speculated that the betalain pathway may be regulated by the anthocyanin regulatory network [42], implying the co-option of regulatory genes. If true, this would be a fascinating evolutionary phenomenon where a regulatory network has switched to regulate new genes and end products, but the new pathway is still responsible for the same functions overall, in this case red pigmentation and all the biological functions that it serves.

Betalain pigments do not consist of a single final compound, but instead include many possible chemical modifications that decorate the standard backbone. These modifications may be a contributing factor causing the countless shades and hues of red/violet/pink or yellow/orange that are seen across the betalain-producing families. So far, there are only a few modification genes/enzymes identified and these are limited to the addition of glucose at two positions on cyclo-DOPA moiety, either before or after condensation with betalamic acid. More work is needed in this arena to identify the genes/enzymes responsible for other types of modifications.

It is not completely understood where exactly in the cell the synthesis of betalains takes place. The DODA appears to be soluble while the identification of the step 3 enzyme as a cytochrome P450 implies membrane localization in the ER. However, the locations of these enzymes have not been reported.

There is no information on how the final products end up in the vacuole. With anthocyanin producing plants there are several transporters that are involved in the movement of pigments. It is expected that betalain pigments will have similar requirements for transport into the vacuole. Thus far, there are no such transporters known.

DODA-like genes are found in nearly all life, but this protein type has evolved a novel function in betalain-producing plants. The biochemical functions of DODA gene homologs in other plant species have a yet to be determine. Sequence analysis of betalain pathway and nonpathway DODAs has identified two divergent motifs at the predicted catalytic site, one conserved in betalain species and the other in nonbetalain species. It has been proposed that this motif change at the catalytic site is responsible for the specific biosynthetic ability of DODAs that produce betalains. However, there has been no experimentation to test this theory [7]. It is possible that the conserved motifs are conserved based on phylogeny and not function. Given the current state of sequencing and transgenic technologies, it should be easy to test this.

Above we have outlined some of the obvious holes in our knowledge of the genes involved in the production of betalains. With the introduction of modern sequencing and analysis techniques, and the generation of more genetic tools and resources it should be much easier to find the undiscovered genes and test their functions. The strange occurrence and isolation of the betalains to a single order remain the most fascinating aspect of this pathway. Once found, discovery of these genes will allow an informed look at how the betalain pigment pathway functions and how it came into being.

References

1. Clement JC, Mabry TJ (1996) Pigment evolution in the *Caryophyllales*: a systematic overview. Bot Acta 109:360–367
2. Strack D, Vogt T, Schliemann W (2003) Recent advances in betalain research. Phytochemistry 62:247–269
3. Cai Y, Sun M, Corke H (2003) Antioxidant activity of betalains from plants of the *Amaranthaceae*. J Agric Food Chem 51:2288–2294

4. Herbach KM, Stintzing FC, Carle R (2006) Betalain stability and degradation—structural and chromatic aspects. J Food Sci 71:41–50
5. Cuenoud P, Savolainen V, Chatrou LW, Powell M, Grayer RJ, Chase MW (2002) Molecular phylogenetics of *Caryophyllales* based on nuclear 18 S rDNA and plastid rbcL, atpB, and matK DNA sequences. Am J Bot 89:132–144
6. Brockington SF, Walker RH, Glover BJ, Soltis PS, Soltis DE (2011) Complex pigment evolution in the *Caryophyllales*. New Phytol 190(4):854–864
7. Christinet L, Burdet FX, Zaiko M, Hinz U, Zrÿd JP (2004) Characterization and functional identification of a novel plant 4,5-extradiol dioxygenase involved in betalain pigment biosynthesis in *Portulaca grandiflora*. Plant Physiol 134:265–274
8. Hatlestad GJ, Sunnadeniya RM, Gonzalez A, Akhavan NA, Goldman IL, McGrath JM, Lloyd AM (2011) The beet R locus encodes a new cytochrome P450 required for red betalain production. Nat Genet 44(7):816–820
9. Schliemann W, Kobayashi N, Strack D (1999) The decisive step in betaxanthin biosynthesis is a spontaneous reaction. Plant Physiol 119:1217–1232
10. Vogt T, Grimm R, Strack D (1999) Cloning and expression of a cDNA encoding betanidin 5-O-glucosyltransferase, a betanidin- and flavonoid-specific enzyme with high homology to inducible glucosyltransferases from the *Solanaceae*. Plant J 19:509–521
11. Isayenkova J, Wray V, Nimtz M, Strack D, Vogt T (2006) Cloning and functional characterisation of two regioselective flavonoid glucosyltransferases from *Beta vulgaris*. Phytochemistry 67:1598–1612
12. Sasaki N, Wada K, Koda T, Kasahara K, Adachi T, Ozeki Y (2005) Isolation and characterization of cDNAs encoding an enzyme with glucosyltransferase activity for cyclo-DOPA from four O'clocks and feather cockscombs. Plant Cell Physiol 46(4):666–670
13. Korner A, Pawelek J (1982) Mammalian tyrosinase catalyzes three reactions in the biosynthesis of melanin. Science 217:1163–1165
14. Steiner U, Schliemann W, Böhm H, Strack D (1999) Tyrosinase involved in betalain biosynthesis of higher plants. Planta 208:114–124
15. Mayer AM (2006) Polyphenol oxidases in plants and fungi: going places? A review. Phytochemistry 67:2318–2331
16. Vaughn KC, Lax AR, Duke SO (1988) Polyphenol oxidase: the chloroplast enzyme with no established function. Physiol Plant 72:659–665
17. Keller W (1936) Inheritance of some major color types in beets. J Agric Res 52:27–38
18. Rheinberger H (2000) Mendelian inheritance in Germany between 1900 and 1910. The case of Carl Correns (1864–1933). C R Acad Sci III 323:1089–1096
19. Showalter HM (1934) Self flower color inheritance and mutation in *Mirabilis jalapa*. Genetics 19:568–580
20. Prakken R (1944) Contribution to the genetics and cytology of *Mirabilis*. Hereditas 30:201–212
21. Engels JMM, Kester WNM, Spitters CJT, Vosselman L, Zeven AC (1975) Investigations of the inheritance of flower variegation in *Mirabilis jalapa* L. 1. General introduction and 2. Inheritance of color in uniformly colored flowers. Euphytica 24:1–5
22. Goldman IL, Austin D (2000) Linkage among the R, Y and BI loci in table beet. Theor Appl Genet 100:337–343
23. Trezzini GF, Zrÿd JP (1990) *Portulaca grandiflora*: a model system for study of the biochemistry and genetics of betalain synthesis. Acta Hortic 280:581–585
24. Mueller LA, Hinz U, Zrÿd JP (1997) The formation of betalamic acid and muscaflavin by recombinant dopa-dioxygenase from *Amanita*. Phytochemistry 44:567–569
25. Shin KS, Murthy HN, Heo JW, Paek KY (2003) Induction of betalain pigmentation in hairy roots of red beet under different radiation sources. Biol Plant 47:149–152
26. Scherer GFE, Holk A (2000) NO donors mimic and NO inhibitors inhibit cytokinin action in betalaine accumulation in *Amaranthus caudatus*. Plant Growth Regul 32:345–350
27. Suresh B, Thimmaraju R, Bhagyalakshmi N, Ravishankar GA (2004) Polyamine and methyl jasmonate-induced enhancement of betalaine production in hairy root cultures of *Beta vulgaris* grown in bubble column reactor and studies on efflux of pigments. Process Biochem 39:2091–2096

28. Savitha BC, Thimmaraju R, Bhagyalakshmi N, Ravishankar GA (2006) Different biotic and abiotic elicitors influence betalain production in hairy root cultures of *Beta vulgaris* in shake-flask and bioreactor. Process Biochem 41:50–60
29. Mukundan U, Bhide V, Dawda H (1999) Production of betalains by hairy root cultures of *Beta vulgaris*. L. In: Fu TJ, Singh G, Curtis WR (eds) Plant cell and tissue culture for the production of food ingredients. Academic/Plenum, New York, pp 121–127. (ISBN 0-306-46100-5)
30. Vogt T, Ibdah M, Schmidt J, Wray V, Nimtz M, Strack D (1999) Light-induced betacyanin and flavonol accumulation in bladder cells of *Mesembryanthemum crystallinum*. Phytochemistry 52:583–592
31. Zhao SZ, Sun H, Chen M, Wang B (2010) Light-regulated betacyanin accumulation in euhalophyte *Suaeda salsa* calli. Plant Cell Tiss Organ Cult 102:99–107
32. Hinz UG, Fivaz J, Girod PA, Zrÿd JP (1997) The gene coding for the DOPA dioxygenase involved in betalain biosynthesis in *Amanita muscaria* and its regulation. Mol Gen Genet 256:1–6
33. Guengerich FP (2001) Common and uncommon cytochrome P450 reactions related to metabolism and chemical toxicity. Chem Res Toxicol 14:611–650
34. Lewellen RT (2004) Registration of Rhizomania resistant, monogerm populations C869 and C869CMS sugarbeet. Crop Sci 44:357–358
35. Vogt T (2002) Substrate specificity and sequence analysis define a polyphyletic origin of betanidin 5- and 6-O-glucosyltransferase from *Dorotheanthus bellidiformis*. Planta 214:492–495
36. Sasaki N, Abe Y, Goda Y, Adachi T, Kasahara K, Ozeki Y (2009) Detection of DOPA 4,5-Dioxygenase (DOD) activity using recombinant protein prepared from *Escherichia coli* cells harboring cDNA encoding DOD from *Mirabilis jalapa*. Plant Cell Physiol 50(5):1012–1016
37. Gandía-Herrero F, García-Carmona F (2012) Characterization of recombinant *Beta vulgaris* 4,5-DOPA-extradiol-dioxygenase active in the biosynthesis of betalains. Planta 236:91–100
38. Harris NN, Javellana J, Davies KM, Lewis KM, Jameson PE, Deroles SC, Calcott KE, Gould KS, Schwinn KE (2012) Betalain production is possible in anthocyanin producing plant species given the presence of DOPA-dioxygenase and L-DOPA. BMC Plant Biol 12:34
39. McGrath J, Trebbi D, Fenwick A, Panella L, Schulz B, Laurent V, Barnes S, Murray S (2007) An open-source first-generation molecular genetic map from a sugarbeet x table beet cross and its extension to physical mapping. Crop Sci 47:S27–S44
40. McGrath JM, Shaw RS, de los Reyes BG, Weiland JJ (2004) Construction of a sugar beet BAC library from a hybrid with diverse traits. Plant Mol Biol Rep 22:23–28
41. Dohm JC, Lange C, Holtgräwe D, Sörensen TR, Borchardt D, Schulz B, Lehrach H, Weisshaar B, Himmelbauer H (2012) Palaeohexaploid ancestry for Caryophyllales inferred from extensive gene-based physical and genetic mapping of the sugar beet genome (*Beta vulgaris*). Plant J 70(3):528–540
42. Stafford HA (1994) Anthocyanins and betalains: evolution of the mutually exclusive pathways. Plant Sci 101:91–98

Chapter 7
Indicaxanthin Dietetics: Past, Present, and Future

Maria A. Livrea and Luisa Tesoriere

Introduction

In the recent 25 years, enormous progresses have been made in understanding the redox biology of cells and tissues. Epidemiological observations have revealed that fruit and vegetable diets have health benefits and protective functions against oxidative stress-based chronic disorders, such as cardiovascular disease and cancer, which lead to broad research on redox-active phytochemicals [1–3]. At first, these compounds were believed to act mostly by providing direct antioxidant protection by trapping radicals and/or chelating redox-active metals. A significant body of evidence, however, now supports the participation of these molecules in the regulation of a complex web of cell signaling as the means by which they affect cell activities and functions [4–6]. The physicochemical characteristics of phytochemicals can play important roles in cell signaling, including the regulation of the cell redox state by interfering with oxidant production and/or antioxidant defenses, interactions with signaling proteins, and the regulation of membrane-associated cell signaling. As a final target of these activities, the epigenetic modulation of gene expression has recently emerged as a finely tuned means to explain the effects of these compounds.

A so-called xenohormesis theory [7] has been put forward to explain why the plant kingdom is such a rich source of redox-active beneficial compounds (Fig. 7.1).

M. A. Livrea (✉) · L. Tesoriere
Department of Stembio, University of Palermo, via M Cipolla, 74, Palermo 90123, Italy
e-mail: maria.livrea@unipa.it

C. Chen (ed.), *Pigments in Fruits and Vegetables,* DOI 10.1007/978-1-4939-2356-4_7

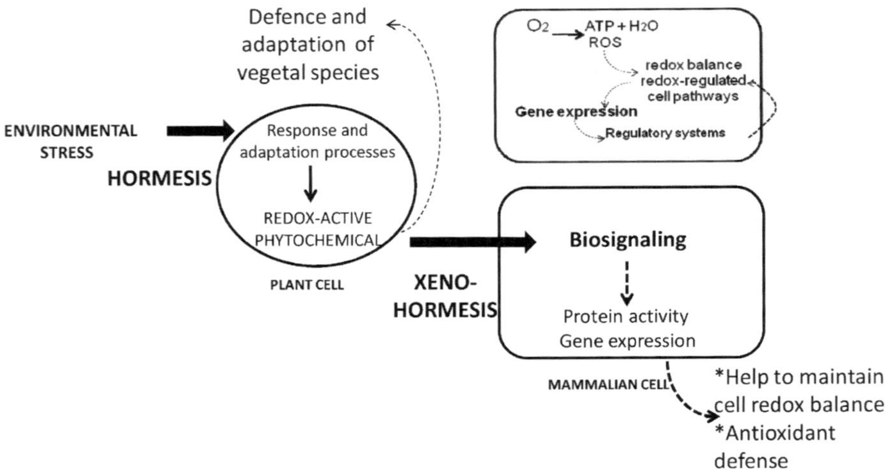

Fig. 7.1 Redox-regulated cell functions in the context of hormesis and xenohormesis. In response to adverse environmental conditions such as dryness, light, heat, or exogenous stressors including wounding and oxidative stress [8–10] plants synthesize their own molecular defense, i.e., phytochemicals, that in turn modulate redox-regulated cellular adaptative responses, a process known as hormesis [11, 12]. Xenohormesis postulates that heterotrophs (animals and fungi) have evolved the ability to sense/interact with these compounds and use them as chemical cues finally inducing cell responses [13, 14]

Indicaxanthin

Chemical and Physicochemical Characteristics

Indicaxanthin is a yellow pigment in the betalain class [15]. These nitrogenous compounds are restricted to 13 plant families of the Cariophyllales order and a few superior fungi in the genus *Amanita* of the Basidiomycetes [16], in which anthocyanins are absent and replaced [17]. Yellow and red beet (*Beta vulgaris*) and cactus pear (*Opuntia* genus) are the main food sources of these phytochemicals [18]. The edible berries of *Rivina humilis* L. were recently identified to be new sources [19]. The main common structure of all betalains is the dihydropyridine moiety from betalamic acid (Fig. 7.2a). This is conjugated with amino acids or the corresponding amines (including dopamine) to generate the yellow betaxanthins (λ_m=480 nm). Indicaxanthin, in particular, is the immonium derivative of proline with betalamic acid (Fig. 7.2b) [20]. The violet-red pigmented betacyanins (λ_m=535 nm) are derivatives of betanidin, the conjugate of betalamic acid with cyclo-DOPA (Fig. 7.2a). The fruit of *Opuntia ficus indica* (yellow variety) is by far the best source of indicaxanthin, with an amount from 8.5 mg/100 g fresh fruit pulp (Sicilian cultivated fruit) [21] to 15 mg/100 g fresh fruit pulp (Sicilian wild fruit, unpublished data from the authors' laboratory). It also contains betanin, around one tenth of indicaxanthin [21]. As a comparison, the *Rivina humilis* fresh berries reportedly contain a minor

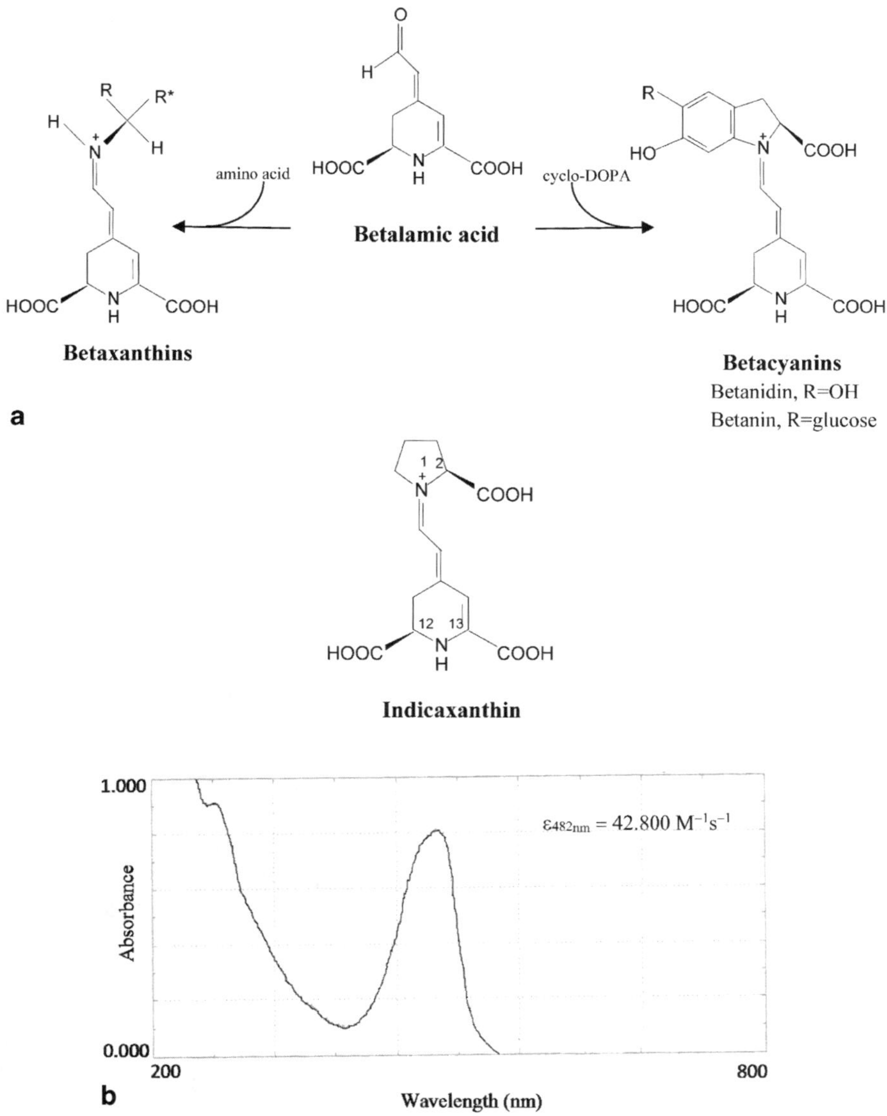

Fig. 7.2 Structure of betalains (**a**), and indicaxanthin and its absorbance spectrum (**b**)

amount of indicaxanthin, but up to 180 mg/100 g eight betaxanthins [19]. In accordance with the hormetic theory [11], it is not surprising that wild Sicilian cactus pear fruits possess higher amounts of betalains than cultivated fruits.

Betacyanins have been known for many years as additives in the food industry, due to their colorant properties, high solubility in water, pH stability in the range 3–7 [22], and the absence of toxicity [22, 23]. Renewed interest in betalains has come in the recent decade to characterize the redox chemistry and radical-scavenging and

Table 7.1 Chemical and physicochemical properties of indicaxanthin. (Adapted from [21, 29])

MW	Partition coefficients		MSA[a] (Å^2)		Peak potential mV	pKa
–	log P^b	log D^c	PSA	NPSA	Ep(a)	COOH[d]
309	0.362	−7.25	126.92	161.11	611; 895	$5.0_{(2)}$; $3.7_{(11)}$; $2.6_{(13)}$

[a] The molecular surface area (MSA) is described by the polar surface area (PSA) and the nonpolar surface area (NPSA)
[b] Octanol/water partition coefficient
[c] Octanol/buffer pH 6.0 partition coefficient
[d] Carbon atom numbers in bracket

antioxidant activities of these molecules [21, 24–26], which has then stimulated a number of studies on their potential activities in biological environments and cell contexts.

The redox potential of indicaxanthin has been measured by cyclic voltammetry [21]. The differential pulse voltammogram showed two anodic waves, with calculated peak potentials of 611 and 895 mV indicating that indicaxanthin is able to donate its electrons to organic radicals commonly formed in cell environments, including alkyl-peroxyl and lipoperoxyl radicals [27], which has been shown [28].

Indicaxanthin is a cationized molecule, with a positive charge localized in proximity of the N1 nitrogen (Fig. 7.2b) and possesses a number of ionizable carboxyl groups. A recent computational analysis predicted the solubility parameters and the polar and nonpolar surface area (NPSA) of indicaxanthin, as well as the dissociation constants of the carboxyl groups [29]. According to these calculations, indicaxanthin mainly exists as a bis-anion within a large range of pH, including physiological pH, and the pH gradient relevant to the intestinal environment during digestion (pH 6.0–7.4) [30, 31], a condition used to investigate the mechanisms of the intestinal absorption [29]. According to the octanol/water partition coefficients ClogP and ClogD (pH 6.0), indicaxanthin has a moderately nonpolar character [29]. However, the calculated NPSA suggested that it has a quite large nonpolar surface, accounting for more than 50 % of the surface area, which appears to substantiate observations on its ability to interact with membranes [25, 32–34] and low-density lipoproteins (LDLs) [28, 35, 36]. Table 7.1 summarizes the chemical and physicochemical properties of indicaxanthin.

Interactions of Indicaxanthin with Membranes

Because it possesses ionizable groups (the charged portions) and lipophilic moieties, indicaxanthin may act as an amphiphilic-like compound at a physiological pH. These chemical properties are critical to drive interaction with cell surface, where bioactive compounds binds or interacts with lipids or/and membrane effectors to initiate cell signaling transduction or transports through the cell membrane to reach inner sites, including nuclear ones, possibly affecting regulation of gene expression.

The location of the pigment in vesicles of either L-R-dipalmitoyl-phosphatidylcholine (DPPC), or L-R-dimyristoyl-phosphatidylcholine (DMPC) has recently been investigated by monitoring the characteristic absorbance spectrum of indicaxanthin in the visible light (Fig. 7.2b) and its variation as a function of temperature and phospholipid concentration [33]. While ruling out that the pigment partitioned at the hydrophobic interior, that study did not unequivocally conclude on the location of indicaxanthin in these vesicles. The partition of indicaxanthin between water and vesicular pseudo phases was investigated with a Gepasi modeling approach [34]. The rate constants of the reaction of indicaxanthin with the water-soluble azo-initiator 2,2′-azobis [2-methylpropionamidine] dihydrochloride (AAPH) in aqueous solution were calculated and compared with the constants from analogous reaction in the presence of DPPC or DMPC bilayers, at varying phospholipid concentration and temperature. Analysis of the data established that indicaxanthin can solubilize in the bilayer between the polar head groups and hydrophobic region, the so-called palisade domain. This location may allow indicaxanthin to scavenge radicals from the water phase, as well as propagating lipoperoxyl radicals floating to the bilayer polar interface ([37]; Fig. 7.3). All this appears consistent with the antioxidant activity observed in simple chemical systems such as soybean phosphatidylcholine (PC) bilayers [28]. In addition, interactions of indicaxanthin with membrane lipids and possibly other components, could elicit cell responses, and account for cell effects of indicaxanthin, including anti-inflammatory and anti-apoptotic effects [38–40].

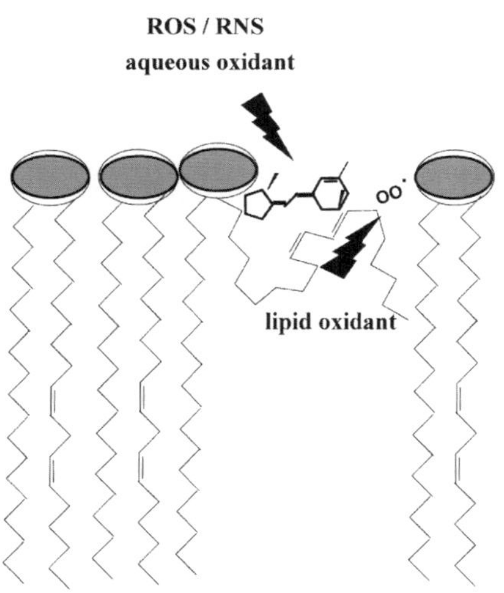

Fig. 7.3 Proposed model for the biophysical properties of Indicaxanthin in which the phytochemical can partition in the interface between hydrophobic core and hydrophilic head groups of lipid bilayers. At this level Indicaxanthin can react with radicals from the water phase as well as chain propagating lipoperoxyl radicals floating to the bilayer polar interface. Only one leaflet of the bilayer is represented

Bioavailability

Bioavailability is considered a critical factor for dietary compounds treatment. Bioactivity and potential effects of dietary phytochemicals on human health can be assessed only on absorbed active molecules and transformed digests. Bioavailability is affected by a number of exogenous and endogenous variables and factors, such as structure of food matrix, processing of food, chemistry and stability of ingested compounds, dietary amount, interactions with other food components, bacterial metabolism, and individual differences. All of these could affect intestinal absorption and/or circulation in plasma and excretion, and ultimately the actual concentration of phytochemicals available to target cells.

Betacyanins appeared to be absorbed, which were detected in the urine of subjects who consumed red beet juice [25, 30], beetroot [41], or cactus pear fruit [36]. Early human studies in the authors' laboratory [36], provided evidence that bioavailability of dietary indicaxanthin was higher than a majority of dietary polyphenol phytochemicals [42], and even than other betalain pigments such as betanin [25, 30, 36]. Data from plasma kinetics and urinary excretion indeed showed that indicaxanthin reached a maximum plasma concentration of 7 µmol/L, 3 h after a cactus pear fruit meal with 28 mg of the pigment, and disappeared from plasma 12 h after the ingestion. The urinary excretion over 12 h represented more than 70% of the amount ingested [36]. Two major points emerged from these studies are: indicaxanthin is absorbed in a quite high amount, and enters circulation in its native form. The evidence that dietary indicaxanthin did not undergo metabolism in enterocytes or hepatocytes allowed a number of studies to investigate effects and mechanistic aspects of indicaxanthin activity in cell models, utilizing the phytochemical concentrations reflecting dietary conditions and post-intestinal blood levels [32, 40, 43]. Data from these studies suggest that dietary indicaxanthin is a potential nutraceutical.

In vitro studies under simulated gastrointestinal conditions shed light on factors that would account for the pigment bioavailability observed in humans. Digestive stability and bioaccessibility of indicaxanthin, that is, the amount of the compound soluble in a post-intestinal digest [44] have been evaluated after a simulated oral, gastric and small intestinal digestion of cactus pear fruit preparations and compared with the stability and bioaccessibility of purified pigment [45]. A minor loss of indicaxanthin was observed at the gastric-like environment only, which was not affected by food matrix, whereas the molecule appeared wholly soluble in the aqueous fraction of post-intestinal digest. Interestingly, the bioaccessibility of indicaxanthin, expressed as percent of the food content, was 77% of the amount in the cactus pear fruit, a finding quite in accordance with the bioavailability in humans [36]. That indicaxanthin bioavailability is mainly determined by stability of the molecule to the digestive environment has also been confirmed by studies on the intestinal *trans*-epithelial transport of the pigment [29].

Like for xenobiotics, the intestinal absorption of phytochemicals may occur passively, through *trans*-cellular permeation or paracellular route in accordance with molecular mass and physicochemical characteristics, and could involve either influx or efflux membrane transporters. Generally, unless utilizing transporters in the epithelial membrane, charged solutes of a suitable molecular mass should diffuse through the paracellular route and be transported passively by solvent drag [46–49]. The intestinal permeation of indicaxanthin was investigated using Caco-2 cell monolayers grown on Transwell® insert, an established model of the intestinal barrier. Originating from a human colorectal carcinoma, these cells spontaneously differentiate into polarized monolayers that exhibit morphological and functional characteristics of the intestinal absorptive epithelium [50, 51]. The *trans*-epithelial transport of dietary-consistent amounts of indicaxanthin was measured in apical-to-basolateral and basolateral-to-apical direction under an inwardly directed pH gradient (pH 6.0–7.4), mimicking luminal and serosal sides of human intestinal cells [29]. According to the apparent bidirectional permeability coefficient (P_{app}), transport kinetics as a function of time and concentration, absence of interference by inhibitors of membrane transporters, and remarkable increase of permeation after ethylene diamine tetra-acetic acid (EDTA) treatment, it was concluded that, quite consistent with its physicochemical features, indicaxanthin crosses the intestinal barrier through paracellular junctions reaching unaltered the basolateral compartment. Moreover, the magnitude of P_{app} ($(4.4 \pm 0.4) \times 10^{-6}$ cm s^{-1}), either pure or food-derived compound, suggests that the *trans*-epithelial gradient of dietary indicaxanthin at the intestinal lumen, with the continuous removal by the bloodstream at the serosal side, would allow a significant intestinal transport and absorption *in vivo*. According to these findings, the bioavailability of dietary indicaxanthin in humans [36] appears as a result of (i) relatively high stability of the molecule to the digestive process [45], (ii) favorable intestinal absorption through paracellular route by solvent drag, and (iii) easy release from food [29]. It is peculiar that indicaxanthin does not undergo metabolic transformations such as glucuronidation, sulfation, or methylation to be released in plasma and circulates as unconjugated molecule. This may be advantageous for the interaction of this amphypathic molecule with membranes and/or membrane effectors leading to cell signaling and response.

Radical-Scavenging and Antioxidant Activity of Indicaxanthin

Structural Implications in the Free Radical-Scavenging Activity of Betalains

Several studies in the latest decade investigated the antiradical activity of betalains [21, 24–26, 52–54]. The cyclic amine of betalamic acid, the building block of the betalain pigments, may be envisaged as the reactive group acting similarly to the amine group of ethoxyquin, a potent antioxidant in lipid systems [55, 56] and *in vivo* [57]. A recent systematic analysis [52, 53] explored the relations between structures, spectroscopic properties and antiradical activity of various betalains, measured by the assay of decolorization of the 2,2′-azino-bis[3-ethylbenzthiazoline-6-sulfonic acid (ABTS) radical cation [58]. The studies considering both natural and synthetic compounds, with or without hydroxyl groups or aromaticity in the pigment structure, ascertained that in betaxanthins there is an "intrinsic activity" not linked to the presence of hydroxyl groups or aromaticity, which might be associated with the common electronic resonance system supported between the two nitrogen atoms, and be general to all betalains that contain a similar resonance system. The presence of phenolic hydroxy group, however, enhances significantly the radical-scavenging activity of betaxanthins [52]. In this context, the proline-derivative of betalamic acid (indicaxanthin) should act as a radical scavenger less effective than betacyanins, e.g., betanin, which has indeed been shown [21].

Chemical Models

The decolorization of the ABTS radical cation is an accurate assay universally used for screening radical-scavenging activity of either food extracts or natural pure compounds, as compared with the hydrophilic vitamin E-analogue Trolox [58]. The reducing activity of indicaxanthin has been evaluated by the reaction with the ABTS radical cation, generated by reacting ABTS with potassium persulfate [21]. When expressed as Trolox equivalents, the radical-scavenging activity of indicaxanthin was one order of magnitude lower than the betacyanin betanin [21], which was in accordance with the structural features [52, 53] and the redox potential of these two betalains [21]. The higher scavenging capacity of betanin can be explained by the ease with which it is possible to withdraw an electron from its phenolic hydroxyl group, and by the stability of the resulting delocalized radical [24]. In contrast, the electron abstracted from the betaxanthins could only be from the π-orbitals, this loss being hindered by the positive charge of the N-atom.

Lipoperoxyl radical-scavenging activity of indicaxanthin has been evaluated by the reaction with radicals generated in methyl linoleate methanol solution by 2,2′-azobis[2,4-dimethylvaleronitrile] (AMVN), and in aqueous soybean PC unilamellar liposomes by AAPH [28]. Indicaxanthin acts as a classical chain-terminating lipoperoxyl radical scavenger in solution, with calculated inhibition constant and stoichiometric factor quite comparable with those reported for the most effective natural lipoperoxyl radical scavenger, α-tocopherol [59], under comparable conditions. According to the data, after H-atom donation, the polyene system of indicaxanthin may allow the formation of a resonance-stabilised aminyl radical, whose reactivity will be affected by the environment. In methanol solution of peroxidizing methyl linoleate, the observed stoichiometry suggests reaction with a second lipoperoxyl radical ($n = 1.98$) [28].

The radical-scavenging capacity is not the only requisite for a compound to act as an antioxidant in a biological context. Kind and site of radicals generated and location of the compound (membrane or solution) may finally decide about the potential of a molecular antioxidant [59–61]. Though effective in preventing lipid

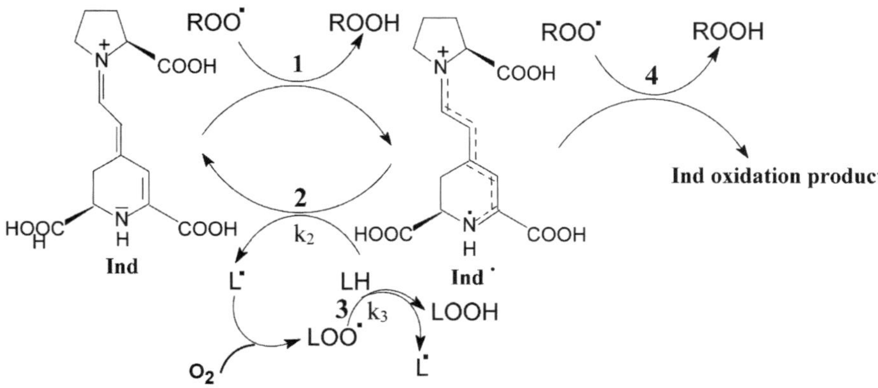

Fig. 7.4 Proposed scheme of inhibition of peroxyl-induced lipid oxidation in aqueous soybean PC liposomes by Indicaxanthin and phytochemical regeneration. Reaction of indicaxanthin (Ind) with lipoperoxyl or aqueous peroxyl radicals (ROO•) generates a resonance stabilized intermediate cation radical (Ind•) and hydroperoxide (ROOH) (reaction 1). It is proposed that Ind• is reduced by reaction with polyunsatured fatty acid (LH) to regenerate Ind and a carbon centered radical (L•) (reaction 2) that in turn reacts with oxygen (O_2) to form LOO• capable of initiating a new oxidation chain (reaction 3). Phytochemical regeneration does not result in a prooxidant event as the rate at which Ind starts oxidation events is much lower than that of the lipid peroxyl radical ($k_2 \ll k_3$). This process competes with termination reactions, which results in consumption of the phytochemical (reaction 4)

oxidation, when incorporated in soybean PC liposomes submitted to aqueous radicals from AAPH, indicaxanthin did not exhibit an antioxidant activity consistent with the activity in solution [28]. The data suggested a mechanism more complex than a classical chain-breaking antioxidant, involving indicaxanthin regeneration and recycling. After reaction of indicaxanthin with a lipid peroxyl radical, reduction of a short-lived indicaxanthin radical intermediate would occur at the expenses of unsaturated lipids, that is a potentially prooxidant insult. Since a clear antioxidant effect is evident, however, the rate at which the indicaxanthin radical starts oxidation events must be lower than that of the lipid peroxyl radical (Fig. 7.4). In this context, the ratio indicaxanthin/unsaturated lipid should be critical in determining the recycling effectiveness, as well as the actual equilibrium between antioxidant and potential prooxidant activity of the molecule.

Indicaxanthin and α-tocopherol, simultaneously incorporated in liposomes, exhibited cooperative antioxidant effects and reciprocal protection [28]. In accordance with the potential "prooxidant" character of indicaxanthin in this system, the extent of synergism decreased at the increase of the ratio (indicaxanthin)/(α -tocopherol).

Biological Models

Ex vivo spiking of human plasma with indicaxanthin, followed by isolation of LDL has provided evidence that the betalain can bind to LDL in a saturable fashion, with a maximum binding of 0.5 nmol/mg LDL protein. The indicaxanthin-enriched LDL appeared more resistant than the homologous native particles to the copper-induced oxidation, as assessed by the elongation of the period required to start lipoperoxide production [35]. Mechanistic aspects of the antioxidant activity of indicaxanthin in LDL appeared comparable with the observations in liposomes [28], including synergistic interactions with the LDL vitamin E and recycling of indicaxanthin at the expenses of unsaturated LDL lipids, with no evidence of prooxidant effects [35].

The oxidation status of the heme iron in proteins such as hemoglobin (Hb), myoglobin (Mb), or myeloperoxidase (MPO) must strictly be controlled in order that the proteins accomplish appropriately their functional roles, and do not become powerful oxidizing agents. Indeed, due to their peroxidase or peroxidase-like activity, in the presence of hydrogen peroxide or organic hydroperoxides these proteins undergo a two-electron oxidation, giving rise to high-valent iron radical species ($X^{\bullet}[Fe^{IV}=O]$) [62–64]. Unless rapidly reduced, the latter can oxidize cell components and damage the protein itself thus impairing its function.

The Hb heme-iron is extensively oxidized under pathological conditions, e.g., β-thalassemia [65], while continuous slow autoxidation of Hb also occurs in healthy individuals resulting in red cell oxidative stress [66, 67]. Kinetic studies of the reaction of indicaxanthin with perferryl-Hb, the two-electron oxidized intermediate in the oxidative degradation of Hb [62], showed that indicaxanthin reduces the hypervalent heme-iron at a rate one order of magnitude higher than that of reductants such as ascorbate and Trolox, on a molar basis [43].

Dietary phytochemicals may act in the gastrointestinal tract [68–71]. Scavenging of highly oxidizing hypervalent-iron myoglobin (perferryl-Mb) formed during meat digestion [68] may preserve oxidable lipids and avoid formation of potentially toxic lipid hydroperoxides. In this context, unpublished observations in the authors' laboratory provided evidence that indicaxanthin concentrations consistent with a dietary intake reduced perferryl-Mb and prevented the lipid oxidation of heated red meat under a simulated gastric digestion.

Despite MPO plays key roles in the defense against invading pathogens by oxygen-dependent antimicrobial action [72, 73], the enzyme has an enormous potential to inflict damage to host tissues, through its ability to catalyze the production of a complex array of reactive oxidants including its product hypochlorous acid (HOCl), and nitrogen dioxide, organic-free radicals and drug metabolites [74–77]. In addition, the harmful potential of MPO in the onset and progression of atherogenetic processes is to be considered because the demonstrated capacity of the enzyme to oxidize LDL [78–81]. According to its halogenation and peroxidase cycles [82], MPO can catalyze both two- and one-electron oxidation reactions, leading to production of the cytotoxic HOCl and generating the hypervalent iron redox intermediates compound I (CI) and compound II (CII). Indicaxanthin has been shown to be a good electron donor for both CI and CII, at very low μmolar concentrations [83]. In addition, it has been reported that the betalain can scavenge HOCl [83].

Table 7.2 reports quantitative parameters of the reaction of indicaxanthin with various oxidants/radicals.

Table 7.2 Scavenging activity of indicaxanthin versus reactive oxidants

ABTS•+a	LOO•b	MPOc compound I	MPOc compound II	HOCld	Perferryl-Hbe
Trolox equivalents	–	Reduction constants (pH 7.0) ($M^{-1}s^{-1}$)		–	Reduction rate (nMs^{-1})
1.76 [21]	3.6×10^{5} [28]	1.1×10^{6} [83]	2.9×10^{5} [83]	7.7×10^{4} [83]	660 [43]

[a] ABTS, [2,2′-azinobis(3-ethylbenzothiazoline- 6-sulfonic acid)] diammonium salt cation radical
[b] LOO•, methylinoleate peroxyl radical
[c] MPO, myeloperoxidase
[d] HOCl, hypochlorous acid
[e] Perferryl-Hb, perferryl-hemoglobin

Effects of Indicaxanthin in Cell, Tissue, and Animal Models

The redox chemistry, physicochemical properties, bioactivity in solution and high bioavailability of indicaxanthin were a promising approach to investigate on the activity and eventual mechanism of action in the cells. A number of experimental setups with various either human or mouse cell cultures showed positive effects of indicaxanthin on the cell response to toxic stimuli modeling pathophysiological conditions, from serious oxidative stress to inflammation and apoptosis. In other studies, evidence has been provided that indicaxanthin can modulate the differentiation of T cells toward an immune-modulatory phenotype, and affect mouse ileal contractility. Anti-inflammatory activity of indicaxanthin was tested in a murine model.

Ex vivo spiking of blood from healthy human subjects with varied indicaxanthin concentrations resulted in a saturable incorporation of the betalain in the RBC, to the extent of approximately 1 nmol/mL packed cell, an amount very close to that measured in human RBCs after consumption of dietary indicaxanthin [32, 36]. Indicaxanthin-enriched erythrocytes resulted more resistant to the cumene hydroperoxide-indicaxanthin-induced oxidative haemolysis than the homologous nonenriched cells, with a significant correlation between the increase of resistance and the amount of the incorporated indicaxanthin [32]. High oxygen tension and large amounts of iron, a transition metal promoting formation of oxygen free radicals [84], make red blood cells (RBCs) highly susceptible to oxidation, leading to impairment of the cell function and premature aging [67]. In addition, the oxidative alterations of the RBC membrane can even cause injury to the endothelial cells [67]. Protective effects in blood of dietary indicaxanthin may be suggested.

An increased generation of reactive oxygen species, first caused by Hb autooxidation and precipitation, characterizes genetic hemolytic disorder such as β-thalassemia. The pathology is associated to depletion of the RBC antioxidant defense resulting in damage to cell components, impairment of morphology and function of cell membrane, and accelerated RBC destruction [65, 85–88]. Protective dose-dependent effects of indicaxanthin have been shown in RBC from β-thalassemia patients submitted to an *in vitro* oxidation by cumene hydroperoxide [43]. The betalain, that entered red blood cells, enhanced the resistance of thalassemia RBC to hemolysis, prevented lipid and Hb oxidation, and retarded vitamin E and GSH depletion. In the same work, spiking of blood from thalassemia patients with indicaxanthin resulted in its incorporation in the RBCs, indicating that the pathological alterations to the membrane do not affect a *trans*-bilayer movement of this phytochemical. Antioxidant vitamins and phytochemicals may be helpful to treat β-thalassemia [89–91]. The finding that indicaxanthin can be incorporated in the redox machinery of β -thalassemia RBCs, may offer novel opportunities of therapeutic interest in this pathology.

Atherosclerosis is a chronic inflammatory condition of the arterial intima. This complex process progresses slowly during a lifetime, with initial dysfunction of the vascular endothelium to a great extent mediated by oxidized LDL, immunocompetent cells, cytokines, growth factors, adhesion molecules, and compounds involved in redox-sensitive regulatory mechanisms underlying inflammatory processes [92–96]. Since atherosclerosis and related cardiovascular problems are a leading cause of death in the western world [97], research on dietary compounds capable of preventing or controlling pathological events is now very active. In this context, the activity of indicaxanthin has been explored in various model systems.

Vascular endothelial cells are a direct target of proinflammatory stimuli that remarkably affect a number of redox-mediated signaling pathways, leading to production of chemotactic factors, lipid mediators, and cytokines [98], as well as adhesion molecules such as the adhesion molecule-1 (ICAM-1) [99, 100]. With an *in vitro* model of inflammation consisting of umbilical vein endothelial cells (HUVEC) stimulated with the proinflammatory cytokine tumor necrosis factor-α (TNF-α), it has been shown that μmolar indicaxanthin concentrations can modulate the expression of ICAM-1 [38].

Apoptosis of macrophages triggered by oxidized cholesterol derivatives, especially 7-ketocholesterol (7-KC), is considered a key event in the development of human atheromas [101, 102]. A recent *in vitro* study showed that dietary consistent amounts of indicaxanthin exert protective effects and prevent the 7-KC-induced pro-apoptotic events in human monocyte/macrophage THP-1 cells. The effects of indicaxanthin have been related to inhibition of the basal activity and overexpression of NADPH oxidase-4 (NOX-4), prevention of the redox-sensitive nuclear factor-*k*B activation, maintaining of cell redox balance and inhibition of the increase of cytosolic calcium, which prevented mitochondrial damage and consequently apoptosis. About the molecular mechanism, though other explanations cannot be ruled out, interactions of indicaxanthin with 7-KC at the level of the THP-1 cell membrane could bring about some stabilizing effect leading to modulate NOX-4 enzyme activity and possibly prevent opening of calcium channels [102].

Immunomodulatory effects of indicaxanthin have been revealed *in vitro* with mouse splenocytes allowed to differentiate in Th1 and Th17 skewing conditions, either in the presence or in the absence of indicaxanthin [103]. Indicaxanthin inhibited both IFN-γ and IL-17 production in the Th17 conditions, accompanied by a marked increase of IL-6 and a significant reduction of IL-10 levels, indicating a modulation of T regulatory differentiation and/or function. The data also showed a reduction of percentage of IL-17- and a parallel increase of IFN-γ- producing cells in indicaxanthin-treated compared to control cells. Consistent with this, indicaxanthin almost doubled the number of IFN-γ^{+}T cells in splenocytes differentiated in Th1 skewing conditions. The data suggest potential activity of indicaxanthin in Th1-mediated protective immunity against pathogens. In addition, positive effects could be hypothesized on Th17-driven autoimmune disorders, e.g., multiple sclerosis or rheumatoid arthritis.

An experimental organ bath technique to record the mechanical activity of the isolated mouse ileum longitudinal muscle was used to investigate the effects of indicaxanthin on the spontaneous and evoked contractility [104], and mechanism of action underlying [105]. Indicaxanthin showed a remarkable spasmolytic effect on the intestinal contractility, with its action ascribable to phosphodiesterase inhibition and elevation of cAMP levels. In view of the capacity of indicaxanthin to interact with and cross cell membranes [32–34, 43], these findings open interesting questions on how this action may be accomplished at a molecular level.

To date, there is only one *in vivo* report on pharmacological activity of indicaxanthin. Remarkable anti-inflammatory effects have been observed in a carrageenin-induced rat pleurisy model [39]. Orally given to the extent of 8 μmol/kg, which is consistent with a human dietary assumption from cactus pear fruits [36], prior to carrageenin injection in the pleural cavity, indicaxanthin caused a 72 % reduction of the inflammatory exudate volume, and 95 % reduction of total leukocyte number. Table 7.3 reports on biological actions of indicaxanthin in various systems.

Epigenetics

Occurrence and frequencies of epigenetic changes throughout the life are now acknowledged crucial for gene activation or silencing [106]. This finally affects the entire cell behavior allowing normal cell functions or, conversely, leading to pathophysiological states, including inflammation, cancer, and aging [107–109]. Differently from definite aberrations in the DNA sequences, epigenetic alterations are potentially reversible [106]. Numerous data pointing out to the influence of food components, especially redox-active phytochemicals, on epigenetic mechanisms emphasize the role of diet as an important environmental factor for maintaining a healthy status, and offer promising evidence for their eventual chemopreventive efficacy [110]. It may be worth noting that steady-state oxidation of cell proteins has been reported to be responsive to diet [111].

Longevity and lower incidence of chronic degenerative disorders and cardiovascular disease among Mediterranean populations have been associated with a peculiar dietary model rich of fruits and vegetables and low of animal fat. As a part of the Mediterranean dietary habit, the cactus pear fruit and its betalain phytochemicals, i.e., indicaxanthin and betanin, may have contributed to these beneficial effects, especially in the past when food was less globalized [112, 113]. Repetitive long-life

Table 7.3 Indicaxanthin bioactivity

Model	Effect	Mechanism	Ref
Cells			
RBCs	↑Resistance to oxidative haemolysis	Radical-scavenging activity	[32]
Thalassemia RBCs	↑Resistance to oxidative haemolysis	Perferryl-Hb reduction	[43]
–	↓Vitamin E and GSH depletion	–	–
–	↓Lipid and Hb oxidation	–	–
HUVEC[a]	↓TNF-α–induced ICAM-1expression	Redox-modulation of cell signaling	[38]
THP-1[b] cells	↓7-ketocholesterol-induced apoptosis	↓NF-kB activation	[40]
–	–	↓NOX-4 expression	–
–	–	↓intracellular [Ca^{2+}]	–
Mouse splenocytes	↑Th1 differentiation	–	[103]
Differentiated in Th1/17 skewing conditions	↓Th17 differentiation	–	–
Tissues			
Mouse ileum muscle	Spasmolytic	Phosphodiesterase inhibition	[105]
–	–	↑cAMP	–
Animals			
Carrageenin-induced rat pleurisy	Anti-inflammatory	↓NF-kB activation	[39]
–	↓Inflammatory exudates	↓COX-2;iNOS expression	–
–	↓Leukocyte recruitment	↓TNF-α;IL-1β;PGE_2;NO· release	–

[a] Human umbilical vein endothelial cells
[b] Human monocyte/macrophage cells

exposure to physiological concentrations of betalains could play a role in maintaining cell functions through epigenetic machinery. The highly bioavailable and bioactive indicaxanthin may represent an important factor in this context, so our current research is going in this direction. Epigenetic activity of indicaxanthin has recently been shown [114]. Hypermethylation of onco-suppressor gene promoters is a well-established epigenetic modification [115]. Among others, the hypermethylation of the p16 gene, an onco-suppressor controlling the cell cycle through cycline inhibition, is associated with tumor development [116–118]. A number of data with a human intestinal cancer (Caco-2) cell line provided evidence that the amount of indicaxanthin consistent with that in the intestinal digests after a dietary consumption of cactus pear fruit pulp [36] could induce demethylation of the p16 gene promoter, reactivation of the p16 synthesis and finally cell apoptosis. Exploring indicaxanthin activity on other onco-suppressor genes in the same as well as other human cancer cell lines and, understanding pathways by which indicaxanthin can modulate epigenetic mechanisms is the near future challenge.

Molecular Toxicology and Biosafety

Betalains, especially red betacyanins, have long been considered important natural pigments for industry, from food to cosmetics and pharma products, then safety of these molecules has been checked long time ago [24, 119, 120]. Studies carried out to determine decomposition and stability [121, 122], mutagenicity [120, 123], and toxicological and toxicokinetic effects [124], showed that these pigments are not harmful. Some *in vivo* studies in rats indicated that a betalain extract from Garambullo, a cactacea fruit (*Myrtillocactus geometrizans*) did not have toxic effects at any of the doses tested, up to 5 g/kg body weight [124]. Nevertheless, in spite of the acute or subacute safety assessment in animals, current and emerging knowledge on (i) the redox regulation of signaling pathways involved in cell survival and death, (ii) the activity of phytochemicals as redox modulators with potential dual behavior of acting (antioxidant vs. prooxidant), and (iii) the cell epigenetic response to changes of its redox milieu, require that actions of redox active phytochemicals of our diet should be checked at a molecular level. In addition, their eventual involvement in the epigenome regulation should at least been established in cellular models under normal and pathophysiological states. Since these substances may have the potential to behave as nutraceutical/ pharmaceutical agents, it should be assured that they do not cause harmful variations of the cell redox homeostasis at recurring dietary-consistent amounts, or eventually at amounts of pharmacological interest. It should be kept in mind that a potential chemoprevention, through modulation of the cell redox state in turn controlling various signaling transduction mechanisms, may require very minute amounts of phytochemicals. Then, in normal cells, eventual prooxidant activity and dangerous effects by high concentrations may occur. Conversely, in tumor cells, where the redox milieu is disturbed [125], prooxidant activity of a given phytochemical could be required [126, 127], to restore homeostatic patterns and cause apoptosis (chemotherapy), which may be toxic for normal cells [14].

So far, with respect to the above mentioned tasks, potential adverse effects of high concentrations of indicaxanthin have been searched in a few experimental conditions in the authors' laboratory, in contexts where low indicaxanthin concentrations exhibited antioxidant, or radical-scavenging effects [28, 35]. Prooxidant activity was not observed neither in the model of copper-stimulated oxidation of human LDL [35], nor in liposomal membranes under the prooxidant action of the azo-initiator AAPH [28], rather indicaxanthin dose-dependently inhibited lipid peroxidation in both systems, even when tested at very high micromolar concentrations (up to 1 mM) [35]. Similarly, no prooxidant activity was evident in human RBC [32]. With its peculiar location in lipid bilayers [33, 34], antioxidant activity of indicaxanthin at the cell membrane could be relevant not only to the integrity of membrane but also to maintain cell homeostasis. Redox-driven signaling pathways may be started by oxidized membrane lipid components, and their soluble derivatives such as reactive aldehydes, being the latter capable of starting oxidation pathways inside the cell [128], damaging proteins, DNA, and also affecting epigenetic mechanisms [129].

The membrane enzymes of the NADPH oxidase (NOX) family are involved in the production of reactive oxygen species (ROS) in all animal cells and their primary role in the control of cell redox homeostasis is widely acknowledged [130–132]. In this context, investigating the activity of phytochemicals at the level of these enzymes may be a strategy to investigate about potentially harmful mechanisms underlying their biological actions in various cells and under different conditions. A surprising pro-senescent activity of resveratrol in human endothelial cells, for instance, has been reported to be mediated by NOX-1 and NOX-4 [133]. Research with monocyte/macrophage THP-1 cells [40] showed that indicaxanthin, at nutritionally relevant amounts, can prevent the NOX-4 dependent ROS formation induced by the toxic 7-keto-cholesterol, however, does not modify the basal activity of the enzyme in the absence of the oxysterol, even at ten-times higher amounts. Then, while protecting THP-1 cells from an oxidant agent, indicaxanthin itself did not modify the basal redox environment of these cells under normal conditions.

Indicaxanthin toxicity has been checked on Caco-2 cells grown at confluence for 15 days, to reach characteristics of normal human colon epithelial cells [50, 51]. The assays were carried out by measuring LDH activity released from the cells exposed to high micromolar concentrations of the phytochemical, in the range 25–100 μM. No significant release of LDH activity was detected after a 24-h incubation, showing that cell integrity had been preserved.

Other studies with a panel of animal cells are necessary to provide a much more complete picture on potentially harmful actions of indicaxanthin and to test influence on epigenome targets. However, to date, adverse effects of the phytochemical have not been observed in any either solution or cell system, even in a range of concentrations widely exceeding the physiological blood levels obtained with food intake [36].

Conclusive Remarks and Perspectives

Dietary redox active phytochemicals are currently studied intensely for basic science and applied research. Understanding how these compounds exert physiological effects when ingested with food and exploiting their nutraceutical/ pharmacological potential are the key goals in the study of molecular nutrition [134]. This hard task requires interdisciplinary collaborations and clinical studies.

There is now a consensus that redox-active phytochemicals are utilized in animal cells as bio-signals to start pathways regulating protective and/or defense activities. How a cell interprets these signals depends on the cell type, the response machinery of that cell, and its history in terms of what signals that cell previously responded to and the proteomic setup the cell had at the time of the arrival of the new signal.

In this picture, environment- induced epigenetic modifications, reversibly silencing or activating certain genes may play an important role in the great variability of cell responses to dietary phytochemicals (which is the field of nutrigenetics), but at the same time may represent molecular targets on which a phytochemical can intervene (which is the field of nutrigenomics). While signal transduction mechanisms evoked by phytochemicals are mostly unknown, the modulation of epigenetic mechanisms in nonmalignant cells has been associated with the positive effects of such compounds (chemoprevention). Once molecular targets for selected dietary compounds are found, dietary strategies could be designed to maintain cell functions and possibly prevent or reduce risk of developing chronic/degenerative diseases.

Almost all fruits and vegetables on our table are now extensively cultivated, so they do not cope with, or receive much less, environmental stress. When considering hormetic production of defensive phytochemicals, this "easy and cozy" condition may possibly decrease the protective value of green food, which deserves to be tested. In this context, cactus, pear, and its betalains, may be an interesting exception. Even cultivated, the plant is not subjected to continuous reproductive cycles; fruits cannot long be stored and are available for some months in a year. In addition, many plants are still growing wild in Mediterranean areas, facing such tough environmental conditions as intense sunlight, dryness, heat.

Since dietary betalains are only a very limited number as compared with thousands phenol/polyphenol phytochemicals, their biological and health-promoting properties have been the object of less, but recently growing, interest and research [135]. However, among so many compounds indicaxanthin, a yellow betalain pigment, holds a special place and appears to be a promising dietary factor to maintain cell functions and positively affect human health. As a result of relatively high stability during the digestive process [45], favorable intestinal absorption through paracellular route without metabolic transformation, and easy release from food [29], dietary indicaxanthin possesses high bioavailability in humans [36]. Moreover, its redox chemistry and physicochemical characteristics [21, 29] allow the molecule to interact with membranes [34], penetrate cells [32] and scavenge both lipid radicals and water-soluble oxidants [28, 35]. So far, evidence of anti-apoptotic [40] and immune-modulator [103] activity has been provided, and capacity of indicaxanthin of reverting epigenetic changes associated with cancer development [111] shown in *in vitro* models, whereas only one study reported on anti-inflammatory effects in rat [39]. Future investigations in animal models, and characterization of indicaxanthin from a "nutri-epigenomic perspective," will provide important information on the potential to act as a chemopreventive agent.

The edible fruits of the cactus *Opuntia ficus indica*, L., a plant spread in the South of Italy, are the only abundant source of indicaxanthin, which may somewhat account for why nutritional studies and basic science investigations, including biological actions, on this molecule have been carried out almost exclusively in the authors' laboratory in Sicily. While feeling this as a great commitment, we are challenged to find out as many information as possible on the potentiality of this pigment and provide scientific and sound evidence of eventual beneficial effects from its chronic use as a nutraceutical agent. The road ahead is still very long, but really exciting.

References

1. Slavin JL, Lloyd B (2012) Health benefits of fruits and vegetables. Adv Nutr 3:506–516
2. Murakami A, Ohnishi K (2012) Target molecules of food phytochemicals: food science bound for the next dimension. Food Funct 3:462–476
3. Obrenovich ME, Li Y, Parvathaneni K, Yendluri BB, Palacios HH, Leszek J, Aliev G (2011) Antioxidants in health, disease and aging. CNS Neurol Disord Drug Targets 10:192–207
4. Halliwell B, Rafter J, Jenner A (2005) Health promotion by flavonoids, tocopherols, tocotrienols, and other phenols: direct or indirect effects? Antioxidant or not? Am J Clin Nutr 81:268S–276S
5. Doré S (2005) Unique properties of polyphenol stilbenes in the brain: more than antioxidant actions; gene/protein regulatory activity. Neurosignals 14:61–70
6. Izzi V, Masuelli L, Tresoldi I, Sacchetti P, Modesti A, Galvano F, Bei R (2012) The effects of dietary flavonoids on the regulation of redox inflammatory networks. Front Biosci 17: 2396–2418
7. Howitz KT, Sinclair DA (2008) Xenohormesis: sensing the chemical cues of other species. Cell 133:387–391
8. Sepúlveda-Jiménez GP, Rueda-Benítez P, Porta H, Rocha-Sosa M (2004) Betacyanin synthesis in red beet (*Beta vulgaris*) leaves induced by wounding and bacterial infiltration is preceded by an oxidative burst. Physiol Mol Plant Pathol 64:125–133 (Erratum in Physiological and Molecular Plant Pathology 66:75)
9. Suzuki N, Kousevitzky S, Mittler R, Miller G (2012) ROS and redox signaling in the response of plants to abiotic stress. Plant Cell Environ 35:259–270
10. Suzuki N, Mittler R (2012) Reactive oxygen species-dependent wound responses in animals and plants. Free Radic Biol Med 53:2269–2276
11. Howitz KT, Bitterman KJ, Cohen HY, Lamming DW, Lavu S, Wood JG et al (2003) Small molecule activators of sirtuins extend *Saccharomyces cerevisiae* lifespan. Nature 425:191–196
12. Calabrese V, Cornelius C, Dinkova-Kostova AT, Iavicoli I, Di Paola R, Koverech A et al (2012) Cellular stress responses, hormetic phytochemicals and vitagenes in aging and longevity. Biochim Biophys Acta 1822:753–783
13. Surh YJ (2011) Xenohormesis mechanisms underlying chemopreventive effects of some dietary phytochemicals. Ann N Y Acad Sci 1229:1–6
14. Nair S, Li W, Kong AT (2007) Natural dietary anti-cancer chemopreventive compounds: redox-mediated differentiatal signaling mechanisms in cytoprotection of normal cells versus cytotoxycity in tumor cells. Acta Pharmacol Sin 28:459–472
15. Piattelli M (1981) The betalains: structure, biosynthesis and chemical taxonomy. In: Conn EE (ed) The biochemistry of plants: a comprehensive treatise, vol 7. Academic, New York, pp 557–575
16. Strack D, Vogt T, Schliemann W (2003) Recent advances in betalain research. Phytochemistry 62:247–269
17. Stafford HA (1994) Anthocynins and betalains: evolution of the mutually exclusive pathways. Plant Sci 101:91–98
18. Stintzing FC, Schieber A, Carle R (2002) Identification of betalains from yellow beet (*Beta vulgaris*, L) and cactus pear (*Opuntia ficus indica*, L Mill) by high performance liquid chromatography-electrospray ionization mass spectrometry. J Agric Food Chem 50:2302–2307
19. Khan MI, Harsha PSC, Giridhar P, Ravishankar GA (2012) Pigment identification, nutritional composition, bioactivity, and in vitro cancer cell cytotoxicity of *Rivina humilis* L. berries, potential source of betalains. LWT Food Sci Technol 47:315–323
20. Piattelli M, Minale L, Prota G (1964) Isolation structure and absolute configuration of indicaxanthin. Tetrahedron 20:2325–2329
21. Butera D, Tesoriere L, Di Gaudio F, Bongiorno A, Allegra M, Pintaudi AM et al (2002) Antioxidant activities of sicilian prickly pear (*Opuntia ficus indica*) fruit extracts and reducing properties of its betalains: betanin and indicaxantin. J Agric Food Chem 50:6895–6901

22. Stintzing F, Carle R (2007) Betalains—emerging prospects for food scientists. Trends Food Sci Technol 18:514–525
23. Schwartz SJ, von Elbe JH, Pariza MW, Goldsworthy T, Pitot HC (1983) Inability of red beet betalain pigments to initiate or promote hepatocarcinogenesis. Food Chem Toxicol 21:531–535
24. Escribano J, Pedreno MA, Garcia-Carmona F, Munoz R (1998) Characterization of the anti-radical activity of betalains from *Beta vulgaris* L. roots. Phytochem Anal 9:124–127
25. Kanner J, Harel S, Granit R (2001) Betalains—a new class of dietary cationized antioxidants. J Agric Food Chem 49:5178–5185
26. Pedreno MA, Escribano J (2000) Studying the oxidation and the antiradical activity of betalain from beetroot. J Biol Educ 35:49–51
27. Buettner GR (1993) The pecking order of free radicals and antioxidants: lipid peroxidation, a-tocopherol, and ascorbate. Arch Biochem Biophys 300:535–543
28. Tesoriere L, Allegra M, Butera D, Gentile C, Livrea MA (2007) Kinetics of the lipoperoxyl radical scavenging activity of indicaxanthin in solution and in unilamellar liposomes. Free Radical Res 41:226–233
29. Tesoriere L, Gentile C, Angileri F, Attanzio A, Tutone M, Allegra M, Livrea MA (2013) Trans-epithelial transport of the betalain pigments indicaxanthin and betanin across Caco-2 cell monolayers and influence of food matrix. Eur J Nutr 52:1077–1088
30. Frank T, Stintzing FC, Carle R, Bitsch I, Quaas D, Strass G et al (2005) Urinary pharmacokinetics of betalains following consumption of red beet juice in healthy humans. Pharmacol Res 52:290–297
31. Gliszczynska-Swiglo A, Szymusiak H, Malinowska P (2006) Betanin, the main pigment of red beet: molecular origin of its exceptionally high free radical-scavenging activity. Food Addit Contam 23:1079–1087
32. Tesoriere L, Butera D, Allegra M, Fazzari M, Livrea MA (2005) Distribution of betalain pigments in red blood cells after consumption of cactus pear fruits and increased resistance of the cells to ex vivo-induced oxidative hemolysis in humans. J Agr Food Chem 53:1266–1270
33. Turco Liveri ML, Sciascia L, Lombardo R, Tesoriere L, Passante E, Livrea MA (2007) Spectrophotometric evidence for the solubilization site of betalain pigments in membrane biomimetic systems. J Agric Food Chem 55:2836–2840
34. Turco Liveri ML, Sciascia L, Allegra M, Tesoriere L, Livrea MA (2009) Partition of indicaxanthin in membrane biomimetic systems. A kinetic and modeling approach. J Agric Food Chem 57:10959–10963
35. Tesoriere L, Butera D, D'Arpa D, Di Gaudio F, Allegra M, Gentile C, Livrea MA (2003) Increased resistance to oxidation of betalain-enriched human low density lipoproteins. Free Radic Res 37:689–696
36. Tesoriere L, Allegra M, Butera D, Livrea MA (2004) Absorption, excretion, and distribution in low density lipoproteins of dietary antioxidant betalains. Potential health effects of betalains in humans. Am J Clin Nutr 80:941–945
37. Barclay LRC (1993) Model biomembranes: quantitative studies of peroxidation, antioxidant action, partitioning, and oxidative stress. Can J Chem 71:1–16
38. Gentile C, Tesoriere L, Allegra M, Livrea MA, D'Alessio P (2004) Antioxidant betalains from cactus pear (*Opuntia ficus-indica*) inhibit endothelial ICAM-1 expression. Ann NY Acad Sci 1028:481–486
39. Allegra M, Ianaro A, Tersigni M, Tesoriere L, Livrea MA (2009) Indicaxanthin exerts anti-inflammatory effects on carrageenin-induced rat pleurisy. Free Radical Res 43 (Suppl 1):S47
40. Tesoriere L, Attanzio A, Allegra M, Gentile C, Livrea MA (2013) Phytochemical indicaxanthin suppresses 7-ketocholesterol-induced THP-1 cell apoptosis by preventing cytosolic Ca2 + increase and oxidative stress. Br J Nutr 1–11 (e-pub), 110:230–240
41. Watts AR, Lennard ML, Tucker GT, Woods HF (1993) Beeturia and the biological fate of beetroot pigments. Pharmacogenetics 3:302–311
42. Manach C, Williamson G, Morand C, Scalbert A, Rémésy C (2005) Bioavailability and bioefficacy of polyphenols in humans. I. Review of 97 bioavailability studies. Am J Clin Nutr 81:230S–242S

43. Tesoriere L, Allegra M, Butera D, Gentile C, Livrea MA (2006) Cytoprotective effects of the antioxidant phytochemical indicaxanthin in beta-thalassemia red blood cells. Free Radic Res 40:753–761
44. McDougall GJ, Dobson P, Smith P, Blake A, Stewart D (2005) Assessing potential bioavailability of raspberry anthocyanins using an in vitro digestion system. J Agric Food Chem 53:5896–5904
45. Tesoriere L, Fazzari M, Angileri F, Gentile C, Livrea MA (2008) In vitro digestion of betalainic foods. Stability and bioaccessibility of betaxanthins and betacyanins and antioxidative potential of food digesta. J Agric Food Chem 56:10487–10492
46. Adson A, Raub TJ, Burton PS, Barshun CL, Hilgers AR, Audus KL, Ho NF (1994) Quantitative approach to delineate paracellular diffusion in cultured epithelial cell monolyer. J Pharm Sci 83:1529–1536
47. Karlsson J, Ungell AL, Artursson P (1994) Effect of an oral rehydration solution on paracellular drug transport in intestinal epithelial cells and tissues: assessment of charge and tissue selectivity. Pharm Res 11:S248
48. Pade V, Stavchansky S (1997) Estimation of the relative contribution of the transcellular and paracellular pathway to the transport of passively absorbed drugs in the Caco-2 cell culture model. Pharm Res 14:1210–1215
49. Knipp GT, Ho NF, Barsuhn CL, Borchardt R (1997) Paracellular diffusion in Caco-2 cell monolayers: effect of perturbation on the transport of hydrophilic compounds that vary in charge and size. J Pharm Sci 86:1105–1110
50. Galijatovic A, Otake Y, Walle UK, Walle T (2001) Induction of UDP-glucuronyltransferase UGT1A1 by the flavonoid chrisin in Caco2 cells- potential role in carcinogen bioactivation. Pharmaceutical Res 18:374–379
51. Sun DX, Lennernas H, Welage LS, Barnett JL, Landowski CP, Foster D et al (2002) Comparison of human duodenum and Caco2 gene expression profiles for 12,000 gene sequence tags and correlation with permeabililty of 26 drugs. Pharm Res 19:1400–1416
52. Gandía-Herrero F, Escribano J, García-Carmona F (2009) The role of phenolic hydroxy groups in the free radical scavenging activity of betalains. J Nat Prod 72:1142–1146
53. Gandía-Herrero F, Escribano J, García-Carmona F (2010) Structural implications on color, fluorescence, and antiradical activity in betalains. Planta 232:449–460
54. Gandía-Herrero F, Escribano J, García-Carmona F (2012) Purification and antiradical properties of the structural unit of betalains. J Nat Prod 75:1030–1036
55. Belitz HD, Grosch W (1987) Food chemistry. Springer-Verlag, Berlin, pp 177
56. Lin JC, Olcott HS (1995) Ethoxyquin nitroxide. J Agric Food Chem 23:798–800
57. Lockart B, Bonhomme N, Roger A, Dorey G, Casara P, Lestage P (2001) Protective effect of the antioxidant 6-ethoxy-2,2-pentamethylen-1,2-dihydroquinoline (S 33113) in models of cerebral neurodegeneration. Eur J Pharmacol 416:59–68
58. Pellegrini N, Re R, Yang M, Rice-Evans C (1999) Screening of dietary carotenoids and carotenoid-rich fruit extracts for antioxidant activities applying 2,2†²-azinobis(3-ethylenebenzothiazoline-6-sulfonic cid) radical cation decolorization assay. Methods Enzymol 299:379–389
59. Niki E, Saito T, Kawakami A, Kamiya Y (1984) Inhibition of oxidation of methyl linoleate in solution by vitamin E and vitamin C. J Biol Chem 259:4177–4182
60. Burton GW, Ingold KU (1981) Autoxidation of biological molecules. 1. The antioxidant activity of vitamin E and related chainbreaking phenolic antioxidants in vitro. J Am Chem Soc 103:6472–6477
61. Barclay LRC, Ingold KU (1981) Autoxidation of biological molecules. 2. Autoxidation of a model membrane. Comparison of the autoxidation of egg lecithin phosphatidylcholine in water and in chlorobenzene. J Am Chem Soc 103:6478–6485
62. Everse J, Hsia N (1997) The toxicities of native and modified hemoglobins. Free Radic Biol Med 22:1075–1099
63. Everse J (1998) The structure of heme proteins compounds I and II: some misconceptions. Free Radic Biol Med 24:1338–1346

64. Furtmuller PG, Obinger C, Hsuanyu Y, Dunford HB (2000) Mechanism of reaction of myeloperoxidase with hydrogen peroxide and chloride ion. Eur J Biochem 267:5858–5864
65. Rund D, Rachmilewitz E (2005) Beta-thalassemia. New Engl J Med 353:1135–1146
66. Vollaard NB, Reeder BJ, Shearman JP, Menu P, Wilson MT, Cooper CE (2005) A new sensitive assay reveals that hemoglobin is oxidatively modified in vivo. Free Radic Biol Med 39:1216–1228
67. Rifkind JM, Nagababu E (2012) Hemoglobin redox reactions and red blood cell aging. Antiox Redox Sign. doi:10.1089/ars.2012.4867
68. Halliwell B, Zhao K, Whiteman M (2001) The gastrointestinal tract: a major site of antioxidant action? Free Radic Res 33:819–880
69. Kanner J, Lapidot T (2001) The stomach as a bioreactor: dietary lipid peroxidation in the gastric fluid and the effects of plant-derived antioxidants. Free Radic Bio Med 31:1388–1395
70. Scalbert A, Morand C, Manach C, Remesy C (2002) Absorption and metabolism of polyphenols in the gut and impact on health. Biomed Pharmacother 56:276–282
71. Ursini F, Sevanian A (2002) Wine polyphenols and optimal nutrition. Ann NY Acad Sci 957:200–209
72. Klebanoff SJ (1980) Oxygen metabolism and the toxic properties of phagocytes. Ann Intern Med 93:480–489
73. Klebanoff SJ (1975) Antimicrobial mechanisms in neutrophilic polymorphonuclear leukocytes. Seminar Haematol 12:117–142
74. Hazell LJ, Arnold L, Flowers D, Waeg G, Malle E, Stocker R (1996) Presence of hypochlorite-modified proteins in human atherosclerotic lesions. J Clin Invest 97:1535–1544
75. Heinecke JW (1998) Oxidants and antioxidants in the pathogenesis of atherosclerosis: implications for the oxidized low density lipoprotein hypothesis. Atherosclerosis 141:1–15
76. Klebanoff SJ (2005) Myeloperoxidase: friend and foe. J Leukoc Biol 77:598–525
77. Leeuwenburgh C, Hardy MM, Hazen SL, Wagner P, Oh-ishi S, Steinbrecher UP, Heinecke JW (1997) Reactive nitrogen intermediates promote low density lipoprotein oxidation in human atherosclerotic intima. J Biol Chem 272:1433–1436
78. Hazell LJ, Arnold L, Flowers D, Waeg G, Malle E, Stocker R (1996) Presence of hypochlorite-modified proteins in human atherosclerotic lesions. J Clin Invest 97:1535–1544
79. Hazell LJ, Stocker R (1993) Oxidation of low-density lipoprotein with hypochlorite causes transformation of the lipoprotein into a high-uptake form for macrophages. Biochem J 290:165–172
80. Hazen SL, Heinecke JW (1997) 3-Chlorotyrosine, a specific marker of myeloperoxidase-catalyzed oxidation, is markedly elevated in low density lipoprotein isolated from human atherosclerotic intima. J Clin Invest 99:2075–2081
81. Kostyuk VA, Kraemer T, Sies H, Schewe T (2003) Myeloperoxidase/nitrite-mediated lipid peroxidation of low-density lipoprotein as modulated by flavonoids. FEBS Lett 537:146–150
82. Dunford HB (ed) (1999) Heme peroxidases. Wiley-VCH: New York and Toronto
83. Allegra M, Furtmüller PG, Jantschko W, Zederbauer M, Tesoriere L, Livrea MA, Obinger C (2005) Mechanism of interaction of betanin and indicaxanthin with human myeloperoxidase and hypochlorous acid. Biochem Biophys Res Commun 332:837–844
84. Halliwell B, Gutteridge JMC (1999) Free radical in biology and medicine, 3rd edn. Oxford University Press: New York
85. Chiu DT, van den Berg J, Kuypers FA, Hung IJ, Wie JS, Liu TZ (1996) Correlation of membrane lipid peroxidation with oxidation of hemoglobin variants: possibly related to the rates of hemin release. Free Radic Biol Med 21:89–95
86. Grinberg LN, Rachmilewitz EA, Kitrossky N, Chevion M (1995) Hydroxyl radical generation in beta-thalassemic red blood cells. Free Radic Biol Med 18:611–615
87. Scott MD, van den Berg JJ, Repka T, Rouyer-Fessard P, Hebbel RP, Beuzard Y, Lubin BH (1993) Effect of excess alpha-hemoglobin chains on cellular and membrane oxidation in model beta-thalassemic erythrocytes. J Clin Invest 91:1706–1712
88. Van Dyke B, Saltman P (1996) Hemoglobin: a mechanism for the generation of hydroxyl radicals. Free Radic Biol Med 20:985–989

89. Das N, Chowdhury TD, Chattopadhyay A, Datta AG (2004) Attenuation of oxidative stress-induced changes in thalassemic erythrocytes by vitamin E. Polish J Pharmacol 56:85–96
90. Grinberg LN, Rachmilewitz EA, Newmark H (1994) Protective effects of rutine against hemoglobin oxidation. Biochem Pharmacol 48:643–49
91. Tesoriere L, D'Arpa D, Butera D, Allegra M, Renda D, Maggio A, Bongiorno A, Livrea MA (2001) Oral supplements of vitamin E improve measures of oxidative stress in plasma and reduce oxidative damage to LDL and erythrocytes in beta-thalassemia intermedia patients. Free Radic Res 34:529–540
92. Stocker R, Keany J Jr (2004) Role of the oxidative modifications in atherosclerosis. Physiol Rev 84:1381–1478
93. Lusis AJ (2000) Atherosclerosis. Nature 407:233–241
94. Moore KJ, Tabas I (2011) Macrophages in the pathogenesis of atherosclerosis. Cell 145: 341–355
95. Lonn ME, Dennis JM, Stocker R (2012) Actions of "antioxidants" in the protection against atherosclerosis. Free Radic Biol Med 53:863–884
96. Badimon L, Martinez-Gonzalez J, Llorente-Cortes V, Rodriguez C, Padro T (2006) Cell biology and lipoproteins in atherosclerosis. Curr Mol Med 5:439–456
97. D'Alessio P (2004) Aging and the endothelium. Exp Gerontol 36:165–171
98. Loyd-Jones D, Adams RJ, Brown TM, Carnethon M, Dai S, De Simone G, Ferguson TB, Ford E, Furie K, Gillespie C (2010) American Heart Association Statistics Committee and Stroke Statistics Subcommittee. 2010. Update. Circulation 121:948–954
99. D'Alessio P (2002) Endothelium as a pharmacological target. Curr Op Invest Drugs 2: 1720–1724
100. Carman CV, Jun CD, Salas A, Springer TA (2003) Endothelial cells proactively form microvilli-like membrane projections upon intracellular adhesion molecule-1 engagement of leukocyte LFA-1. J Immunol 171:6135–6144
101. Brown AJ, Jessup W (1999) Oxysterols and atherosclerosis. Atherosclerosis 142:1–28
102. Larsson DA, Baird S, Nyhalah JD, Yuan XM, Li W (2006) Oxysterol mixtures, in atheroma-relevant proportions, display synergistic and proapoptotic effects. Free Radic Biol Med 41:902–910
103. Allegra M, Rattazzi A, Attanzio A, Tesoriere L, Livrea MA, D'Acquisto F (2012) Modulation of TH1/TH17 equilibrium in vitro by indicaxanthin from *Opuntia ficus indica* (L. Mill). Free Radic Biol Med 53 (Suppl 1):S47
104. Baldassano S, Tesoriere L, Rotondo A, Serio R, Livrea MA, Mulè F (2010) Inhibition of the mechanical activity of mouse ileum by cactus pear (*Opuntia Ficus Indica*, L, Mill.) fruit extract and its pigment indicaxanthin. J Agric Food Chem 58:7565–7571
105. Baldassano S, Rotondo A, Serio R, Livrea MA, Tesoriere L, Mulè F (2011) Inhibitory effects of indicaxanthin on mouse ileal contractility: analysis of the mechanism of action. Eur J Pharmacol 658:200–205
106. Lin HJ, Zuo T, Chao JR, Peng Z, Asamoto LK, Yamashita SS, Huang TH (2009) Seed in soil, with an epigenetic view. Biochim Biophys Acta 1790:920–924
107. Dolinoy DC Weidman JR, Jirtle RL (2007) Epigenetic gene regulation: linking early developmental environment to adult disease. Reprod Toxicol 23:297–307
108. Esteller M (2008) Epigenetics in cancer. N Engl J Med 358:1148–1159
109. Huang J, Plass C, Gerhauser C (2011) Cancer chemoprevention by targeting epigenome. Curr Drug Targets 12:1925–1956
110. Malireddy S, Kotha SR, Secor JD, Gurney TO, Abbott JL, Maulik G et al (2012) Phytochemical antioxidants modulate mammalian cellular epigenome:implications in health and disease. Antiox Redox Signal 17:327–339
111. Moriarty-Craige SE, Jones DP (2004) Extracellular thiols and tiol/disulfide redox in metabolism. Annu Rev Nutr 24:481–509
112. Martínez-González MA, García-López M, Bes-Rastrollo M, Toledo E, Martínez-Lapiscina EH, Delgado-Rodriguez M, Vazquez Z, Benito S, Beunza JJ (2011) Mediterranean diet and the incidence of cardiovascular disease: a Spanish cohort. Nutr Metab Cardiovasc Dis 21:237–244

113. Sofi F, Cesari F, Abbate R, Gensini GF, Casini A (2008) Adherence to Mediterranean diet and health status: meta-analysis. BMJ 337:a1344
114. Naselli F, Modica M, Attanzio A, Caradonna F, Gentile C, Tesoriere L, Livrea MA (2012) Pro-apoptotic activity of the phytochemical Indicaxanthin on colorectal carcinoma cells (Caco-2) and epigenetic CpG demethylation of the promoter and reactivation of the expression of p16. 56th National Meeting of the Italian Society of Biochemistry and Molecular Biology, Chieti
115. Draht MX, Riedl RR, Niessen H, Carvalho B, Meijer GA, Herman JG, van Engeland M, Melotte V, Smits KM (2012) Promoter CpG island methylation markers in colorectal cancer: the road ahead. Epigenomics 4:179–194
116. Matsuda Y (2008) Molecular mechanism underlying the functional loss of cyclin-dependent kinase inhibitors p16 and p27 in hepatocellular carcinoma. World J Gastroenterol 14:1734–1740
117. Zang JJ, Xie F, Xu JF, Qin YY, Shen RX, Yang JM, He J (2011) P16 gene hypermethylation and hepatocellular carcinoma: a systematic review and meta-analysis. World J Gastroenterol 17:3043–3048
118. Shima K, Nosho K, Baba Y, Cantor M, Meyerhardt JA, Giovannucci EL, Fuchs CS, Ogino S (2011) Prognostic significance of CDKN2A (p16) promoter methylation and loss of expression in 902 colorectal cancers: cohort study and literature review. Int J Cancer 128:1080–1094
119. von Elbe JH, Klement JT, Amundson CH, Cassens RG, Lindsay RC (1974) Evaluation of betalain pigments as sausage colorants. J Food Sci 39:128–132
120. von Elbe JT, Schwartz SJ (1981) Absence of mutagenic activity and a short term toxicity study of beet pigments as food colorants. Arch Toxicol 49:93–98
121. Patkai G, Barta J (1997) Decomposition of anticarcinogen factors of the beetroot during juice and nectar productions. Canc Lett 114:105–106
122. Reynoso R, Garcia FA, Morales D, de Mejia Gonzales E (1997) Stability of betalain pigments from a Cactacea fruit. J Agric Food Chem 45:2884–2889
123. Haveland-Smith RB (1981) Evaluation of the genotoxicity of some natural food colours using bacterial assays. Mutat Res 91:285–289
124. Reynoso RC, Giner TV, de Mejia Gonzales E (1999) Safety of a filtrate of fermented Garambullo fruit: biotransformation and toxicity studies. Food Chem Toxicol 37:825–830
125. Trachootham D, Lu W, Ogasawara MA, Rivera-Del Valle N, Huang P (2008) Redox regulation of cell survival. Antiox Redox Signal 1:1343–1347
126. Fruehauf JP, Meyskens FL Jr (2007) Reactive oxygen species: a breath of life or death? Clin Cancer Res 13:789–794
127. Lambert JD, Elias RJ (2010) The antioxidant and pro-oxidant activities of green tea polyphenols: a role in cancer prevention. Arch Biochem Biophys 501:65–72
128. Dianzani U (2003) 4-Hydroxynonenal from pathology to physiology. Mol Aspects Med 24:213–218
129. Higdon A, Diers AR, Oh JY, Landar A, Darley-Usmar VM (2012) Cell signaling by reactive lipid species: new concepts and molecular mechanisms. Biochem J 442:453–464
130. Jiang F, Zhang Y, Dusting GJ (2011) NADPH-mediated redox signaling: roles in cellular stress response, stress tolerance, and tissue repair. Pharmacol Rev 63:218–242
131. Bedard K, Krause K-H (2007) The NOX family of ROS-generating NADPH oxidases: physiology and pathophysiology. Physiol Rev 87:245–313
132. Leto TL, Morand S, Hurt D, Ueyama T (2009) Targeting and regulationof reactive oxygen species generation by Nox family NADPH oxidases. Antioxid Redox Signal 11:2607–2619
133. Schilder Y, Heiss EH, Schachner D, Ziegler J, Reznicek G, Sorescu D, Dirsch VM (2009) NADPH oxidases 1 and 4 mediate cellular senescence induced by resveratrol in human endothelial cells. Free Radic Biol Med 46:1598–1606
134. Konishi T (2011) From herb to kitchen and bedside: food factors are pharmacological molecules with antioxidant activity. Free Radic Res 45:863–864
135. Moreno DA, Garcia-Viguera C, Gil JI, Gil-Izquierdo A (2008) Betalains in the era of global agri-food science, technology and nutritional health. Phytochem Rev 7:261–280

Chapter 8
Pigments in Citrus

Chunxian Chen, PhD, Angela Roberta Lo Piero, and Fred Gmitter Jr.

Introduction

Citrus fruits are rich with carotenoid and flavonoid pigments, vitamin C, sugars, acids, volatiles, microelements, and other nutrients [1–6]. The characteristic aromatic flavors and healthy nutraceutical compounds in various citrus fruits promote regular consumption of fresh fruits or juice, and special use in beverage blending, culinary seasoning, or herb medication as well [4, 7–12]. As one of the highest-valued fruit crops, usually evergreen citrus trees have small white fragrant five-petaled flowers and are grown commercially in around 140 countries. The *Citrus* genus belongs to the Rutaceae family and its taxonomy has been very controversial [13, 14]. Domesticated *Citrus* cultivars have evolved into diverse edible fruit types with distinct characteristics and origins (Fig. 8.1), including sweet orange (*Citrus sinensis*), mandarin (*Citrus reticulata*), pummelo (*Citrus maxima*), grapefruit (*Citrus paradisi*), sour (bitter) orange (*Citrus aurantium*), lime (*Citrus aurantifolia*), lemon (*Citrus limon*), and citron (*Citrus medica*), to name a few. Most of them are well-domesticated and widely planted, and all classified into one subgenus, also called *Citrus*, which is distinguished from *Papeda,* the other less noticed wild subgenus in the *Citrus* genus. Additionally, citrus as a general term also includes species in *Fortunella* (kumquat, a small fruit whose peel is sweet and edible), *Poncirus* (trifoliate orange, a widely used citrus rootstock), and other relatively distant genera in the family. Sweet oranges account for about 60–70% of the total citrus production worldwide and are mostly used for juice processing [13]. From both taxonomic

C. Chen (✉)
Citrus Research and Education Center, University of Florida,
Lake Alfred, Florida, USA
e-mail: chencx2k@gmail.com

F. Gmitter Jr.
Citrus Research and Education Center, University of Florida, Lake Alfred, FL 33850, USA

A. R. Lo Piero
Department of Agricultural and Food Science, University of Catania, Catania, Italy

C. Chen (ed.), *Pigments in Fruits and Vegetables,* DOI 10.1007/978-1-4939-2356-4_8

Fig. 8.1 Mature citrus fruits of various types, sizes, colors, and seediness, using selected cultivars as examples and displayed in full (**a**) and crosscut (**b**) in the same order. *Top*: "Siamese" pummelo (*C. maxima*); "Duncan" grapefruit (*C. paradisi*); "Flame" grapefruit; *middle*: "Eureka" lemon (*C. limon*); "Tahiti" lime (*C. aurantifolia*); "Sunburst" tangerine (*C. reticulata*); *bottom*: "Cara Cara" *navel orange* (*C. sinensis*); "Glen" *navel orange*; "Valencia" (common) *orange*; "Sanguinelli" *blood orange*. All the colors depend on the concentration of various carotenoids. High levels of lycopene yield deep red flesh color in "Flame" and "Cara Cara", whereas the *red* color in "Sanguinelli" is from anthocyanins, which in Florida are barely induced and accumulated to the "blood" levels commonly seen in other climatically suitable regions such as Sicily, Italy. (Courtesy of Chunxian Chen)

and genetic perspectives, it was widely believed that only one or a few ancestors each from mandarin, pummelo, and citron might be the true biological species of these cultivated *Citrus*. The others were evidently natural ancient hybrids or introgressions, according to morphological traits and molecular markers [14–16]. For example, sweet orange is believed to be a natural ancient backcross introgression of mandarin and pummelo, grapefruit as a natural F1 hybrid of pummelo and sweet orange, and varied lemons/limes as natural offspring of citron and other accessions. Each citrus type includes many cultivars. Except for a few conventional hybrids, most cultivars are spontaneous bud sports (mutants), chance seedlings, or induced mutants with at least one distinctly improved fruit trait, such as red flesh color, late (or early) maturity, seedlessness, and so on. Such improvements of certain traits via the latter three routes are demonstrated in the selection of grapefruit cultivars [17], for example, from pale white to dark red flesh color, and from many to zero seeds (Fig. 8.1b). Deepening orange to red colors, particularly the red of accumulated lycopene or anthocyanins, are one of most important quality attributes of citrus fruits and juice, generally representing better market appeal and value, more potential health benefits, and greater research interests as well.

Other categorizations of citrus and citrus fruits are also used, according to commonly shared characteristics and/or origin within a fruit type. For example, navel oranges (a second fruit grown inside at the apex, like a navel), red oranges (lycopene accumulated), blood oranges (anthocyanins accumulated), acidless oranges (extremely low acidity), and Valencia oranges (aka. summer oranges, ripening during spring and summer) are the mutant types of common sweet oranges; blond oranges is a term sometimes used as a counterpart to blood oranges to refer to all oranges not containing lycopene or anthocyanin. Clementine, satsuma, and other mandarins may be derived from distinct parentages and/or domestications. These diverse mandarins are relatively easy to peel and primarily for fresh consumption, and therefore also called "loose-skin citrus." The other types of citrus fruits, including sweet oranges, grapefruits, pummelos, lemons, limes, and so on, are generally difficult to peel (together called "tight-skin citrus") and may be eaten fresh, squeezed to juice, added to beverages, or used for culinary purposes. Due to the acid content, limes and lemons and the like are called "acid citrus," which are primarily used for beverage flavoring and culinary seasoning.

Unlike most well-domesticated and widely planted *Citrus* spp., *Papeda* spp. and their hybrids are at more wild status, and their fruits are barely edible due to unpleasant off-flavors (bitter, sour, etc.). They are mostly used as localized culinary flavoring and herbal medicine [4]. Kumquat fruits are small, attractive, and edible, for fresh consumption or culinary use, and very popular in certain Asian countries. Interestingly, the rind of kumquat fruits is the primary tissue to eat (the whole fruit can be eaten) because sugar is predominantly accumulated there, different from oranges and mandarins in which sugar accumulates primarily in the juice vesicles. *Poncirus* fruits are inedible but the seedlings from the species and its hybrid selections have been among the most popular rootstocks upon which citrus scions are grafted, because they possess many unique resistance or tolerance genes not found in *Citrus* [13, 14]. Such rootstocks not only can help protect more vulnerable *Citrus* scions from many biotic and abiotic stresses, including citrus tristeza virus (CTV), *Phytophthora* root rot, citrus nematode, and cold hardiness but also can impact the yield and quality of fruits produced. These genes and traits are also the targets for genetic characterization and manipulation, albeit difficult, to improve citrus varieties through conventional introgression and/or genetic engineering.

Carotenoids and flavonoids are the two rich pigment families in citrus fruits [2, 5, 18–22]. In citrus fruits, many carotenoids are characteristically colorful while most flavonoids are colorless except anthocyanins [6, 20, 23, 24]. Therefore, carotenoids are the primary pigments to produce a wide spectrum of characteristic colors, from pale yellow, orange, pink to deep red in different varieties, except in blood oranges where anthocyanins are cold induced and synthesized for additional "blood" color [18, 25, 26]. The actual colors of mature citrus fruits vary greatly among fruit types and depend on the composition and concentration of various colorful carotenoids (Figs. 8.1 and 8.2). In general, phytoene is colorless, while α-carotene, β-cryptoxanthin, and zeaxanthin may contribute more toward yellow, β-carotene, lutein, and violaxanthin toward orange, and lycopene toward red in a range of light pink to dark red depending on its accumulated concentrations [2, 5, 24, 27–31].

Pigments, particularly lycopene and anthocyanins that are able to deepen citrus fruit color into red, are most widely studied from phytochemical, biosynthetic, genetic, genomic, and dietetic perspectives. A great deal of progress has been made over the past decades to characterize the diverse pigment profiles among different citrus types and mutants, to uncover the genes and mechanisms involved in the accumulation of different colored pigments via biosynthesis pathways, and to validate the health benefits or medicinal functions of these pigments in the human diet.

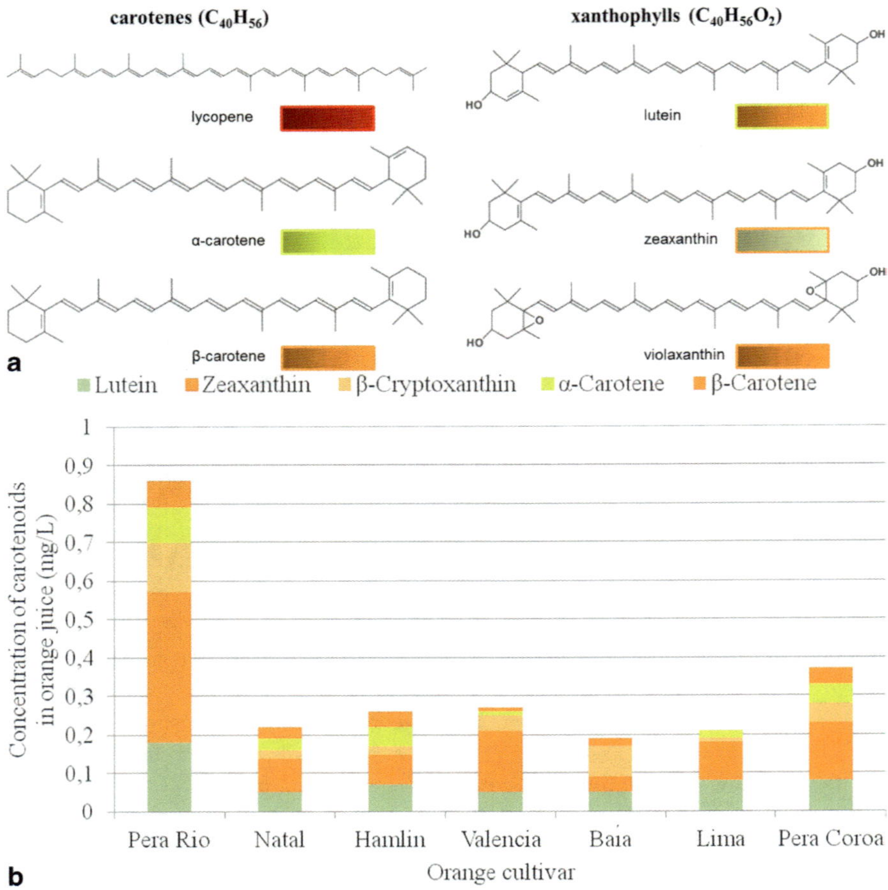

Fig. 8.2 The chemical structures of some representative carotenoids in the carotene (molecular formula, $C_{40}H_{56}$) and xanthophyll ($C_{40}H_{56}O_2$) subgroups (**a**), in which the six chemical diagrams are adapted from a review of Melendez-Martinez et al. [136], and the concentration range of main carotenoids in orange juice from selected cultivars (**b**), in which the data are based from a table of Pupin et al. [2] and re-presented in the current form. The color used for each type of carotenoid is only symbolic and may not be relevant to their respective color in a purified state. (**a**: Adapted from Melendez-Martinez AJ, Vicario IM, Heredia FJ 2007. With permission from Elsevier)

Carotenoids in Citrus Fruits

The Basics of Carotenoids

Carotenoids in plants, algae, and some bacteria are water-insoluble pigments primarily working with chlorophylls in photosynthesis [32–35], but in humans, marines, birds, and other animals they are food-derived functional secondary metabolites [36–38]. As a family of more than 600 chemicals, most carotenoids are structurally tetraterpenoids (containing four terpene units) that can be categorized into two subgroups: carotenes ($C_{40}H_{56}$) and their oxygenated derivatives, xanthophylls ($C_{40}H_{56}O_2$) [34, 37]. They generally share very similar polyene hydrocarbon chain structures but differ in the terminal rings and oxygen atoms (Fig. 8.2a). The most common carotenes responsible for fruit and flower colors are α-carotene, β-carotene, and lycopene; and common xanthophylls are lutein, zeaxanthin, and violaxanthin [2, 6, 24, 27, 28]. In plants, carotenoids also serve as potent antioxidants against photooxidation, as essential precursors for biosynthesis of certain hormones, as attractive colorants for flower pollination and fruit/seed dispersal, and/or as responsive substances for environmental adaptation [32–34]. In humans and animals, some dietary carotenoids are essential precursors of vitamin A, and others may function as most potent antioxidants and nutraceuticals to enhance cell immune response, inhibit induced mutagenesis, and reduce various neoplastic events [36–42].

Carotenoids have been quantified within many citrus fruit types in various locations and conditions. The quantity and composition of different carotenoids, i.e., the characteristic colors of each citrus fruit type, are primarily determined by genotype and fruit maturity, although they are also affected to some extent by other environmental factors, such as physical injury, insect damage, temperature, horticultural practices, storage, and so on. More than 115 carotenoids were found in citrus, and their content and composition vary greatly among citrus varieties and mutants, resulting in diverse mature fruit colors [2, 5, 19, 21, 24, 27, 29, 43–46].

Carotenoid Biosynthesis Genomics in Citrus Fruits

Most carotenoid biosynthesis genes have been identified and the pathway well-characterized in model plants and major crops [32, 35], and in citrus as well [28, 43, 47]. All the genes encoding catalytic enzymes in plants are from nuclear genomes, while the biosynthesis process takes place in plastids, as shown in a schematic diagram of the main steps in the carotenoid biosynthesis pathway (Fig. 8.3; [28, 48–50]). The messenger RNA (mRNA) levels of phytoene synthase (PSY), phytoene desaturase (PDS), ζ-carotene desaturase (ZDS), β-ring hydroxylase (CHYB) increased in flavedo and juice sacs during the massive accumulation of chromoplast-specific carotenoids [43, 48]. The copy numbers and allele polymorphisms of carotenoid biosynthesis genes appear to be related to the variability of carotenoid content and composition among citrus cultivars and species [46, 48, 50–52]. Moreover, citrus apparently possesses more and diverse carotenoid biosynthesis gene members, ranging from 2 to 14, compared to other plants [50], which might be relevant to the fact that citrus contain much more diverse types of carotenoids than any other fruits [53]. Such large expansions of gene families with roles in the carotenoid biosynthesis might occur and result from its long domestication and cultivation history, during which selection driven by fruit visual attractiveness and internal flavor/color quality has never ceased [13, 17]. Intriguingly, most upstream genes (e.g., PSY, PDS) have fewer expressed copies than downstream genes (e.g, lycopene β-cyclase (LCYB); CHYB; capsorubin–capsanthin synthase (CCS)) in citrus [50], suggesting

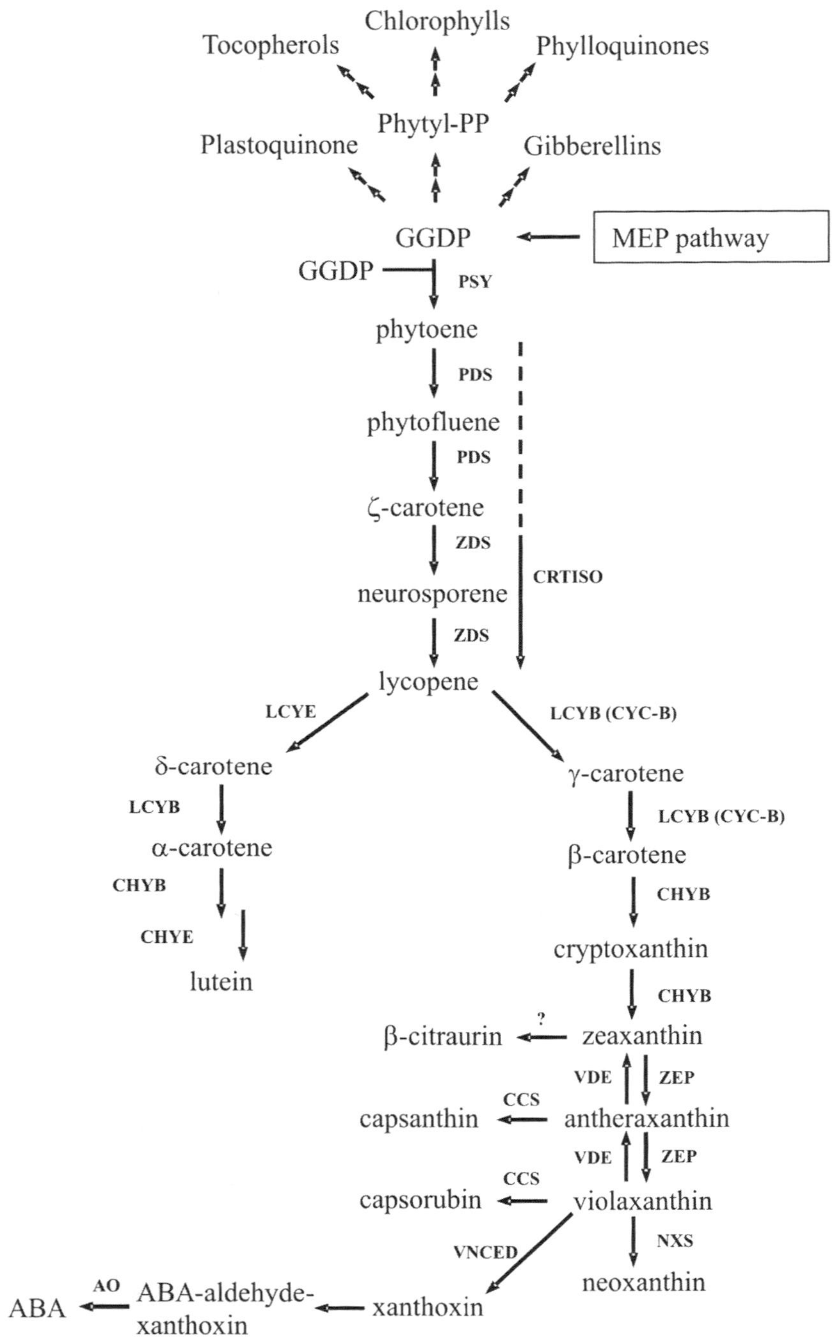

Fig. 8.3 The carotenoid biosynthesis pathway. The abbreviated genes/enzymes are: *PSY* phytoene synthase, *PDS* phytoene desaturase, *ZDS* ζ-carotene desaturase, *CRTISO* carotenoid isomerase, *LCYE* lycopene ε-cyclase, *LCYB* (CYC-B) lycopene β-cyclase, *CHYE* ε-ring hydroxylase, *CHYB* β-ring hydroxylase, *ZEP* zeaxanthin epoxidase, *VDE* violaxanthin de-epoxidase, *CCS* capsorubin-capsanthin synthase, *NXS* neoxanthin synthase, *VNCED* 9-cis-epoxycarotenoid dioxygenase, *AO* aldehyde oxidase. The other abbreviations are *MEP* methylerythritol 4-phosphate, *GGDP* geranyl-geranyl diphosphate, *ABA* abscisic acid

that downstream loci tend to duplicate and evolve more quickly (nucleotide changes at higher rates) because they are under more relaxed constraints and less pleiotropic consequences [54].

Comparative Genomics in Lycopene-Enriched (Red-Fleshed) Fruits

Many citrus cultivars are selected as a consequence of spontaneous mutations leading to altered carotenoid profiles. Some of these cultivars have unusually high lycopene accumulation that results in deep red flesh, and these have been top choices to study the molecular mechanisms of the flux of the relevant pigments, including "Cara Cara" navel orange [30, 31, 55–57], "Hong Anliu" sweet orange [58–62], and "Flame" and other red grapefruits [19, 63, 64]. They are also widely cultivated in certain regions primarily for fresh consumption because of visual appeal and the presumed health benefits of lycopene and sweeter taste in red grapefruits [65, 66]. "Cara Cara" accumulates higher amounts of lycopene and substantial phytoene, phytofluene, and β-carotene in the pulp, which are likely associated with higher expression of some upstream isoprenoid genes in the methyl erythritol phosphate (MEP) pathway and downstream phytoene desaturase (PDS) in the carotenoid pathway. Increased levels of isoprenoid precursors in the mutant likely are relevant to lycopene accumulation [30]. A newly isolated gene of LCYB, *CsLCYB2*, showed an allele-differential, chromoplast-specific expression, and a marked induction in both peel and pulp of oranges and grapefruits in parallel with the accumulation of β,β-xanthophylls. It was different from *CsLCYB1* that was expressed at low levels and remained relatively constant during fruit ripening [67]. One allele of *CsLCYB2*, expressed only in normally pigmented fruits, has normal LCYB activity, while the other, preferentially expressed in red-fleshed grapefruit, loses that activity and likely leads to the lycopene accumulation [63, 67]. "Hong Anliu" has a 1000-fold higher lycopene accumulation in the albedo, segment membranes, and juice sacs than its wild type, but no change of carotenoid composition in leaves. Lycopene accumulation in juice sacs is regulated by coordinate expression of carotenoid biosynthetic genes, and likely provides induced feedback to synthesize in and/or transport to albedo and segment membranes [58].

Recently, different gene expression patterns were observed in "Cara Cara" and "Hong Anliu" fruits, indicating different metabolic mechanisms involved in their respective carotenoid pathways [60]. Further global transcriptional and small RNA comparisons suggest that enhanced photosynthesis and partial impairment of lycopene downstream flux are critical for the lycopene accumulation trait in the mutant, and nine microRNAs (miRNAs) are differentially expressed between the mutant and wild types [61, 62]. Differences in genomic, transcriptional, post-transcriptional, and proteomic levels have been found between these red and their normal types in many reports, suggesting lycopene accumulation may be controlled and regulated in more complex manners and mechanisms; further investigations may provide a more decisive final conclusion [30, 31, 56, 57, 59–63, 66, 67].

Flavonoids and Anthocyanins in Citrus Fruits

The Basics of Flavonoids and Anthocyanins

Flavonoids are a much more functionally and structurally diverse family of more than 9000 water-soluble polyphenolic derivatives [23, 68]. Flavonoids share a basic 15-carbon benzo-γ-pyrone (C6–C3–C6) flavonoid skeleton structure, i.e., one heterocyclic benzopyran ring (ring C), one fused aromatic ring (A), and one phenyl constituent (B), as illustrated in Fig. 8.4 [69]. Hydroxyl, methoxyl, O-glycoside, or other substitutional groups added in various positions and combinations result in at least nine major subgroups: colorless chalcones, aurones, isoflavonoids, flavones, flavonols, and flavandiols, and colored anthocyanins (anthocyanidins), condensed tannins (proanthocyanidins), and phlobaphenes [23, 68]. Along with their natural ubiquity and structural diversity, flavonoids have various biological functions in plants, humans, and other organisms [23, 70]. In plants, for examples, colorful flavonoids in flowers/fruits attract insects and other animals to pollinate flowers and eventually to disperse seeds. Flavonoids with antioxidant potency play an essential role in protection against oxidation and ultraviolet (UV) damage [23]. Flavonoids with antimicrobial or defensive functions help plants to defend against or adapt to biotic and abiotic stresses such as UV light damage, cold, and pathogen attacks [71–73]. Flavonoids secreted by the host help to signal the plant–pathogen interaction, such as to establish symbiotic relationship between Rhizobia and legumes at the infection stage to initiate the nitrogen fixation [72]. Flavonoids, rich in many plant foods and beverages, are also among the most beneficial dietary compounds with their primary function as potent antioxidants in humans and animals, from which great benefits have been demonstrated in many studies [74–82].

Anthocyanins are glycosylated or acylglycosylated anthocyanidins differing in hydroxyl or methoxyl substitutions in the basic C6–C3–C6 structure [83, 84]. Both the hue and color stability are directly affected by the hydroxylation and methylation pattern of the B ring. Blueness is enhanced with the increasing of free hydroxyl groups, whereas redness intensifies with the methylation of the hydroxyl groups. In

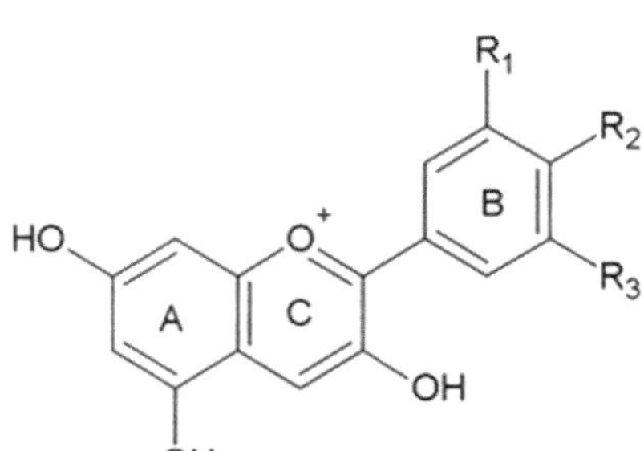

Anthocyanidin	R1	R2	R3
Delphinidin	OH	OH	OH
Cyanidin	OH	OH	H
Petunidin	OH	OH	OCH_3
Pelargonidin	H	OH	H
Peonidin	OCH_3	OH	H
Malvidin	OCH_3	OH	OCH_3

Fig. 8.4 Structure of 3,5,7,4′-tetrahydroxylflavilium ion and of the six basic anthocyanins. (Reprinted with permission from Dugo et al. [69]. Copyright 2003, American Chemical Society)

most plants, O-glycosylation of anthocyanidins occurs to produce anthocyanins in which the sugar moiety is typically glucose, or occasionally other sugars such as galactose, rhamnose, arabinose, or xylose.

Structural Diversity of Anthocyanins in Blood Oranges

Four subgroups of flavonoids (flavanones, flavones, flavonols, and anthocyanins) are found in citrus fruits [85]. Anthocyanins are cold induced and only present in several blood orange cultivars, which are mutants derived from blond common sweet oranges independently occurring in China, Italy, and other Mediterranean regions at least several hundred years ago [86]. For example, "Doppio Sanguigno" was reportedly the first mutant found in Sicily, Italy, and produced several relatively modern varieties, such as "Tarocco" and "Moro," which generally contain higher levels of anthocyanins. Anthocyanins, primarily present in juice sac vesicles and/or rind, vary in the amount and composition due to genotype, maturity, growing regions, and other environmental factors [69, 87–89]. All blood orange varieties require a broad day–night thermal range for deep red pigmentation in fruit flesh [86], which geographically limits reliable quality commercial production to only a few countries/regions, such as around Mount Etna in Sicily [87, 89].

Blood orange juices contain structurally diverse anthocyanins with seasonal concentration variations among varieties ([26, 69]; Fig. 8.5). The two major anthocyanins in blood oranges (ca. 90%) were identified as cyanidin 3-glucoside [90–93] and cyanidin 3-(6′-malonylglucoside) [93], and confirmed in later studies as well [18, 26, 69]. Several minor anthocyanin components were also identified by Maccarone and coworkers' pioneering work [91, 93], and later other contributors [26, 69]. Interestingly, cyanidin 3-(6′-dioxalylglucoside) and four other anthocyanin-derived pigments are formed through a direct reaction between anthocyanins and hydroxycinnamic acids only during prolonged storage of blood orange juice; these are novel pyranoanthocyanins with an additional pyran ring between the C4 and the hydroxyl group attached to C5 [26].

Biosynthesis of Anthocyanins in Blood Oranges

Anthocyanins are synthesized via the ubiquitous and well-described flavonoid pathway in plants [23, 94]. Both enzymatic and regulatory genes are required for anthocyanin biosynthesis, most of which have been cloned and characterized in various species [23]. The former encode enzymes directly catalyzing metabolic reactions, and the latter are transcription factors controlling the expression of enzymatic genes [95–97]. In grape, upstream enzymes, involved in diverse metabolic functions in the flavonoid pathway, are usually encoded by larger gene families whereas downstream enzymes are encoded by single genes [98]. This seems to be applicable to blood orange, according to relevant studies on these genes involved in the

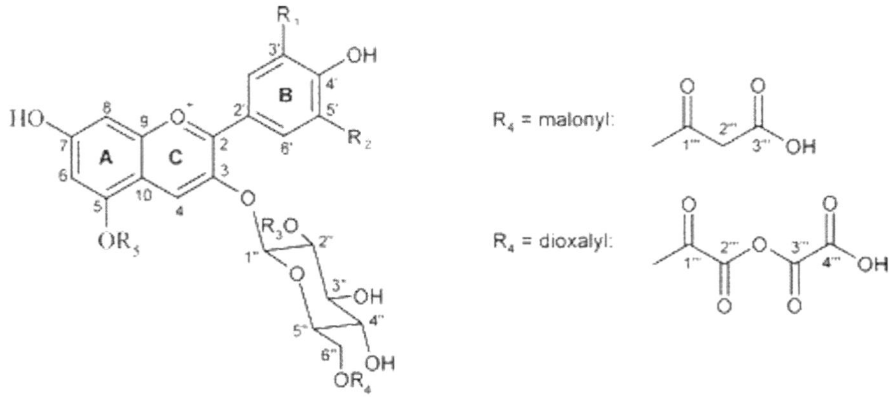

compd	R_1	R_2	R_3	R_4	R_5
1	OH	H	H	H	glucosyl
2	OH	OH	H	H	H
3	OH	H	glucosyl	H	H
4	OH	H	H	H	H
5	OH	OH	H	malonyl	H
6	OH	H	H	malonyl	H
7	OH	H	H	dioxalyl	H
8	OCH_3	H	H	malonyl	H

For compound labeling cf. Figure 1

Fig. 8.5 The structures of anthocyanins from blood orange juice: cyanidin 3,5-diglucoside (*1*), delphinidin 3-glucoside (*2*), cyanidin 3-sophoroside (*3*), cyanidin 3-glucoside (*4*), delphinidin 3-(6″-malonylglucoside) (*5*), cyanidin 3-(6″-malonylglucoside) (*6*), cyanidin 3-(6″-dioxalylglucoside) (*7*), and peonidin 3-(6″-malonylglucoside) (*8*). (Reprinted with permission from Hillebrand et al. [26]. Copyright 2004, American Chemical Society)

flavonoid and anthocyanin biosynthesis in blood oranges and other citrus cultivars [87, 99–103].

Anthocyanins are synthesized by an extremely complex network in the pathway, in which a multienzyme complex is formed and regulated by transcriptional factors to carry out the biosynthesis, as demonstrated in some model plants [104]. Phenylalanine is a direct precursor for the synthesis of anthocyanidins. The conversion of phenylalanine to anthocyanins requires a series of reactions, as illustrated in Fig. 8.6. First, transformation of phenylalanine to *trans*-cinnamic acid through the elimination of ammonia is catalyzed by phenylalanine ammonia lyase (**PAL**) that has been extensively studied for its involvement in the responses to various plant stresses [87, 100]. Hydroxylation of cinnamic acid by cinnamate 4-hydroxylase (**C4H**) then generates *p*-coumaric acid, which is activated by 4-coumarate:CoA ligase (**4CL**) to the respective CoA ester. Chalcone synthase (**CHS**) condenses three malonyl-CoA molecules and one *p*-coumaroyl-CoA to produce the naringenin chalcone for the eventual biosynthesis of anthocyanin and other phenolic compounds [105]. As a consequence, different CHS genes may act in different pathways producing distinct secondary metabolites [106].

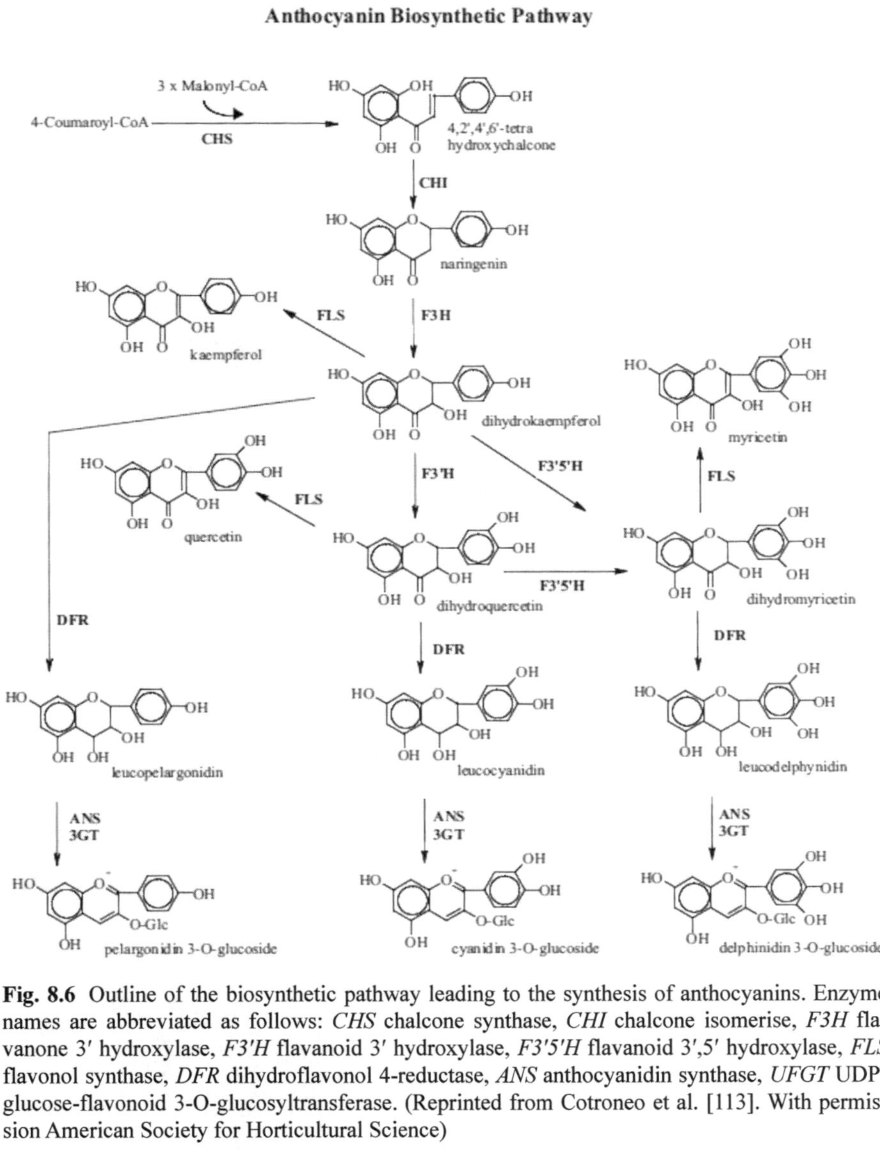

Fig. 8.6 Outline of the biosynthetic pathway leading to the synthesis of anthocyanins. Enzyme names are abbreviated as follows: *CHS* chalcone synthase, *CHI* chalcone isomerise, *F3H* flavanone 3′ hydroxylase, *F3′H* flavanoid 3′ hydroxylase, *F3′5′H* flavanoid 3′,5′ hydroxylase, *FLS* flavonol synthase, *DFR* dihydroflavonol 4-reductase, *ANS* anthocyanidin synthase, *UFGT* UDP-glucose-flavonoid 3-O-glucosyltransferase. (Reprinted from Cotroneo et al. [113]. With permission American Society for Horticultural Science)

Naringenin chalcone is isomerized by chalcone isomerase (**CHI**) to flavanone naringenin, which is subsequently converted to dihydrokaempferol by flavanone 3′-hydroxylase (**F3H**) [99, 107]. The crucial role of both CHS and F3H in the anthocyanin pigmentation was confirmed by proteomic profiling of ripening "Moro" (blood orange) and "Cadenera" (blond or common orange), which are differentially expressed and could be used to distinguish blood and common oranges [108].

Dihydroquercetin and dihydromyricetin, two dihydroflavonols, are synthesized from dihydrokaempferol by hydroxylation reactions catalyzed by flavonoid

3′-hydroxylase (**F3′H**) and flavonoid 3′,5′-hydroxylase, respectively [109]. Dihydroflavonol 4-reductase (**DFR**) can reduce the dihydroflavonols to their corresponding leucoanthocyanidins. The gene encoding DFR was isolated from both cDNA library (accession no. AY519363) and genomic DNA (DQ084722). The successful *in vitro* expression of the DFR complementary DNA (cDNA), along with relevant motifs found in its promoter region, confirmed its involvement in the biosynthesis of anthocyanins [102]. One of the motifs is a putative unit I, the homologous binding site for the G box factor and the downstream Myb homologous binding site involved in the transcriptional activation of DFR [110]. There are no differences in the *dfr* gene coding and promoter sequences between blood and blond oranges, but DFR transcripts are normally detected only in blood oranges. Therefore, its expression should be under strict control of transcriptional regulators [102].

Anthocyanidin synthase (**ANS**), an oxoglutarate-dependent Fe^{2+}/Fe^{3+} dioxygenase, converts colorless leucoanthocyanidins into colored anthocyanidins that, once formed, must be immediately modified as they are inherently unstable under physiological conditions (Fig. 8.6). Therefore, the uridine diphosphate (UDP)-glucose flavonoid glucosyl-transferase (**UFGT**) catalyzes the addition of one glucose moiety in the 3-OH positions of anthocyanidins increasing their hydrophilicity and stability. ANS and UFGT genes in Tarocco blood orange have been cloned by Lo Piero and coworkers (accession no. ANS, AY581048; UFGT, AY519364). The presence of di-glycosylated anthocyanins in blood orange juice [91, 93] indicates that other glucosylating enzymes might be involved in anthocyanin modification as it occurs in grape [111], although homologous genes have not yet identified in *Citrus*. Acylation is one of the most common modifications of blood orange anthocyanins, resulting in greatly increased structural diversity of anthocyanins from the addition of aromatic and/or aliphatic constituents linked to the C6″ positions of the glucosyl groups [93]. Anthocyanins, without the protection of acylation can be easily and quickly decolorized in neutral or weakly acidic aqueous solutions. The acylation of anthocyanins in plants is catalyzed by the action of anthocyanin acyltransferases (ACTs, also known as AATs), which have high substrate specificity for both the anthocyanin acceptors and the acyl group donors. In plants, there are mainly two types of ACTs that are classified based on the acyl group donors: the BAHD family using acyl-CoA and the serine carboxypeptidase-like (SCPL) group using acyl-activated sugars [112]. Although acylated anthocyanins can account for more than 40 % of the total anthocyanin content in some blood oranges, until now there has been no report about the exact genes required in blood orange anthocyanin acylation.

Regulation of Anthocyanin Biosynthesis in Blood Oranges

Anthocyanins are synthesized and accumulated only in blood orange flesh and rind, not in blond ones. Comparison of six anthocyanin biosynthetic genes (CHS, CHI, F3H, DFR, ANS, UFGT) revealed that CHS, CHI, and F3H, the genes involved in the early steps in the pathway, are expressed in juice vesicles of both oranges. On the contrary, DFR, ANS, and UFGT are at much lower or no detected expression in blond cultivars [103, 113]. Anthocyanin content increases during fruit ripening but varies greatly among blood orange cultivars [88, 113] and upon other genetic factors and environmental conditions such as temperature [87, 89, 114].

Several different families of regulatory genes in plants, such as Myb, Myc (encoding basic helix-loop-helix proteins (bHLH)) and WD40-like transcription factors, generally control the spatial and temporal expression of enzymatic genes in the anthocyanin biosynthesis pathway [23, 115, 116]. They can form a ternary complex (MYB-bHLH-WD40, MBW complex) in a hierarchical network to regulate the expression of multiple genes, such as CHI, CHS, F3H, and DFR, and play crucial roles in the regulation of anthocyanins and other flavonoid products, such as flavonols and proanthocyanidins [111]. In blood oranges, several anthocyanin biosynthetic genes show increased expression compared with blond oranges [101–103, 114, 117]. Variation in anthocyanin content and/or tissue specificity is largely governed by the activity of the R2R3 Myb transcription factor, named *Ruby*, in the complex. The *Ruby* cDNA encodes a 262-amino acid protein containing an R2R3 Myb domain with a signature motif for interaction with bHLH proteins from the clade 3f (DLX2RX3LX6LX3R) and it also has a conserved motif KPXPR(S/T)F in its C-terminal domain found in other R2R3 Myb regulators of anthocyanin biosynthesis. The ability of *Ruby* to activate anthocyanin biosynthesis was verified by its ectopic expression under the control of the constitutive cauliflower mosaic virus (CaMV) 35S promoter in tobacco (*Nicotiana tabacum*), where it resulted in visible purple-red pigmentation in undifferentiated callus and regenerated plant tissues [86]. *CsMyc*, the gene encoding the bHLH partner in the MWB complex [118], is expressed at detectable levels in both blond and blood oranges, supporting the hypothesis that *Ruby* regulates anthocyanin biosynthesis [86]. *Ruby* is only expressed in blood orange fruits, primarily flesh and rind at high levels, but not in blond oranges. Sequence analysis of the *Ruby* genes cloned from three blood ("Sanguinelli," "Maltaise Sanguine," and "Moro") and three blond ("Navelina," "Salustiana," and "Cadenera") varieties revealed 100% nucleotide identity in the three exons and two introns. The results suggested that *Ruby* transcription is strictly regulated by another element, as was later confirmed as a complete *copia*-type retrotransposon (Tcs1) inserted at the site of the 551 bp upstream from the initiating ATG of *Ruby* [86].

Fruit ripening stages affect transcriptional levels of biosynthetic genes and hence the amount of anthocyanins in blood oranges. The expression of CHS, ANS, and UFGT increased during the entire period and the levels of all three transcripts appeared to vary in a similar manner. Total anthocyanins were undetectable at the beginning but their levels rose at the end of November and showed a sharp increase between late January and middle February [113]. The anthocyanin levels also varied among varieties or selections grown under the same environmental conditions ranging from 34 mg/100 g in OTA 9 to 0.25 mg/100 g in "Tarocco Messina" [113].

Cool temperature during blood orange ripening is the key to induction and accumulation of anthocyanins. Lo Piero and coworkers have shown that post-harvest exposure to cold induces up to an eightfold increase of anthocyanins in blood orange flesh depending on the storage time, which is accomplished by the transcriptional stimulation of the genes involved in the anthocyanin biosynthesis including PAL, CHS, DFR, ANS, UFGT, and glutathione S-transferase (GST) [87, 101, 114]. Further transcriptomic study of Tarocco Sciara blood orange after exposure at 4 °C for 77 days revealed cold-enhanced transcriptions of genes involved in the defense mechanisms against oxidative damage, osmoregulating processes, and lipid desaturation as well as those implicated in the primary and secondary metabolisms, particularly toward the flavonoid biosynthesis pathway in blood oranges. It included those reactions involved in anthocyanin biosynthesis (CHS and ANS), and some upstream metabolic pathways such as the shikimate pathway leading to the biosynthesis of phenylalanine, the flavonoid "mother" molecule [89]. On the basis of the metabolic reactions in which some upregulated genes such as citrate lyase, phosphoenolpyruvate/phosphate translocator (PEP/Pi translocator), and phosphoenolpyruvate carboxykinase (PEPCK) are involved, it seems that they may act in a coordinated manner. In the cytosol, citrate lyase gives rise to oxaloacetate and acetyl CoA. This latter molecule might be further utilized to synthesize fatty acids, sterol, and isoprenoid compounds, or becomes part of the flavonoid skeleton through the reaction catalyzed by CHS (Fig. 8.6). Interestingly, among the genes encoding for regulatory proteins, several transcription factors have been identified for the first time as cold responsive genes in plants. In particular, a transcription factor belonging to the no apical meristem, *Arabidopsis* transcription activation factor, cup-shaped cotyledon (NAC) transcription factor family was found specifically induced in blood orange, but not in blond orange, thus proposing it as candidate gene involved in the signaling cascade leading to the response to cold [114]. However, the cold dependency of anthocyanin production in blood oranges might depend upon the cold induction of retroelement transcription as elevated levels of Tcs1 transcripts were observed in "Tarocco" and "Moro" blood varieties following storage of fruit in the cold, indicating that Tcs1 and Tcs1-like elements are activated by cold and ultimately might be involved in the temperature-dependent anthocyanin accumulation in blood oranges [86]. Interestingly, by comparing the blood orange response to cold stress with those of other plant sources, such as grapefruit [119], it seems to be similar to that of the chilling acclimated species. Therefore, it is likely that the rise of anthocyanin levels occurred in the oranges subjected to cold stress might contribute to the control of cell osmotic potential thus coping with the imposed stressful conditions [89].

Additional intrinsic and environmental factors have been linked to anthocyanin accumulation in several plant species [120, 121]. Nutrient deficiencies, especially of phosphorus (P) and nitrogen (N), commonly induce the accumulation of anthocyanins in many plant species. Also, exposure to lowered pH, methyl jasmonate, wounding, pathogen infection, and fungal elicitors may trigger a significant increase in the anthocyanin levels [120, 121]. Both auxins and/or cytokinins have been shown to induce anthocyanins in cell cultures or whole plants [120]. Abscisic acid (ABA) and ethylene promotes biosynthesis of anthocyanin in certain fruit tissues [121].

Dietetics of Citrus Fruits Pigments

Carotenoids and flavonoids, due to their antioxidant properties and nutritional benefits, have been some of the most studied dietary and therapeutic components in the dietetics and preventive medicine areas, yielding hundreds of publications and reviews [36, 39, 41, 65, 66, 122–126]. Much anecdotal or scientific evidence indicate that some of these components may possess potent antioxidant, anti-inflammatory, antiallergic, hypolipidaemic, vasoprotective, and anticarcinogenic properties [127] and protect against oxidative stress, coronary heart diseases, certain cancers, and other age-related diseases. Orange juice, enriched in both carotenoids and flavonoids, as one of the most popular drinks, has therefore been used in many dietetics and/or pharmaceutics studies. Results suggest it can not only improve diet quality and nutrient adequacy but also substantially reduce the risk of obesity and/or cardiovascular diseases, and help prevent and suppress some cancers [122–125, 128, 129].

Studies during 2003–2006 revealed that children of 2–18 years old who consumed 100 % orange juice had a significantly higher percentage of the population meeting the estimated average requirement (EAR) for vitamin A, vitamin C, folate, and magnesium, and the healthy eating index (HEI) in 2005 was significantly higher in the consumers [129]. In the adult group, compared to nonconsumers, consumers had higher percentage of the population meeting the EAR for the abovementioned nutrients, and lower mean body mass index, total cholesterol level, and low-density lipoprotein cholesterol level. As a result, the subgroup consuming 100 % orange juice on a regular basis was 21 % less likely to be obese, and male consumers were 36 % less likely to have metabolic syndrome (MetS). These improved biomarkers imply the risk of relevant or induced diseases is decreased accordingly [123].

Combinational use of natural carotenoids from mandarin citrus fruit juice (enriched with β-cryptoxanthin) with capsules of a carotenoid mixture resulted in statistically significant suppression of liver cancer during 2.5-year treatment period, compared to the group treated with carotenoid mixture capsules alone or the control group without treatment. The group treated with carotenoid mixture capsules alone also had lower, but not statistically significant, cumulative cancer incidences than the control group [126].

The evaluation of the pharmacokinetics of hesperetin and naringenin, two flavanones in blood orange juice, suggested that the plasma concentrations of these compounds in human subjects were dose dependent and the peak concentration was reached in 5.1 h after juice consumption. The conjugated forms of these flavanones represent more than 95 % of the plasma concentration [127]. Some health-promoting features of these components, with chemical structures that appear ideal for free radical scavenging, can be attributed to their antioxidant properties [130]. The antioxidant properties and/or activities have been reported in many carotenoids and flavonoids, though they may differ in potency, strength, and capacity [128, 131–133].

Conclusions and Perspectives

Carotenoids and flavonoids in citrus fruits are pigmentation agents for fruit coloration and organoleptic attraction, and functional nutrients in the human diet as well. Lycopene and anthocyanin enriched mutants, both resulting in red-fleshed fruits, are the two leading research targets to understand their biosynthesis genomics and molecular regulation mechanisms, through comparisons with their respective progenitor counterparts. Most genes in the two pathways are cloned and characterized, and the regulatory genes and mechanisms are being gradually uncovered. Such knowledge will greatly facilitate continued varietal improvement for these beneficial components. The available citrus genome sequences [134, 135], along with the advances in any new "omics" biotechnologies, will accelerate the progress in these pigment related areas, such as the regulation of various carotenoids and lycopene accumulation; the coordination of the carotenoid relevant pathways and networks; genes involved in anthocyanin modifications; the mechanism of anthocyanin vacuolar compartmentation; and the role of phytohormones and/or biotic and abiotic factors in inducing anthocyanin accumulation.

Another most interesting area is the dietetics of citrus fruits and juice (primarily orange juice), in which various carotenoids and flavonoids showed both nutritional and preventive medicinal benefits. Some emphasis was on lycopene and anthocyanins primarily due to the special red color and enhanced health benefits. The preventive roles include higher quality and availability of nutrients, substantially improved health biomarkers for lipid profiles, lower risk of obesity and cardiovascular disease, and potential suppression of some cancers. According to hundreds of reports on the beneficial effects of orange juice and/or its functional components, moderate regular consumption of 100 % orange juice and/or fresh citrus fruits should be encouraged to help children and adults meet the USDA daily recommendation for fruit intake and to provide them a vital component of a healthy diet [123, 129]. The science-based information should be more broadly broadcast to consumers via multiple media to increase their awareness of the potential benefits and encourage consumption of more fruit or juice.

References

1. Kawaii S, Tomono Y, Katase E, Ogawa K, Yano M (1999) Quantitation of flavonoid constituents in citrus fruits. J Agric Food Chem 47(9):3565–3571
2. Pupin AM, Dennis MJ, Toledo MCF (1999) HPLC analysis of carotenoids in orange juice. Food Chem 64(2):269–275
3. Moufida S, Marzouk B (2003) Biochemical characterization of blood orange, sweet orange, lemon, bergamot and bitter orange. Phytochemistry 62(8):1283–1289
4. Lu Y, Zhang C, Bucheli P, Wei D (2006) Citrus flavonoids in fruit and traditional Chinese medicinal food ingredients in China. Plant Foods Hum Nutr 61(2):57–65
5. Melendez-Martinez AJ, Vicario IM, Heredia FJ (2007) Carotenoids, color, and ascorbic acid content of a novel frozen-marketed orange juice. J Agric Food Chem 55(4):1347–1355

6. Fuzfai Z, Molnar-Perl I (2007) Gas chromatographic-mass spectrometric fragmentation study of flavonoids as their trimethylsilyl derivatives: analysis of flavonoids, sugars, carboxylic and amino acids in model systems and in citrus fruits. J Chromatogr A 1149(1):88–101
7. Atkinson W, Downer P, Lever M, Chambers ST, George PM (2007) Effects of orange juice and proline betaine on glycine betaine and homocysteine in healthy male subjects. Eur J Nutr 46(8):446–452
8. Guarnieri S, Riso P, Porrini M (2007) Orange juice vs vitamin C: effect on hydrogen peroxide-induced DNA damage in mononuclear blood cells. Br J Nutr 97(4):639–643
9. Rapavi E, Kocsis I, Feher E, Szentmihalyi K, Lugasi A, Szekely E, Blazovics A (2007) The effect of citrus flavonoids on the redox state of alimentary-induced fatty liver in rats. Nat Prod Res 21(3):274–281
10. Roza JM, Xian-Liu Z, Guthrie N (2007) Effect of citrus flavonoids and tocotrienols on serum cholesterol levels in hypercholesterolemic subjects. Altern Ther Health Med 13(6):44–48
11. Titta L, Trinei M, Stendardo M, Berniakovich I, Petroni K, Tonelli C, Riso P, Porrini M, Minucci S, Pelicci PG et al (2010) Blood orange juice inhibits fat accumulation in mice. Int J Obesity 34(3):578–588
12. Assini JM, Mulvihill EE, Huff MW (2013) Citrus flavonoids and lipid metabolism. Curr Opin Lipidol 24(1):34–40
13. Gmitter FG Jr, Chen C, Rao M, Soneji J (2007) Citrus fruits. In: Kole CR (ed) Fruits and nuts, genome mapping & molecular breeding in plants, vol 4. Springer, Heidelberg, pp 265–279
14. Barrett HC, Rhodes AM (1976) A numerical taxonomic study of affinity relationships in cultivated citrus and its close relatives. Syst Bot 1:105–136
15. Nicolosi E, Deng ZN, Gentile A, La Malfa S, Continella G, Tribulato E (2000) Citrus phylogeny and genetic origin of important species as investigated by molecular markers. Theor Appl Genet 100:1155–1166
16. Nicolosi E (2007) Origin and taxonomy. In: Khan IA (ed) Citrus genetics, breeding and biotechnology. CABI, Wallingford, pp 19–43
17. Gmitter FG Jr (1995) Origin, evolution, and breeding of the grapefruit. Plant Breed Rev 13:345–363
18. Lee HS (2002) Characterization of major anthocyanins and the color of red-fleshed Budd Blood orange (*Citrus sinensis*). J Agric Food Chem 50(5):1243–1246
19. Xu CJ, Fraser PD, Wang WJ, Bramley PM (2006) Differences in the carotenoid content of ordinary citrus and lycopene-accumulating mutants. J Agric Food Chem 54(15):5474–5481
20. Nogata Y, Sakamoto K, Shiratsuchi H, Ishii T, Yano M, Ohta H (2006) Flavonoid composition of fruit tissues of *Citrus* species. Biosci Biotechnol Biochem 70(1):178–192
21. Dhuique-Mayer C, Fanciullino AL, Dubois C, Ollitrault P (2009) Effect of genotype and environment on citrus juice carotenoid content. J Agric Food Chem 57(19):9160–9168
22. Roowi S, Crozier A (2011) Flavonoids in tropical *Citrus* species. J Agric Food Chem 59(22):12217–12225
23. Winkel-Shirley B (2001) Flavonoid biosynthesis. A colorful model for genetics, biochemistry, cell biology, and biotechnology. Plant Physiol 126(2):485–493
24. Matsumoto H, Ikoma Y, Kato M, Kuniga T, Nakajima N, Yoshida T (2007) Quantification of carotenoids in citrus fruit by LC-MS and comparison of patterns of seasonal changes for carotenoids among citrus varieties. J Agric Food Chem 55(6):2356–2368
25. Mondello L, Cotroneo A, Errante G, Dugo G, Dugo P (2000) Determination of anthocyanins in blood orange juices by HPLC analysis. J Pharm Biomed Anal 23(1):191–195
26. Hillebrand S, Schwarz M, Winterhalter P (2004) Characterization of anthocyanins and pyranoanthocyanins from blood orange [*Citrus sinensis* (L.) Osbeck] juice. J Agric Food Chem 52(24):7331–7338
27. Lee HS, Castle WS (2001) Seasonal changes of carotenoid pigments and color in Hamlin, Earlygold, and Budd blood orange juices. J Agric Food Chem 49(2):877–882
28. Kato M (2012) Mechanism of carotenoid accumulation in citrus fruit. J Jpn Soc Hortic Sci 81(3):219–233

29. Xu JA, Liu BZ, Liu X, Gao HJ, Deng XX (2011) Carotenoids synthesized in citrus callus of different genotypes. Acta Physiol Plant 33(3):745–753
30. Alquezar B, Rodrigo MJ, Zacarias L (2008) Regulation of carotenoid biosynthesis during fruit maturation in the red-fleshed orange mutant Cara Cara. Phytochemistry 69(10):1997–2007
31. Tao N, Hu Z, Liu Q, Xu J, Cheng Y, Guo L, Guo W, Deng X (2007) Expression of phytoene synthase gene (Psy) is enhanced during fruit ripening of Cara Cara navel orange (*Citrus sinensis* Osbeck). Plant Cell Rep 26(6):837–843
32. Cunningham FX, Gantt E (1998) Genes and enzymes of carotenoid biosynthesis in plants. Ann Rev Plant Physiol Plant Mol Biol 49:557–583
33. Cazzonelli CI (2011) Carotenoids in nature: insights from plants and beyond. Funct Plant Biol 38(11):833–847
34. Walter MH, Strack D (2011) Carotenoids and their cleavage products: biosynthesis and functions. Nat Prod Rep 28(4):663–692
35. Ruiz-Sola MA, Rodriguez-Concepcion M (2012) Carotenoid biosynthesis in Arabidopsis: a colorful pathway. Arabidopsis Book 10:e0158
36. Bendich A (1989) Carotenoids and the immune response. J Nutr 119(1):112–115
37. Bendich A, Olson JA (1989) Biological actions of carotenoids. FASEB J 3(8):1927–1932
38. Maoka T (2011) Carotenoids in marine animals. Mar Drugs 9(2):278–293
39. Garewal H, Meyskens F Jr, Friedman S, Alberts D, Ramsey L (1993) Oral cancer prevention: the case for carotenoids and anti-oxidant nutrients. Prev Med 22(5):701–711
40. Hix LM, Lockwood SF, Bertram JS (2004) Bioactive carotenoids: potent antioxidants and regulators of gene expression. Redox Rep 9(4):181–191
41. Sharoni Y, Linnewiel-Hermoni K, Khanin M, Salman H, Veprik A, Danilenko M, Levy J (2012) Carotenoids and apocarotenoids in cellular signaling related to cancer: a review. Mol Nutr Food Res 56(2):259–269
42. Stahl W, Sies H (2005) Bioactivity and protective effects of natural carotenoids. Biochim et Biophys Acta 1740(2):101–107
43. Kato M, Ikoma Y, Matsumoto H, Sugiura M, Hyodo H, Yano M (2004) Accumulation of carotenoids and expression of carotenoid biosynthetic genes during maturation in citrus fruit. Plant Physiol 134(2):824–837
44. Gao HJ, Xu J, Liu X, Liu BZ, Deng XX (2011) Light effect on carotenoids production and expression of carotenogenesis genes in citrus callus of four genotypes. Acta Physiol Plant 33(6):2485–2492
45. Matsumoto H, Ikoma Y, Kato M, Nakajima N, Hasegawa Y (2009) Effect of postharvest temperature and ethylene on carotenoid accumulation in the flavedo and juice sacs of Satsuma Mandarin (*Citrus unshiu* Marc.) fruit. J Agric Food Chem 57(11):4724–4732
46. Fanciullino AL, Dhuique-Mayer C, Luro F, Casanova J, Morillon R, Ollitrault P (2006) Carotenoid diversity in cultivated citrus is highly influenced by genetic factors. J Agric Food Chem 54(12):4397–4406
47. Zhang L, Ma G, Kato M, Yamawaki K, Takagi T, Kiriiwa Y, Ikoma Y, Matsumoto H, Yoshioka T, Nesumi H (2012) Regulation of carotenoid accumulation and the expression of carotenoid metabolic genes in citrus juice sacs in vitro. J Exp Bot 63(2):871–886
48. Fanciullino AL, Cercos M, Dhique M, Froelicher Y, Talon M, Ollitrault P, Morillon R (2008) Changes in carotenoid content and biosynthetic gene expression in juice sacs of four orange varieties (*Citrus sinensis*) differing in flesh fruit color. J Agric Food Chem 56(10):3628–3638
49. Hirschberg J (2001) Carotenoid biosynthesis in flowering plants. Curr Opin Plant Biol 4(3):210–218
50. Chen CX, Costa MGC, Yu QB, Moore GA, Gmitter FG (2010) Identification of novel members in sweet orange carotenoid biosynthesis gene families. Tree Genet Genom 6(6):905–914
51. Fanciullino AL, Dhuique-Mayer C, Luro F, Morillon R, Ollitrault P (2007) Carotenoid biosynthetic pathway in the *Citrus* genus: number of copies and phylogenetic diversity of seven genes. J Agric Food Chem 55(18):7405–7417

52. Wei X, Chen C, Yu Q, Gady A, Yu Y, Liang G, Gmitter F Jr (2014) Novel expression patterns of carotenoid pathway-related genes in citrus leaves and maturing fruits. Tree Genet Genom 10:439–448
53. Gross J (1987) Pigments in fruits. Academic, London
54. Livingstone K, Anderson S (2009) Patterns of variation in the evolution of carotenoid biosynthetic pathway enzymes of higher plants. J Hered 100(6):754–761
55. Lee HS (2001) Characterization of carotenoids in juice of red navel orange (Cara Cara). J Agr Food Chem 49(5):2563–2568
56. Tao NG, Xu J, Cheng YJ, Deng XX (2005) Lycopene-epsilon-cyclase pre-mRNA is alternatively spliced in Cara Cara navel orange (*Citrus sinensis* Osbeck). Biotechnol Lett 27(11):779–782
57. Tao N, Wang C, Xu J, Cheng Y (2012) Carotenoid accumulation in postharvest "Cara Cara" navel orange (*Citrus sinensis* Osbeck) fruits stored at different temperatures was transcriptionally regulated in a tissue-dependent manner. Plant Cell Rep 31(9):1667–1676
58. Liu Q, Xu J, Liu Y, Zhao X, Deng X, Guo L, Gu J (2007) A novel bud mutation that confers abnormal patterns of lycopene accumulation in sweet orange fruit (*Citrus sinensis* L. Osbeck). J Exp Bot 58(15–16):4161–4171
59. Pan Z, Liu Q, Yun Z, Guan R, Zeng W, Xu Q, Deng X (2009) Comparative proteomics of a lycopene-accumulating mutant reveals the important role of oxidative stress on carotenogenesis in sweet orange (*Citrus sinensis* [L.] osbeck). Proteomics 9(24):5455–5470
60. Pan Z, Zeng Y, An J, Ye J, Xu Q, Deng X (2012) An integrative analysis of transcriptome and proteome provides new insights into carotenoid biosynthesis and regulation in sweet orange fruits. J Proteomics 75(9):2670–2684
61. Xu Q, Liu Y, Zhu A, Wu X, Ye J, Yu K, Guo W, Deng X (2010) Discovery and comparative profiling of microRNAs in a sweet orange red-flesh mutant and its wild type. BMC Genomics 11:246–2164-2111–2246
62. Xu Q, Yu K, Zhu A, Ye J, Liu Q, Zhang J, Deng X (2009) Comparative transcripts profiling reveals new insight into molecular processes regulating lycopene accumulation in a sweet orange (*Citrus sinensis*) red-flesh mutant. BMC Genomics 10:540–2164-2110–2540
63. Mendes AF, Chen C, Gmitter FG, Jr., Moore GA, Costa MG (2011) Expression and phylogenetic analysis of two new lycopene beta-cyclases from *Citrus paradisi*. Physiol Plantarum 141(1):1–10
64. Costa MG, Moreira CD, Melton JR, Otoni WC, Moore GA (2012) Characterization and developmental expression of genes encoding the early carotenoid biosynthetic enzymes in *Citrus paradisi* Macf. Mol Biol Rep 39(2):895–902
65. Gunasekera RS, Sewgobind K, Desai S, Dunn L, Black HS, McKeehan WL, Patil B (2007) Lycopene and lutein inhibit proliferation in rat prostate carcinoma cells. Nutr Cancer 58(2):171–177
66. van Breemen RB, Pajkovic N (2008) Multitargeted therapy of cancer by lycopene. Cancer Lett 269(2):339–351
67. Alquezar B, Zacarias L, Rodrigo MJ (2009) Molecular and functional characterization of a novel chromoplast-specific lycopene beta-cyclase from *Citrus* and its relation to lycopene accumulation. J Exp Bot 60(6):1783–1797
68. Wang Y, Chen S, Yu O (2011) Metabolic engineering of flavonoids in plants and microorganisms. Appl Microbiol Biotechnol 91(4):949–956
69. Dugo P, Mondello L, Morabito D, Dugo G (2003) Characterization of the anthocyanin fraction of sicilian blood orange juice by micro-HPLC-ESI/MS. J Agr Food Chem 51(5):1173–1176
70. Ververidis F, Trantas E, Douglas C, Vollmer G, Kretzschmar G, Panopoulos N (2007) Biotechnology of flavonoids and other phenylpropanoid-derived natural products. Part I: chemical diversity, impacts on plant biology and human health. Biotechnol J 2(10):1214–1234
71. Winkel-Shirley B (2002) Biosynthesis of flavonoids and effects of stress. Curr Opin Plant Biol 5(3):218–223
72. Hassan S, Mathesius U (2012) The role of flavonoids in root-rhizosphere signalling: opportunities and challenges for improving plant-microbe interactions. J Exp Bot 63(9):3429–3444

73. Straney D, Khan R, Tan R, Bagga S (2002) Host recognition by pathogenic fungi through plant flavonoids. Adv Exp Med Biol 505:9–22
74. Claudie DM, Alexandrine D, Bertrand C, Franck T, Marie-Josephe A (2013) Citrus flavanones enhance carotenoid uptake by intestinal Caco-2 cells. Food Funct 4(11):1625–1631
75. Fuhrman B, Aviram M (2001) Flavonoids protect LDL from oxidation and attenuate atherosclerosis. Curr Opin Lipidol 12(1):41–48
76. Hwang SL, Shih PH, Yen GC (2012) Neuroprotective effects of citrus flavonoids. J Agr Food Chem 60(4):877–885
77. Kandaswami C, Lee LT, Lee PP, Hwang JJ, Ke FC, Huang YT, Lee MT (2005) The antitumor activities of flavonoids. In Vivo (Athens, Greece) 19(5):895–909
78. Kawaii S, Tomono Y, Ogawa K, Sugiura M, Yano M, Yoshizawa Y, Shahidi F, Weerasinghe DK (2004) Anticancer activity and flavonoid content of various citrus juices. Nutraceutical Beverages 871:76–89
79. Manthey JA, Grohmann K, Guthrie N (2001) Biological properties of citrus flavonoids pertaining to cancer and inflammation. Curr Med Chem 8(2):135–153
80. Middleton E, Jr (1998) Effect of plant flavonoids on immune and inflammatory cell function. Adv Exp Med Biol 439:175–182
81. Nguyen TD, Canada AT (1993) Citrus flavonoids stimulate secretion by human colonic T84 cells. J Nutr 123(2):259–268
82. Zielinska-Przyjemska M, Ignatowicz E (2008) Citrus fruit flavonoids influence on neutrophil apoptosis and oxidative metabolism. Phytother Res 22(12):1557–1562
83. Mazza G (1995) Anthocyanins in grapes and grape products. Crit Rev Food Sci Nutr 35(4):341–371
84. Bordignon-Luiz MT, Gauche C, Gris EF, Falcao LD (2007) Colour stability of anthocyanins from Isabel grapes (*Vitis labrusca* L.) in model systems. Lwt-Food Sci Technol 40(4):594–599
85. Kong JM, Chia LS, Goh NK, Chia TF, Brouillard R (2003) Analysis and biological activities of anthocyanins. Phytochemistry 64(5):923–933
86. Butelli E, Licciardello C, Zhang Y, Liu J, Mackay S, Bailey P, Reforgiato-Recupero G, Martin C (2012) Retrotransposons control fruit-specific, cold-dependent accumulation of anthocyanins in blood oranges. Plant Cell 24(3):1242–1255
87. Lo Piero AR, Puglisi I, Rapisarda P, Petrone G (2005) Anthocyanins accumulation and related gene expression in red orange fruit induced by low temperature storage. J Agric Food Chem 53(23):9083–9088
88. Rapisarda P, Bellomo SE, Fabroni S, Russo G (2008) Juice quality of two new mandarin-like hybrids (*Citrus clementina* Hort. ex Tan x *Citrus sinensis* L. Osbeck) containing anthocyanins. J Agric Food Chem 56(6):2074–2078
89. Crifo T, Puglisi I, Petrone G, Reforgiato Recupero G, Lo Piero AR (2011) Expression analysis in response to low temperature stress in blood oranges: implication of the flavonoid biosynthetic pathway. Gene 476(1–2):1–9
90. Chandler BV (1958) Anthocyanins of blood oranges. Nature 182(4640):933–933
91. Maccarone E, Maccarrone A, Perrini G, Rapisarda P (1983) Anthocyanins of the Moro Orange Juice. Ann Chim-Rome 73(9–10):533–539
92. Maccarone E, Maccarrone A, Rapisarda P (1985) Acylated anthocyanins from oranges. Ann Chim-Rome 75(1–2):79–86
93. Maccarone E, Rapisarda P, Fanella F, Arena E, Mondello L (1998) Cyanidin-3-(6″-malonyl)-beta-glucoside. One of the major anthocyanins in blood orange juice. Ital J Food Sci 10(4):367–372
94. Dixon RA, Steele CL (1999) Flavonoids and isoflavonoids—a gold mine for metabolic engineering. Trends Plant Sci 4(10):394–400
95. Dooner HK, Robbins TP, Jorgensen RA (1991) Genetic and developmental control of anthocyanin biosynthesis. Annu Rev Genet 25:173–199
96. Goodrich J, Carpenter R, Coen ES (1992) A common gene regulates pigmentation pattern in diverse plant-species. Cell 68(5):955–964

97. Quattrocchio F, Wing JF, Leppen HTC, Mol JNM, Koes RE (1993) Regulatory genes-controlling anthocyanin pigmentation are functionally conserved among plant-species and have distinct sets of target genes. Plant Cell 5(11):1497–1512
98. Sparvoli F, Martin C, Scienza A, Gavazzi G, Tonelli C (1994) Cloning and molecular analysis of structural genes involved in flavonoid and stilbene biosynthesis in grape (*Vitis-Vinifera* L). Plant Mol Biol 24(5):743–755
99. Moriguchi T, Kita M, Tomono Y, Endo-Inagaki T, Omura M (2001) Gene expression in flavonoid biosynthesis: correlation with flavonoid accumulation in developing citrus fruit. Physiol Plantarum 111(1):66–74
100. Sanchez-Ballesta MT, Lafuente MT, Zacarias L, Granell A (2000) Involvement of phenylalanine ammonia-lyase in the response of Fortune mandarin fruits to cold temperature. Physiol Plantarum 108(4):382–389
101. Lo Piero AR, Puglisi I, Petrone G (2006) Gene isolation, analysis of expression, and in vitro synthesis of glutathione S-transferase from orange fruit [*Citrus sinensis* L. (Osbeck)]. J Agric Food Chem 54(24):9227–9233
102. Lo Piero AR, Puglisi I, Petrone G (2006) Gene characterization, analysis of expression and in vitro synthesis of dihydroflavonol 4-reductase from [*Citrus sinensis* (L.) Osbeck]. Phytochemistry 67(7):684–695
103. Lo Piero AR, Puglisi I, Orestano G, Reforgiato Recupero G, Petrone G, Consoli A (2005) Anthocyaninless cultivars of sweet orange lack to express the UDP-glucose flavonoid 3-O-Glucosyl transferase. J Plant Biochem Biotech 14:1–6
104. Springob K, Nakajima J, Yamazaki M, Saito K (2003) Recent advances in the biosynthesis and accumulation of anthocyanins. Nat Prod Rep 20(3):288–303
105. Holton TA, Cornish EC (1995) Genetics and biochemistry of anthocyanin biosynthesis. Plant Cell 7(7):1071–1083
106. Moriguchi T, Kita M, Tomono Y, Endo-Inagaki T, Omura M (1999) One type of chalcone synthase gene expressed during embryogenesis regulates the flavonoid accumulation in citrus cell cultures. Plant Cell Physiol 40(6):651–655
107. Lucheta AR, Silva-Pinhati ACO, Basilio-Palmieri AC, Berger IJ, Freitas-Astua J, Cristofani M (2007) An in silico analysis of the key genes involved in flavonoid biosynthesis in *Citrus sinensis*. Genet Mol Biol 30(3):819–831
108. Muccilli V, Licciardello C, Fontanini D, Russo MP, Cunsolo V, Saletti R, Recupero GR, Foti S (2009) Proteome analysis of *Citrus sinensis* L. (Osbeck) flesh at ripening time. J Proteomics 73(1):134–152
109. Doostdar H, Shapiro JP, Niedz R, Burke MD, Mccollum TG, Mcdonald RE, Mayer RT (1995) A cytochrome-P450 mediated naringenin 3′-hydroxylase from sweet orange cell-cultures. Plant Cell Physiol 36(1):69–77
110. Grotewold E, Drummond BJ, Bowen B, Peterson T (1994) The myb-homologous P-gene controls phlobaphene pigmentation in maize floral organs by directly activating a flavonoid biosynthetic gene subset. Cell 76(3):543–553
111. He F, Mu L, Yan GL, Liang NN, Pan QH, Wang J, Reeves MJ, Duan CQ (2010) Biosynthesis of anthocyanins and their regulation in colored grapes. Molecules 15(12):9057–9091
112. D'Auria JC (2006) Acyltransferases in plants: a good time to be BAHD. Curr Opin Plant Biol 9(3):331–340
113. Cotroneo PS, Russo MP, Ciuni M, Recupero GR, Lo Piero AR (2006) Quantitative real-time reverse transcriptase-PCR profiling of anthocyanin biosynthetic genes during orange fruit ripening. J Am Soc Hortic Sci 131(4):537–543
114. Crifo T, Petrone G, Lo Cicero L, Lo Piero AR (2012) Short cold storage enhances the anthocyanin contents and level of transcripts related to their biosynthesis in blood oranges. J Agric Food Chem 60(1):476–481
115. Koes R, Verweij W, Quattrocchio F (2005) Flavonoids: a colorful model for the regulation and evolution of biochemical pathways. Trends Plant Sci 10(5):236–242
116. Winkel-Shirley B (1999) Evidence for enzyme complexes in the phenylpropanoid and flavonoid pathways. Physiol Plantarum 107(1):142–149

117. Bernardi J, Licciardello C, Russo MP, Luisa Chiusano M, Carletti G, Recupero GR, Marocco A (2010) Use of a custom array to study differentially expressed genes during blood orange (*Citrus sinensis* L. Osbeck) ripening. J Plant Physiol 167(4):301–310
118. Cultrone A, Cotroneo PS, Recupero GR (2010) Cloning and molecular characterization of R2R3-MYB and bHLH-MYC transcription factors from *Citrus sinensis*. Tree Genet Genom 6(1):101–112
119. Maul P, McCollum GT, Popp M, Guy CL, Porat R (2008) Transcriptome profiling of grapefruit flavedo following exposure to low temperature and conditioning treatments uncovers principal molecular components involved in chilling tolerance and susceptibility. Plant Cell Environ 31:752–768
120. Chalker-Scott L (1999) Environmental significance of anthocyanins in plant stress responses. Photochem Photobiol 70(1):1–9
121. de Pascual-Teresa S (2008) S-BMT: anthocyanins: from plant to health. Phytochem Rev 7:281–299
122. Stookey JD, Hamer J, Espinoza G, Higa A, Ng V, Tinajero-Deck L, Havel PJ, King JC (2012) Orange juice limits postprandial fat oxidation after breakfast in normal-weight adolescents and adults. Adv Nutr (Bethesda, MD) 3(4):629S–635S
123. O'Neil CE, Nicklas TA, Rampersaud GC, Fulgoni VL III (2012) 100 % Orange juice consumption is associated with better diet quality, improved nutrient adequacy, decreased risk for obesity, and improved biomarkers of health in adults: national health and nutrition examination survey, 2003–2006. Nutr J 11:107–2891-2811–2107
124. Niv E, Shapira Y, Akiva I, Rokhkind E, Naor E, Arbiv M, Vaisman N (2012) Effect of levan supplement in orange juice on weight, gastrointestinal symptoms and metabolic profile of healthy subjects: results of an 8-week clinical trial. Nutrients 4(7):638–647
125. Buscemi S, Rosafio G, Arcoleo G, Mattina A, Canino B, Montana M, Verga S, Rini G (2012) Effects of red orange juice intake on endothelial function and inflammatory markers in adult subjects with increased cardiovascular risk. Am J Clin Nutr 95(5):1089–1095
126. Nishino H, Murakoshi M, Tokuda H, Satomi Y (2009) Cancer prevention by carotenoids. Arch Biochem Biophys 483(2):165–168
127. Gardana C, Guarnieri S, Riso P, Simonetti P, Porrini M (2007) Flavanone plasma pharmacokinetics from blood orange juice in human subjects. Br J Nutr 98(1):165–172
128. Sanchez-Moreno C, Cano MP, de Ancos B, Plaza L, Olmedilla B, Granado F, Martin A (2003) Effect of orange juice intake on vitamin C concentrations and biomarkers of antioxidant status in humans. Am J Clin Nutr 78(3):454–460
129. O'Neil CE, Nicklas TA, Rampersaud GC, Fulgoni VL 3rd (2011) One hundred percent orange juice consumption is associated with better diet quality, improved nutrient adequacy, and no increased risk for overweight/obesity in children. Nutr Res (New York, NY) 31(9):673–682
130. Amorini AM, Lazzarino G, Galvano F, Fazzina G, Tavazzi B, Galvano G (2003) Cyanidin-3-O-beta-glucopyranoside protects myocardium and erythrocytes from oxygen radical-mediated damages. Free Radic Res 37(4):453–460
131. Riso P, Visioli F, Gardana C, Grande S, Brusamolino A, Galvano F, Galvano G, Porrini M (2005) Effects of blood orange juice intake on antioxidant bioavailability and on different markers related to oxidative stress. J Agr Food Chem 53(4):941–947
132. Navarro P, Melendez-Martinez AJ, Heredia F, Gabaldon JA, Carbonell-Barrachina AA, Soler A, Perez-Lopez AJ (2011) Effects of beta-cyclodextrin addition and farming type on vitamin C, antioxidant activity, carotenoids profile, and sensory analysis in pasteurised orange juices. Int J Food Sci Tech 46(10):2182–2190
133. Franke AA, Cooney RV, Henning SM, Custer LJ (2005) Bioavailability and antioxidant effects of orange juice components in humans. J Agr Food Chem 53(13):5170–5178
134. Wu GA, Prochnik S, Jenkins J, Salse J, Hellsten U, Murat F, Perrier X, Ruiz M, Scalabrin S, Terol J, Takita MA, Labadie K, Poulain J, Couloux A, Jabbari K, Cattonaro F, Del Fabbro C, Pinosio S, Zuccolo A, Chapman J, Grimwood J, Tadeo FR, Estornell LH,

Munoz-Sanz JV, Ibanez V, Herrero-Ortega A, Aleza P, Perez-Perez J, Ramon D, Brunel D, Luro F, Chen C, Farmerie WG, Desany B, Kodira C, Mohiuddin M, Harkins T, Fredrikson K, Burns P, Lomsadze A, Borodovsky M, Reforgiato G, Freitas-Astua J, Quetier F, Navarro L, Roose M, Wincker P, Schmutz J, Morgante M, Machado MA, Talon M, Jaillon O, Ollitrault P, Gmitter F, and Rokhsar D (2014) Sequencing of diverse mandarin, pummelo and orange genomes reveals complex history of admixture during citrus domestication. Nature Biotechnol 32(7):656–662

135. Xu Q, Chen LL, Ruan X, Chen D, Zhu A, Chen C, Bertrand D, Jiao WB, Hao BH, Lyon MP et al (2012) The draft genome of sweet orange (*Citrus sinensis*). Nat Genet 45(1):59–66
136. Melendez-Martinez AJ, Vicario IM, Heredia FJ (2007) Review: analysis of carotenoids in orange juice. J Food Compos Anal 20(7):638–649

Chapter 9
Pigments in Grape

Christopher L. Owens

Introduction

Grapevine is the most valuable horticultural crop in the world. Significant grape acreage exists on all continents, save for Antarctica. Approximately 8 million ha of grapevine are currently planted and 60 million metric t of fruit are produced annually worldwide (Food and Agriculture Organization of the United Nations (FAO) production statistics). Spain, France, and Italy are the largest grape producers in the world, followed by many other European countries, the USA, Argentina, Chile, Australia, South Africa, and China. The majority of the fruit, in terms of yield and area, is used to produce wine, but the remaining is destined for fresh consumption, dried into raisins, processed into nonalcoholic juice, or distilled into spirits. The quality of wine and other grape products is the key to the crop's value, so sustainably maximizing quality is the primary goal of grape producers.

Most grape cultivars are used specifically in one market, but some cultivars may be used in several market classes. Premium wine and table grape cultivars are more specialized in their utilization than are raisin, juice, and concentrate varieties. For example, "Cabernet Sauvignon" is primarily used for wine but is not desirable as a table or raisin grape. "Sultanina" (known as "Thompson Seedless" in the USA) is the predominant raisin cultivar worldwide and also is an important table grape, wine grape, and concentrate cultivar. Wine grape cultivars usually have relatively small-seeded berries. Important wine cultivars include "Cabernet Sauvignon" and "Pinot Noir," used for red wine production, and "Chardonnay" and "sauvignon blanc," used for white wine production. Table grapes are consumed fresh. Table grape cultivars have relatively large berries, and seedlessness is valued by many consumers. Most dried grapes, usually called raisins, are made from seedless grapes. Unfermented juice is manufactured from cultivars with distinctive flavors and aromas. Varieties

C. L. Owens (✉)
Grape Genetics Research Unit, United States Department of Agriculture,
Agricultural Research Service, 630 W. North St., Geneva, NY 14456, USA
e-mail: chris.owens@ars.usda.gov

C. Chen (ed.), *Pigments in Fruits and Vegetables,* DOI 10.1007/978-1-4939-2356-4_9

with relatively heat-stable flavors and aromas, such as "Concord" and "Niagara," are used in the production of pasteurized juices. Cultivars such as "Chasselas" with flavors and aromas that are noticeably altered by pasteurization are processed for unfermented juice production using ultrafiltration for juice sterilization. Jams, jellies, and other spreads are made from juice grape cultivars. Grape concentrate is juice with some water removed; it is used as a natural sweetener and coloring agent for beverages and foods. The concentrate market is an outlet for excess grapes in all market classes and is a target market for certain cultivars; "Rubired," a highly pigmented cultivar, is used in red concentrate.

The grape is a member of the Vitaceae, commonly called the grape family. The genus *Vitis* consists of about 60 species, plus some natural interspecific hybrids [1]. Nearly all grapes cultivated for fruit production are of the species *Vitis vinifera* or are hybrids that include *V. vinifera* in their parentage. *Vitis* species are found across the temperate zones of the Northern Hemisphere. The genus has the highest species diversity in east Asia and in eastern and southern North America, with about 30 species in each region. *Vitis* is separated into two subgenera, *Euvitis* and *Muscadinia*; some authorities treat the sections as the genera *Vitis* and *Muscadinia*. The subgenera are separated by morphological, anatomical, and cytological characteristics. Subgenus *Euvitis* species have $2n=2x=38$ chromosomes, forked tendrils, striate bark, pyriform seeds, and nodal diaphragms. These species and their hybrids are called "bunch grapes." Subgenus *Muscadinia* species have $2n=2x=40$ chromosomes, unforked tendrils, stellate bark, naviform seeds, and lack diaphragms at the nodes; they are known as muscadine grapes. Within a subgenus, species are maintained in nature by range and flowering time and can be considered ecospecies. Hybrids between species within a subgenus are typically fully fertile and many interspecific hybrids between *Euvitis* species have been developed as scion and rootstock cultivars. Hybrids between the subgenera are usually sterile due to the difference in chromosome number; two have been commercialized as rootstocks [2], and backcrossing with partially fertile intersubgeneric hybrids has led to the introduction of disease resistance from *Vitis rotundifolia* into bunch grape gene pools [3].

Subgenus *Euvitis* species (about 57 species) are the most important in viticulture. Most grape cultivars belong to the species *V. vinifera,* which is a native of the Mediterranean basin, southern and central Europe, northern Africa, and southwest and central Asia. *V. vinifera* cultivars are grown worldwide and account for the overwhelming majority of cultivated areas and grapes produced. Interspecific hybrid cultivars selected from crosses of *V. vinifera* with other species, including *Vitis labrusca, Vitis amurensis, Vitis riparia, Vitis rupestris,* and *Vitis aestivalis*, are locally important and account for a minor portion of world viticulture and enology. Rootstocks are used exclusively for bunch grape varieties, which are mostly interspecific hybrids or selections of North American *Euvitis* species.

The subgenus *Muscadinia* includes only three species. The range of the subgenus is limited to the southeastern USA and eastern Mexico. Muscadine grape cultivars, primarily *V. rotundifolia* and a few interspecific hybrids, are grown commercially only in the native region of *V. rotundifolia* in the southeastern USA.

Pigments in Grapes

The primary pigments in grapes are anthocyanins. The distribution and classification of anthocyanins within grape berries have been extensively studied. Anthocyanins are specific to red cultivars and localized in the berry skin of most cultivars. A small number of red-fleshed cultivars (also known as teinturier, Fig. 9.1) play a key role in producing highly pigmented juice for wine making. Anthocyanin profiles of many *V. vinifera* cultivars as well as several other *Vitis* species have been reported and previously reviewed [4–6]. Structurally, anthocyanins are glycosides and acylglycosides of anthocyanidins, and the aglycones and flavyliums differ in different hydroxyl or methoxyl substitutions in their basic structures [4]. The proportion of the primary individual anthocyanins is the 3-*O*-monoglucosides of delphinidin, cyanidin, peonidin, petunidin, and malvidin, with the proportion of 3′ and 3′, 5′ forms showing variation among cultivars as well as the contribution of acetylated forms [7]. In cultivars of *V. vinifera,* the glucose molecules can only be linked to the anthocyanidin through glycosidic bonds at the C_3 position to form 3-*O*-monoglucoside anthocyanins. In the other non-*V. vinifera* species, the 3,5-*O*-diglucoside anthocyanins can be present. Most cultivars possess acylated forms of the anthocyanins, but some have a much simpler profile, with the cultivar "Pinot Noir" being a notable example, possessing only the five basic anthocyanidin 3-glucosides. Some grape species, such as *V. rotundifolia,* are also known not to produce acylated forms of anthocyanins [8, 9]. It has recently been shown that the majority of anthocyanins within a cross section of North American grape species show a predominance of nonacylated forms of anthocyanins [10].

Further diversity in anthocyanin content is possible due to the presence of two distinct forms at vacuolar pH, the neutral quinonoidal anhydro base and the flavylium cation form [11]. The wide range in possible anthocyanin forms has led to a large amount of diversity in fruit color among cultivated grapes (Fig. 9.2). Because of this high level of diversity, grape cultivars have distinct anthocyanin profiles and it has been proposed as a method for the authentication of grape varieties and wine [12–14]. *V. vinifera* cultivars are characterized by the presence of only the monoglucoside forms of anthocyanins, while several of the wild *Vitis* species possess diglucosides. A double mutation occurring in an anthocyanin 5-*O*-glucosyltransferase within *V. vinifera* appears to be the cause for the loss of this enzymatic activity within this species [15].

Fig. 9.1 Anthocyanin accumulation within berry flesh of the teinturier cultivar "Bailey Alicante" during berry maturation

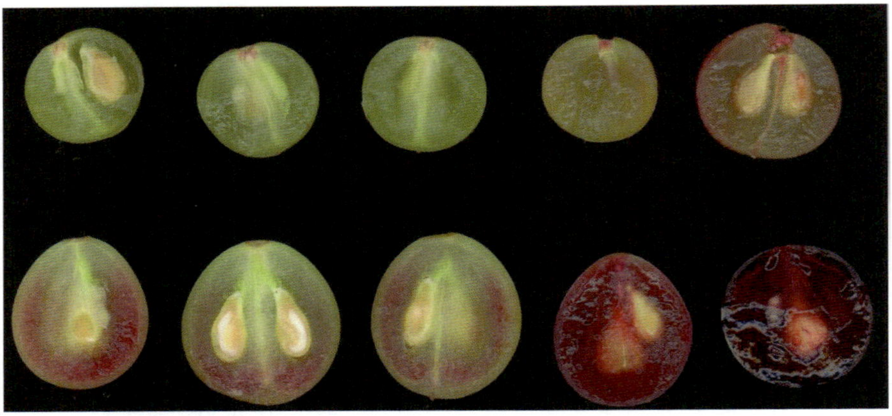

Fig. 9.2 Berry color diversity present within the cultivated grapevine *Vitis vinifera*

The production of red wine requires the maceration of the berries in order to extract the pigments from the berry skins. The concentration of anthocyanins begins to decrease after a few days of skin contact within the fermenting must [16]. During the course of wine ageing, polymeric pigments are known to form from reactions of anthocyanins with tannins [17], or from reactions with aldehydes [18]. Additionally, anthocyanin reactions that do not involve tannins are known to occur [19]. The resulting pigments include both anthocyanin polymers [20, 21] and small molecules such as pyranoanthocyanins [16, 22, 23], that are derived from the addition of yeast metabolites to the anthocyanins as well as caftaric acid–anthocyanin adducts that are formed through enzymatic reactions [24].

Anthocyanins are usually represented in the red flavylium cation form, the color of which shifts toward higher wavelengths (from orange to purple) as the number of substituents on the B ring increases. However, when dissolved in water, flavylium cations undergo proton transfer and hydration reactions, generating blue quinonoidal bases and colorless hemiketals, respectively [16]. At wine pH, grape anthocyanins occur mostly as the colorless, hydrated hemiketal form.

Red wine color is ensured through two stabilization processes: conversion of grape anthocyanins to other pigments, and association mechanism collectively called copigmentation. Anthocyanin-derived pigments show a wide range of colors from orange to blue. Tannin–anthocyanin adducts are red, like their anthocyanin precursors (λ_{max} 515–526 nm). Other reaction products include orange pyranoanthocyanins (λ_{max} 480–510 nm) [25, 26], purple ethyl-linked species (λ_{max} 528–540 nm) [18, 27], and blue flavanyl-vinylpyranoanthocyanins (λ_{max} 575 nm) [28]. These compounds may also change colors with the alternation of pH.

Color stabilization may also occur through copigmentation. The phenomenon of copigmentation is due to the molecular association between pigments and other nonpigmented organic molecules, leading to the exhibition of far greater color than would be expected by the pigment concentration [29]. Copigmentation can account for between 30 and 50% of the color in young wines and is vitally important for determining wine color, the variation in color and pigment concentration between wines. Copigmentation both increases color intensity (the hyperchromic effect) and shifts the color toward purple (the bathochromic effect). Colored anthocyanins are planar structures that can interact with other planar species (copigments) to form molecular stacks from which water is excluded. The flavylium ion is thus trapped and protected from hydration. This is particularly important at wine pH, where hydrated forms predominate.

Genetic Control and Variability in Anthocyanin Accumulation

The genetic control and inheritance of fruit color or anthocyanin production in grapevine are not fully understood despite evidence that the primary determination of anthocyanin production in berries appears to be controlled by a single dominant locus in *V. vinifera* [30, 31] with white fruit being a recessive character. This observation is supported by numerous reports showing that controlled crosses between white-fruited vines universally result in white-fruited progeny [32–37]. Although fruit color in grape is frequently characterized as a qualitative trait, quantitative variation does exist within segregating populations for total anthocyanin and for anthocyanin content. Broad-sense heritability is typically high for total anthocyanin content, and the majority of the phenotypic variation localizes to a genetic loci containing a cluster of myeloblastosis (MYB) transcription factors [38–41]. A genetic association analysis with candidate genes from the anthocyanin biosynthetic pathway revealed many polymorphisms within five regulatory and ten structural genes to be positively associated with anthocyanin content [42].

The genetic regulation of the flavonoid biosynthetic pathways has been extensively studied. Work conducted primarily in *Arabidopsis* and maize has shown that the basic flavonoid pathway upstream of anthocyanin biosynthesis is under the control of several different families of regulatory genes, consisting of complexes of MYB, basic helix-loop-helix (bHLH), and WD40. These regulatory genes are connected to form a network to regulate the expression of the structural genes involved in flavonoid synthesis. Similarities in sequence homology have been used to isolate and characterize many of the flavonoid biosynthetic structural genes in grape [43–47].

In grapes, a series of R2R3-Myb transcription factors has been demonstrated to be involved in the control of different aspects of the phenylpropanoid pathway [48]. Despite the characterization of several MYB proteins and their role in grapevine anthocyanin regulation, characterization of additional members of the protein complexes is not as well described. One bHLH gene, *VvMYC1*, was characterized as a component of the transcriptional complex controlling anthocyanin biosynthesis in grapevine [49]. The first *Myb* transcription factors identified in grape were isolated from *V. labrusca* hybrids [50], with subsequent orthologs identified in *V. vinifera* [51].

White-fruited grapes are linked to the homozygous presence of *Gret1* in the promoter region of *Vvmyba1* as well as mutations in the tightly linked gene *Vvmyba2* [52, 53]. *Gret1* is inserted in the 5′ flanking region of *VvmybA1,* resulting in the loss of function of this transcription factor. In pigmented somatic mutants of white-fruited cultivars, the full *Gret1* insertion is absent from the *VvymbA1* gene, but a solo long terminal repeat (LTR) has been shown to remain [54, 55]. Pigmented cultivars possess at least one allele at the *VvmybA1* locus not containing this large insertion [56]. In pigmented cultivars, *VvmybA1* is expressed only after veraison and has been shown to regulate anthocyanin biosynthesis by controlling the expression of anthocyanin biosynthetic genes, particularly flavonoid-3-*O*-glucosyltransferase (UFGT). Further evidence suggested that *VvmybA1* is not expressed in the young leaves, tendrils, or stem of grape vines and is tissue-specific to the berry [54, 57], while ectopic expression of *Vlmyba1–2* was shown to induce the expression of many of the genes within the flavonoid biosynthetic pathway [58].

VvmybA1 belongs to a family of linked regulatory genes which have related sequence homology, *VvmybA2* and *VvmybA3*. Sequence analysis of *VvmybA2* in 55 white cultivars of *V. vinifera* shows that the existence of rare mutations in two adjacent regulatory genes, the insertion of *Gret1* in *VvmybA1* and two nonconservative mutations in *VvmybA2* are present in many white-fruited cultivars [59]. Although white-fruited somatic mutations appear to be independent developments of white-fruited cultivars [55], evidence shows that *VvmybA1* co-segregates with the morphological marker for berry color [60] and that mutations in *VvmybA1* are associated with the vast majority of white-fruited *V. vinifera* accessions and many pink and red accessions as well [53, 60]. Additional polymorphisms within this cluster of closely related *myb* genes are also significantly associated with quantitative variation in anthocyanin content of berries [61, 62]. Recent work, mapping gene expression quantitative trait loci (QTL), also shows that the majority of the phenotypic variation within a population derived from a cross between two pigmented cultivars, “Syrah” and “Grenache,” colocated with the *VvmybA* gene cluster, while an additional large eQTL co-localized with *VvUFGT* [63].

Cellular Transport of Anthocyanins

The distribution of anthocyanins in the different branches of a grape cluster is highly variable depending on many environmental and physiological factors. Anthocyanins primarily accumulate in the hypodermal cell layers of the berry skin post-veraison. Anthocyanins are synthesized in the cytoplasm but accumulate in the vacuoles. Several glutathione S-transferase (GST) genes have been identified in grapes, which could function as anthocyanin transporters [64, 65]. In *Arabidopsis* and maize, GST proteins function with a glutathione S-conjugate pump and belong to the multidrug resistance-associated protein (MRP) family. To date, no MRP has been identified in grape involved with anthocyanin transport.

Two genes that encode multidrug and toxic compound extrusion (MATE) transporters have been identified in grape [66]. These genes were specifically expressed in fruit and showed expression patterns that correlated with anthocyanin accumulation. These genes have also been shown in vivo to be involved in vesicular anthocyanin transport in addition to GST [67]. Additionally, putative flavonoid translocators, which may be responsible for anthocyanin transport, and are similar to mammalian bilitranslocase, have been identified in ripening red- and white-fruited cultivars [68, 69].

Factors Affecting Variation in Anthocyanins

Variability in anthocyanin concentration and composition is known to be influenced by genotypes, berry developmental stages, environmental conditions, and cultural factors [7, 70].

The developmental stages of ripening have been shown to affect anthocyanin accumulation, with a two-stage accumulation, including a rapid increase of anthocyanins at early stages of development, closely correlated with sugar accumulation, followed by a second stage with slower accumulation [71]. These stages of ripening-related anthocyanin accumulation have been shown to be associated with vine vegetative conditions and climatic conditions [7]. The 3′-subsituted forms of anthocyanins have been observed to form earlier in the berry development than the 3′,5′ forms [72, 73].

Mineral nutrition is also known to influence variability in anthocyanin accumulation [70]. Higher concentrations of malvidin-3-*O*-glucoside and malvidin-3-*O*-coumarate glucoside in "Cabernet Sauvignon" vines that had been grafted on the high-vigor rootstock 1103P compared to those grafted on SO4 [74]. Heavy nitrogen application earlier in the season has been shown to delay ripening and to affect anthocyanin accumulation and composition [75].

Light and temperature also affect anthocyanin accumulation, with divergent effects depending on the ambient temperature [76–78]. Compositional changes in the constituent anthocyanins have also been observed in response to increases in solar radiation [76, 79, 80], while higher temperatures have also been reported to increase the proportion of acetylated to nonacetylated forms of anthocyanins [72, 76, 80]. Water deficits have also been showed to result in changes to anthocyanin concentrations as well as composition [81–83]. High nighttime temperatures have been shown to reduce the expression of several of the structural genes within the flavonoid biosynthetic pathway and also result in a reduction in the accumulation of anthocyanins [84–86]. Conversely, low storage temperatures have been shown to lead to an increase in expression of some of the flavonoid biosynthetic genes as well as an increase in anthocyanin accumulation [87, 88].

Considering the high value of the crop, extensive research has been conducted on utilizing viticultural practices to enhance fruit and wine quality, including the production of grape pigments. Cluster thinning, either by hand or mechanically, early in the season, prior to fruit set, has been shown to have many effects on fruit quality, including an increase in anthocyanin accumulation [89]. Conversely, cluster thinning at veraison has been shown to have much less of an effect on fruit quality [90]. Numerous additional cultural factors are known to affect berry quality, including the accumulation of anthocyanins, such as girdling of stems, disease infection, or the use of partial root-zone drying [91–93].

Ultraviolet irradiation is known to increase the expression of anthocyanin biosynthetic genes in several plant species and also lead to an increase in the accumulation of anthocyanins [70, 94]. Research examining the structure of the promoter sequences of two of the structural genes involved in flavonoid biosynthesis in grapes revealed that these genes could be induced by exposure to ultraviolet-A light [95, 96]. Similarly, it has been shown that exposure of grape berries to ultraviolet irradiation leads to the accumulation of anthocyanins in the cultivar "Gros Colman" [97].

Phytohormones are also known to influence anthocyanin accumulation in grape berries and tissues. Abscisic acid (ABA) is a phytohormone involved in stress responses, especially in response to water stress. It is known that the application of ABA can increase the anthocyanin content in grape berry skin and that application of ABA can greatly enhance the color of grapes [98]. Addition of ABA to grape cell suspension cultures can also promote anthocyanin accumulation and increase the expression of chalcone isomerase [99]. Additional work has shown exogenous application of ABA leads to enhance the expression of several structural genes involved in flavonoid biosynthesis as well as the regulatory gene VvmybA1 [100, 101].

Ethylene is a phytohormone known to influence many aspects of fruit ripening [70]. Application of the ethylene-releasing compound, 2-chloroethylphosphonic acid (2-CEPA), can hasten the accumulation of anthocyanins in grape skin, and application of 2-CEPA has been shown to increase the expression of several genes involved in flavonoid biosynthesis [102]. Similarly, the postharvest application of ethylene and/or 1-methylcyclopropene (1-MCP) to grape berries has been shown to improve the stability of anthocyanins during storage [103].

The application of some phytohormones has been shown to have a negative impact on anthocyanin accumulation. Specifically, application of the auxins 2,4-dichlorophenoxyacetic acid (2,4-D) or 1-napthaleneacetic acid (1-NAA) to grape berries has been shown to reduce the expression of several structural genes in the flavonoid biosynthetic pathways, and decrease the expression of *Vvmyba1* as well [100, 101].

Several nonphytohormone chemicals have also been shown to influence the accumulation of anthocyanins in grape berries and other tissues. The accumulation of sugars is closely correlated with the accumulation of anthocyanins in developing grape berries, and some evidence shows that sugars can stimulate the expression of some of the structural genes involved in flavonoid biosynthesis as well as lead to an increase in the accumulation of anthocyanins [104, 105]. Similarly, the application of a 5% ethanol solution has been shown to enhance the accumulation of anthocyanins in grape berries at veraison as well as increase the expression of several of the structural genes involved in the flavonoid biosynthetic pathway [106].

Also, eutypine [4-hydroxy-3-(3-methyl-3-butene-1-ynyl)benzaldehyde], the toxin produced by *Eutypa lata,* the pathogen causing Eutypa dieback in grapevines, has been shown to reduce the gene expression of UFGT, but not additionally tested flavonoid structural genes, and also inhibits the accumulation of anthocyanins in developing grape berries [107].

Anthocyanin Production Using Cell Suspension Culture

There is significant interest in using cell suspension cultures as a means for producing anthocyanins as natural colorants in the food industry [108, 109]. The primary source of cells for suspension cultures for the production of grape anthocyanins come from the two intensely pigmented teinturier cultivars, "Gamay Freaux" or "Bailey Alicante," in which high osmotic potential has been shown to enhance anthocyanin production [110, 111]. Elevating the sucrose content in the cell culture medium can increase the external osmotic potential and lead to an increase in anthocyanin accumulation, presumably through a mechanism similar to the osmotic stress induced under water deficit. Similar increases in anthyocyanin accumulation are observed through the addition of other osmolytes, such as mannitol. The addition of nitrate has also been shown to increase anthocyanin levels in cell suspension cultures, presumably by removing an inhibitory effect on production and transport [111]. In contrast, high ammonium concentrations in the culture medium can lead to a reduction in the accumulation of anthocyanins [112]. Combined, the effect of high sugar and low nitrogen can lead to an increase in anthocyanin accumulation [111]. Similarly, phosphate deprivation has been shown to increase the expression of dihydroflavonol 4-reductase (DFR) and additional increase in the accumulation of anthocyanins in cell suspension culture [113]. Polysaccharide elicitors have also been shown to enhance the accumulation of anthocyanins in cell suspension culture [114].

Conclusions

Grapes are one of the world's most important horticultural crops, in which the quality of wine and other grape products is the chief determinant of its value and importance. Wine, juice, and grape color are critical in determining the maximum quality of grape products, and substantial research has been conducted into the diversity of grape anthocyanins, regulation, physiology, and environmental/cultural factors that affect the production of anthocyanins. Although the anthocyanin pigments are widely distributed in the plant kingdom and have been extensively studied in many plant species, including many cultivated fruit crops, grapevine anthocyanins are of extreme agricultural significance due to the value and worldwide distribution of the crop. Additionally, the diversity of anthocyanins present within cultivated grape suggests the importance of these compounds for this crop and presents exciting opportunities to study the underlying processes leading to diversification of grape pigments over time.

References

1. Wen J (2007) Vitaceae. In: Kubitzki K (ed) Flowering plants-eudicots, vol 9. Springer-Verlag, Berlin, pp 467–479
2. Walker MA, Lider LA, Goheen AC, Olmo HP (1991) VR 039-16 grape rootstock. Hortscience 26(9):1224–1225. PubMed PMID: WOS:A1991GH45000049
3. Pauquet J, Bouquet A, This P, Adam-Blondon AF (2001) Establishment of a local map of AFLP markers around the powdery mildew resistance gene Run1 in grapevine and assessment of their usefulness for marker assisted selection. Theor Appl Genet 103(8):1201–1210. PubMed PMID: WOS:000173261300011
4. Mazza G (1995) Anthocyanins in grapes and grape products. Crit Rev Food Sci Nutr 35(4):341–371. PubMed PMID: WOS:A1995RM36300003
5. Liang Z, Owens CL, Zhong G-Y, Cheng L (2011) Polyphenolic profiles detected in the ripe berries of Vitis vinifera germplasm. Food Chem 129(3):940–950. PubMed PMID: WOS:000293726500034
6. Pomar F, Novo M, Masa A (2005) Varietal differences among the anthocyanin profiles of 50 red table grape cultivars studied by high performance liquid chromatography. J Chromatogr A 1094(1–2):34–41. PubMed PMID: WOS:000233150600005
7. Dai ZW, Ollat N, Gomes E, Decroocq S, Tandonnet J-P, Bordenave L et al (2011) Ecophysiological, genetic, and molecular causes of variation in grape berry weight and composition: a review. Am J Enol Vitic 62(4):413–425. PubMed PMID: WOS:000298125500001
8. Goldy RG, Maness EP, Stiles HD, Clark JR, Wilson MA (1989) Pigment quantity and quality characteristics of some native vitis-rotundifolia michx. Am J Enol Viti 40(4):253–258. PubMed PMID: WOS:A1989CC66100005
9. Pastrana-Bonilla E, Akoh CC, Sellappan S, Krewer G (2003) Phenolic content and antioxidant capacity of muscadine grapes. J Agr Food Chem 51(18):5497–5503. PubMed PMID: WOS:000184941500048
10. Liang Z, Yang Y, Cheng L, Zhong G-Y (2012) Polyphenolic composition and content in the ripe berries of wild Vitis species. Food Chem 132(2):730–738. PubMed PMID: WOS:000300475800008
11. Adams DO (2006) Phenolics and ripening in grape berries. Am J Enol Viti 57(3):249–256. PubMed PMID: WOS:000240934100002
12. Mattivi F, Guzzon R, Vrhovsek U, Stefanini M, Velasco R (2006) Metabolite profiling of grape: flavonols and anthocyanins. J Agr Food Chem 54(20):7692–7702. PubMed PMID: WOS:000240795400041
13. Edelmann A, Diewok J, Schuster KC, Lendl B (2001) Rapid method for the discrimination of red wine cultivars based on mid-infrared spectroscopy of phenolic wine extracts. J Agr Food Chem 49(3):1139–1145. PubMed PMID: WOS:000168967400011
14. Etievant P, Schlich P, Bertrand A, Symonds P, Bouvier JC (1988) Varietal and geographic classification of french red wines in terms of pigments and flavonoid compounds. J Sci Food Agr 42(1):39–54. PubMed PMID: WOS:A1988L932000005
15. Janvary L, Hoffmann T, Pfeiffer J, Hausmann L, Toepfer R, Fischer TC et al (2009) A double mutation in the anthocyanin 5-O-Glucosyltransferase gene disrupts enzymatic activity in *Vitis vinifera* L. J Agr Food Chem 57(9):3512–3518. PubMed PMID: WOS:000265896600018
16. Cheynier V, Prieur C, Guyot S, Rigaud J, Moutounet M (1997) The structures of tannins in grapes and wines and their interactions with proteins. In: Watkins TR (ed) Wine: nutritional and therapeutic benefits. ACS Symposium Series 661, pp 81–93

17. Salas E, Atanasova V, Poncet-Legrand C, Meudec E, Mazauric JP, Cheynier V (2004) Demonstration of the occurrence of flavanol-anthocyanin adducts in wine and in model solutions. Analytica Chimica Acta 513(1):325–332. PubMed PMID: WOS:000221718000043
18. Timberlake CF, Bridle P (1976) Interactions between anthocyanins, phenolic compounds, and acetaldehyde and their significance in red wines. Am J Enol Viti 27(3):97–105. PubMed PMID: WOS:A1976CM12400001
19. Cheynier V (2005) Polyphenols in foods are more complex than often thought. Am J Clin Nutr 81(1):223S–229S. PubMed PMID: WOS:000226401200003
20. Salas E, Duenas M, Schwarz M, Winterhalter P, Cheynier W, Fulcrand H (2005) Characterization of pigments from different high speed countercurrent chromatography wine fractions. J Agr Food Chem 53(11):4536–4546. PubMed PMID: WOS:000229405000035
21. Vidal S, Meudec E, Cheynier V, Skouroumounis G, Hayasaka Y (2004) Mass spectrometric evidence for the existence of oligomeric anthocyanins in grape skins. J Agr Food Chem 52(23):7144–7151. PubMed PMID: WOS:000225080100047
22. Fulcrand H, Benabdeljalil C, Rigaud J, Cheynier V, Moutounet M (1998) A new class of wine pigments generated by reaction between pyruvic acid and grape anthocyanins. Phytochemistry 47(7):1401–1407. PubMed PMID: WOS:000073714200040
23. Fulcrand H, Duenas M, Salas E, Cheynier V (2006) Phenolic reactions during winemaking and aging. Am J Enol Viti 57(3):289–297. PubMed PMID: WOS:000240934100006
24. Sarni-Manchado P, Cheynier V, Moutounet M (1997) Reactions of polyphenoloxidase generated caftaric acid o-quinone with malvidin 3-O-glucoside. Phytochemistry 45(7):1365–1369. PubMed PMID: WOS:A1997XL54600008
25. Benabdeljalil C, Cheynier V, Fulcrand H, Hakiki A, Mosaddak M, Moutounet M (2000) Evidence of new pigments resulting from reaction between anthocyanins and yeast metabolites. Sci Aliment 20(2):203–219. PubMed PMID: WOS:000088101400002
26. Sarni-Manchado P, Fulcrand H, Souquet JM, Cheynier V, Moutounet M (1996) Stability and color of unreported wine anthocyanin-derived pigments. J Food Sci 61(5):938–941. PubMed PMID: WOS:A1996VN17500018
27. Atanasova V, Fulcrand H, Le Guerneve C, Cheynier W, Moutounet M (2002) Structure of a new dimeric acetaldehyde malvidin 3-glucoside condensation product. Tetrahedron Lett 43(35):6151–6153. PubMed PMID: WOS:000177467500015
28. Mateus N, Silva AMS, Rivas-Gonzalo JC, Santos-Buelga C, De Freitas V (2003) A new class of blue anthocyanin-derived pigments isolated from red wines. J Agr Food Chem 51(7):1919–1923. PubMed PMID: WOS:000181706300027
29. Boulton R (2001) The copigmentation of anthocyanins and its role in the color of red wine: a critical review. Am J Enol Viti 52(2):67–87. PubMed PMID: WOS:000170693800001
30. Doligez A, Bouquet A, Danglot Y, Lahogue F, Riaz S, Meredith CP et al (2002) Genetic mapping of grapevine (*Vitis vinifera* L.) applied to the detection of QTLs for seedlessness and berry weight. Theor Appl Genet 105:780–795
31. Riaz S, Dangl GS, Edwards KJ, Meredith CP (2004) A microsatellite marker based framework linkage map of *Vitis vinifera* L. Theor Appl Genet 108(5):864–872. PubMed PMID: ISI:000220424400012
32. Madero E, Boubals D, Truel P (1986) Transmission hereditaire des principaux caracters des cepages Cabernet Franc, Cabernet Sauvignon et Merlot (*V. vinifera* L.). Vigne Vini 13(Suppl. 12):209–219
33. Barritt BH, Einset J (1969) Inheritance of 3 major fruit colors in grapes. J Amer Soc Hortic Sci 94(2):87–89. PubMed PMID: ISI:A1969D534400002
34. Hedrick UP, Anthony RD (1915) Inheritance of certain characters of grapes. New York State Agricultural College Technical Bulletin No 45, pp 3–19
35. Snyder E, Harmon FN (1939) Grape progenies of self-pollinated vinifera varieties. Proc Am Soc Hortic Sci 37:625–626
36. Snyder E, Harmon FN (1952) Grape breeding summary 1923–1951. Proc Am Soc Hortic Sci 60:243–246. PubMed PMID: ISI:A1952YB31100040
37. Wellington R (1939) The Ontario grape and its seedlings as parents. Proc Am Soc Hortic Sci 37:630–634

38. Liang Z, Sang M, Wu B, Ma A, Zhao S, Zhong G-Y et al (2012) Inheritance of anthocyanin content in the ripe berries of a tetraploid x diploid grape cross population. Euphytica 186(2):343–356. PubMed PMID: WOS:000304698100007
39. Fournier-Level A, Le Cunff L, Gomez C, Doligez A, Ageorges A, Roux C et al (2009) Quantitative genetic bases of anthocyanin variation in grape (*Vitis vinifera* L. ssp *sativa*) berry: a quantitative trait locus to quantitative trait nucleotide integrated study. Genetics 183(3):1127–1139. PubMed PMID: WOS:000272295800030
40. Bayo-Canha A, Ignacio Fernandez-Fernandez J, Martinez-Cutillas A, Ruiz-Garcia L (2012) Phenotypic segregation and relationships of agronomic traits in Monastrell x Syrah wine grape progeny. Euphytica 186(2):393–407. PubMed PMID: WOS:000304698100011
41. Viana AP, Riaz S, Walker MA (2013) Genetic dissection of agronomic traits within a segregating population of breeding table grapes. Genet Mol Res 12(2):951–964. PubMed PMID: WOS:000320030100009
42. Cardoso S, Lau W, Dias JE, Fevereiro P, Maniatis N (2012) A candidate-gene association study for berry colour and anthocyanin content in *Vitis vinifera* L. Plos ONE 7(9):e46021. PubMed PMID: WOS:000309973900073
43. Boss PK, Davies C, Robinson SP (1996) Analysis of the expression of anthocyanin pathway genes in developing *Vitis vinifera* L cv Shiraz grape berries and the implications for pathway regulation. Plant Physiol 111(4):1059–1066. PubMed PMID: WOS:A1996VD68800012
44. Bogs J, Ebadi A, McDavid D, Robinson SP (2006) Identification of the flavonoid hydroxylases from grapevine and their regulation during fruit development. Plant Physiol 140(1):279–291. PubMed PMID: WOS:000234492100025
45. Boss PK, Davies C, Robinson SP (1996) Expression of anthocyanin biosynthesis pathway genes in red and white grapes. Plant Mol Biol 32(3):565–569. PubMed PMID: WOS:A1996VY66300019
46. Ford CM, Boss PK, Hoj PB (1998) Cloning and characterization of *Vitis vinifera* UDP-glucose: flavonoid 3-O-glucosyltransferase, a homologue of the enzyme encoded by the maize Bronze-1 locus that may primarily serve to glucosylate anthocyanidins in vivo. J Biol Chem 273(15):9224–9233. PubMed PMID: WOS:000072990800091
47. Hall D, Yuan XX, Murata J, De Luca V (2012) Molecular cloning and biochemical characterization of the UDP-glucose: flavonoid 3-O-glucosyltransferase from Concord grape (*Vitis labrusca*). Phytochemistry 74:90–99. PubMed PMID: WOS:000300815700009
48. Deluc L, Bogs J, Walker AR, Ferrier T, Decendit A, Merillon J-M et al (2008) The transcription factor VvMYB5b contributes to the regulation of anthocyanin and proanthocyanidin biosynthesis in developing grape berries. Plant Physiol 147(4):2041–2053. PubMed PMID: WOS:000258184800051
49. Hichri I, Heppel SC, Pillet J, Leon C, Czemmel S, Delrot S et al (2010) The basic helix-loop-helix transcription factor MYC1 Is involved in the regulation of the flavonoid biosynthesis pathway in grapevine. Mol Plant 3(3):509–523. PubMed PMID: WOS:000278819900004
50. Kobayashi S, Ishimaru M, Hiraoka K, Honda C (2002) Myb-related genes of the Kyoho grape (*Vitis labruscana*) regulate anthocyanin biosynthesis. Planta 215(6):924–933. PubMed PMID: WOS:000178995100005
51. Kobayashi S, Goto-Yamamoto N, Hirochika H (2004) Retrotransposon-induced mutations in grape skin color. Science 304(5673):982. PubMed PMID: WOS:000221383300039
52. Walker AR, Lee E, Bogs J, McDavid DAJ, Thomas MR, Robinson SP (2007) White grapes arose through the mutation of two similar and adjacent regulatory genes. Plant J 49:772–785
53. This P, Lacombe T, Cadle-Davidson M, Owens CL (2007) Wine grape (*Vitis vinifera* L.) color associates with allelic variation in the domestication gene *VvmybA1*. Theor Appl Genet 114:723–730
54. Kobayashi S, Yamamoto NG, Hirochika H (2005) Association of VvmybA1 gene expression with anthocyanin production in grape (*Vitis vinifera*) skin—color mutants. J Jpn Soc Hortic Sci 74(3):196–203. PubMed PMID: WOS:000229191200003
55. Vezzulli S, Leonardelli L, Malossini U, Stefanini M, Velasco R, Moser C (2012) Pinot blanc and Pinot gris arose as independent somatic mutations of Pinot noir. J Exp Bot 63(18):6359–6369. PubMed PMID: WOS:000311645700006

56. Kobayashi S, Goto-Yamamoto N, Hirochika H (2004) Retrotransposon-induced mutations in grape skin color. Science 304:982
57. Jeong ST, Goto-Yamamoto N, Hashizume K, Esaka M (2006) Expression of the flavonoid 3†²-hydroxylase and flavonoid 3†²,5†²-hydroxylase genes and flavonoid composition in grape (*Vitis vinifera*). Plant Sci 170(1):61–69. PubMed PMID: WOS:000233340800008
58. Cutanda-Perez M-C, Ageorges A, Gomez C, Vialet S, Terrier N, Romieu C et al (2009) Ectopic expression of VlmybA1 in grapevine activates a narrow set of genes involved in anthocyanin synthesis and transport. Plant Mol Biol 69(6):633–648. PubMed PMID: WOS:000263899400001
59. Walker AR, Lee E, Bogs J, McDavid DAJ, Thomas MR, Robinson SP (2007) White grapes arose through the mutation of two similar and adjacent regulatory genes. Plant J 49(5):772–785. PubMed PMID: WOS:000244280300001
60. Lijavetzky D, Ruiz-Garcia L, Cabezas JA, De Andres MT, Bravo G, Ibanez A et al (2006) Molecular genetics of berry colour variation in table grape. Mol Genet Genom 276:427–435
61. Fournier-Level A, Le Cunff L, Gomez C, Doligez A, Ageorges A, Roux C et al (2009) Quantitative genetic bases of anthocyanin variation in grape (*Vitis vinifera* L. ssp *sativa*) berry: a quantitative trait locus to quantitative trait nucleotide integrated study. Genetics 183(3):1127–1139. PubMed PMID: ISI:000272295800030
62. Azuma A, Udo Y, Sato A, Mitani N, Kono A, Ban Y et al (2011) Haplotype composition at the color locus is a major genetic determinant of skin color variation in *Vitis xlabruscana* grapes. Theor Appl Genet 122(7):1427–1438. PubMed PMID: WOS:000289478500015
63. Huang Y-F, Bertrand Y, Guiraud J-L, Vialet S, Launay A, Cheynier V et al (2013) Expression QTL mapping in grapevine-revisiting the genetic determinism of grape skin colour. Plant Sci 207:18–24. PubMed PMID: WOS:000319488200004
64. Ageorges A, Fernandez L, Vialet S, Merdinoglu D, Terrier N, Romieu C (2006) Four specific isogenes of the anthocyanin metabolic pathway are systematically co-expressed with the red colour of grape berries. Plant Sci 170(2):372–383. PubMed PMID: WOS:000234732700020
65. Conn S, Curtin C, Bezier A, Franco C, Zhang W (2008) Purification, molecular cloning, and characterization of glutathione S-transferases (GSTs) from pigmented *Vitis vinifera* L. cell suspension cultures as putative anthocyanin transport proteins. J Exp Bot 59(13):3621–3634. PubMed PMID: WOS:000259973800012
66. Gomez C, Terrier N, Torregrosa L, Vialet S, Fournier-Level A, Verries C et al (2009) Grapevine MATE-type proteins act as vacuolar H^+-dependent acylated anthocyanin transporters. Plant Physiol 150(1):402–415. PubMed PMID: WOS:000265661700033
67. Gomez C, Conejero G, Torregrosa L, Cheynier V, Terrier N, Ageorges A (2011) In vivo grapevine anthocyanin transport involves vesicle-mediated trafficking and the contribution of anthoMATE transporters and GST. Plant J 67(6):960–970. PubMed PMID: WOS:000294866100002
68. Bertolini A, Peresson C, Petrussa E, Braidot E, Passamonti S, Macri F et al (2009) Identification and localization of the bilitranslocase homologue in white grape berries (*Vitis vinifera* L.) during ripening. J Exp Bot 60(13):3861–3871. PubMed PMID: WOS:000269609800021
69. Braidot E, Petrussa E, Bertolini A, Peresson C, Ermacora P, Loi N et al (2008) Evidence for a putative flavonoid translocator similar to mammalian bilitranslocase in grape berries (*Vitis vinifera* L.) during ripening. Planta 228(1):203–213. PubMed PMID: WOS:000256083200017
70. He F, Mu L, Yan G-L, Liang N-N, Pan Q-H, Wang J et al (2010) Biosynthesis of anthocyanins and their regulation in colored grapes. Molecules 15(12):9057–9091. PubMed PMID: WOS:000285709000036
71. Guidoni S, Ferrandino A, Novello V (2008) Effects of seasonal and agronomical practices on skin anthocyanin profile of Nebbiolo grapes. Am J Enol Viti 59(1):22–29. PubMed PMID: WOS:000253927900003
72. Downey MO, Dokoozlian NK, Krstic MP (2006) Cultural practice and environmental impacts on the flavonoid composition of grapes and wine: a review of recent research. Am J Enol Viti 57(3):257–268. PubMed PMID: WOS:000240934100003

73. Fournand D, Vicens A, Sidhoum L, Souquet J-M, Moutounet M, Cheynier V (2006) Accumulation and extractability of grape skin tannins and anthocyanins at different advanced physiological stages. J Agr Food Chem 54(19):7331–7338. PubMed PMID: WOS:000240465000056
74. Koundouras S, Hatzidimitriou E, Karamolegkou M, Dimopoulou E, Kallithraka S, Tsialtas JT et al (2009) Irrigation and rootstock effects on the phenolic concentration and aroma potential of *Vitis vinifera* L. cv. cabernet sauvignon grapes. J Agr Food Chem 57(17):7805–7813. PubMed PMID: WOS:000269376000028
75. Keller M, Arnink KJ, Hrazdina G (1998) Interaction of nitrogen availability during bloom and light intensity during veraison. I. Effects on grapevine growth, fruit development, and ripening. Am J Enol Viti 49(3):333–340. PubMed PMID: WOS:000076793300015
76. Tarara JM, Lee J, Spayd SE, Scagel CF (2008) Berry temperature and solar radiation alter acylation, proportion, and concentration of anthocyanin in Merlot grapes. Am J Enol Viti 59(3):235–247. PubMed PMID: WOS:000259247500002
77. Azuma A, Yakushiji H, Koshita Y, Kobayashi S (2012) Flavonoid biosynthesis-related genes in grape skin are differentially regulated by temperature and light conditions. Planta 236(4):1067–1080. PubMed PMID: WOS:000309229300010
78. Dokoozlian NK, Kliewer WM (1996) Influence of light on grape berry growth and composition varies during fruit development. J Am Soc Hortic Sci 121(5):869–874. PubMed PMID: WOS:A1996VF06300019
79. Chorti E, Guidoni S, Ferrandino A, Novello V (2010) Effect of different cluster sunlight exposure levels on ripening and anthocyanin accumulation in nebbiolo grapes. Am J Enol Viti 61(1):23–30. PubMed PMID: WOS:000275805400003
80. Spayd SE, Tarara JM, Mee DL, Ferguson JC (2002) Separation of sunlight and temperature effects on the composition of *Vitis vinifera* cv. Merlot berries. Am J Enol Viti 53(3):171–182. PubMed PMID: WOS:000178782800001
81. Castellarin SD, Pfeiffer A, Sivilotti P, Degan M, Peterlunger E, Di Gaspero G (2007) Transcriptional regulation of anthocyanin biosynthesis in ripening fruits of grapevine under seasonal water deficit. Plant Cell Environ 30(11):1381–1399. PubMed PMID: WOS:000249826400004
82. Olle D, Guiraud JL, Souquet JM, Terrier N, Ageorges A, Cheynier V et al (2011) Effect of pre- and post-veraison water deficit on proanthocyanidin and anthocyanin accumulation during Shiraz berry development. Aust J Grape Wine Res 17(1):90–100. PubMed PMID: WOS:000286626100011
83. Esteban MA, Villanueva MJ, Lissarrague JR (2001) Effect of irrigation on changes in the anthocyanin composition of the skin of cv Tempranillo (*Vitis vinifera* L) grape berries during ripening. J Sci Food Agr 81(4):409–420. PubMed PMID: WOS:000167217000006
84. Mori K, Goto-Yamamoto N, Kitayama M, Hashizume K (2007) Loss of anthocyanins in red-wine grape under high temperature. J Exp Bot 58(8):1935–1945. PubMed PMID: WOS:000248447500003
85. Mori K, Sugaya S, Gemma H (2005) Decreased anthocyanin biosynthesis in grape berries grown under elevated night temperature condition. Scientia Horticulturae 105(3):319–330. PubMed PMID: WOS:000229728500004
86. Yamane T, Jeong ST, Goto-Yamamoto N, Koshita Y, Kobayashi S (2006) Effects of temperature on anthocyanin biosynthesis in grape berry skins. Am J Enol Viti 57(1):54–59. PubMed PMID: WOS:000236618800006
87. Romero I, Sanchez-Ballesta MT, Maldonado R, Escribano MI, Merodio C (2008) Anthocyanin, antioxidant activity and stress-induced gene expression in high CO2-treated table grapes stored at tow temperature. J Plant Physiol 165(5):522–530. PubMed PMID: WOS:000254944100006
88. Sanchez-Ballesta MT, Romero I, Bernardo Jimenez J, Orea JM, Gonzalez-Urena A, Isabel Escribano M et al (2007) Involvement of the phenylpropanoid pathway in the response of table grapes to low temperature and high CO2 levels. Postharvest Biol Technol 46(1):29–35. PubMed PMID: WOS:000250077000004

89. Petrie PR, Clingeleffer PR (2006) Crop thinning (hand versus mechanical), grape maturity and anthocyanin concentration: outcomes from irrigated Cabernet Sauvignon (Vitis vinifera L.) in a warm climate. Aust J Grape Wine Res 12(1):21–29. PubMed PMID: WOS:000236875800003
90. Pena-Neira A, Caceres A, Pastenes C (2007) Low molecular weight phenolic and anthocyanin composition of grape skins from cv. syrah (Vitis vinifera L.) in the maipo valley (Chile): effect of clusters thinning and vineyard yield. Food Sci Technol Int 13(2):153–158. PubMed PMID: WOS:000248730900009
91. Bindon K, Dry P, Loveys B (2008) Influence of partial rootzone drying on the composition and accumulation of anthocyanins in grape berries (Vitis vinifera cv. Cabernet Sauvignon). Aust J Grape Wine Res 14(2):91–103. PubMed PMID: WOS:000257632600004
92. Parrado J, Escudero-Gilete ML, Friaza V, Garcia-Martinez A, Gonzalez-Miret ML, Bautista JD et al (2007) Enzymatic vegetable extract with bio-active components: influence of fertiliser on the colour and anthocyanins of red grapes. J Sci Food Agr 87(12):2310–2318. PubMed PMID: WOS:000249177500018
93. Brar HS, Singh Z, Swinny E, Cameron I (2008) Girdling and grapevine leafroll associated viruses affect berry weight, colour development and accumulation of anthocyanins in 'Crimson Seedless' grapes during maturation and ripening. Plant Sci 175(6):885–897. PubMed PMID: WOS:000261081300019
94. Park J-S, Choung M-G, Kim J-B, Hahn B-S, Kim J-B, Bae S-C et al (2007) Genes up-regulated during red coloration in UV-B irradiated lettuce leaves. Plant Cell Rep 26(4):507–516. PubMed PMID: WOS:000245896300013
95. Gollop R, Even S, Colova-Tsolova V, Perl A (2002) Expression of the grape dihydroflavonol reductase gene and analysis of its promoter region. J Exp Bot 53(373):1397–1409. PubMed PMID: WOS:000175955600003
96. Gollop R, Farhi S, Perl A (2001) Regulation of the leucoanthocyanidin dioxygenase gene expression in Vitis vinifera. Plant Sci 161(3):579–588. PubMed PMID: ISI:000170510400021
97. Kataoka I, Sugiyama A, Beppu K (2003) Role of ultraviolet radiation in accumulation of anthocyanin in berries of 'Gros Colman' grapes (Vitis vinifera L.). J Jpn Soc Hortic Sci 72(1):1–6. PubMed PMID: WOS:000180482600001
98. Kataoka I, Sugiura A, Utsunomiya N, Tomana T (1982) Effect of abscisic-acid and defoliation on anthocyanin accumulation in kyoho grapes (vitis-vinifera l x v labruscana bailey). Vitis 21(4):325–332. PubMed PMID: WOS:A1982QA03500004
99. Hiratsuka S, Onodera H, Kawai Y, Kubo T, Itoh H, Wada R (2001) ABA and sugar effects on anthocyanin formation in grape berry cultured in vitro. Sci Hortic 90(1–2):121–130. PubMed PMID: WOS:000172093200010
100. Ban T, Ishimaru M, Kobayashi S, Shiozaki S, Goto-Yamamoto N, Horiuchi S (2003) Abscisic acid and 2,4-dichlorophenoxyacetic acid affect the expression of anthocyanin biosynthetic pathway genes in 'Kyoho' grape berries. J Hortic Sci Biotechnol 78(4):586–589. PubMed PMID: WOS:000184479900025
101. Jeong ST, Goto-Yamamoto N, Kobayashi S, Esaka A (2004) Effects of plant hormones and shading on the accumulation of anthocyanins and the expression of anthocyanin biosynthetic genes in grape berry skins. Plant Sci 167(2):247–252. PubMed PMID: WOS:000222280600007
102. El-Kereamy A, Chervin C, Roustan JP, Cheynier V, Souquet JM, Moutounet M et al (2003) Exogenous ethylene stimulates the long-term expression of genes related to anthocyanin biosynthesis in grape berries. Physiol Plantarum 119(2):175–182. PubMed PMID: ISI:000185225200002
103. Bellincontro A, Fardelli A, De Santis D, Botondi R, Mencarelli F (2006) Postharvest ethylene and 1-MCP treatments both affect phenols, anthocyanins, and aromatic quality of Aleatico grapes and wine. Aust J Grape Wine Res 12(2):141–149. PubMed PMID: WOS:000239405400008

104. Zheng Y, Tian L, Liu H, Pan Q, Zhan J, Huang W (2009) Sugars induce anthocyanin accumulation and flavanone 3-hydroxylase expression in grape berries. Plant Growth Regul Jul;58(3):251–260. PubMed PMID: WOS:000266455600004
105. Gonzalez-sanjose ML, Diez C (1992) Relationship between anthocyanins and sugars during the ripening of grape berries. Food Chem 43(3):193–197. PubMed PMID: WOS:A1992HF29800004
106. El Kereamy A, Chervin C, Souquet JM, Moutounet M, Monje MC, Nepveu F et al (2002) Ethanol triggers grape gene expression leading to anthocyanin accumulation during berry ripening. Plant Sci 163(3):449–454. PubMed PMID: WOS:000178419800008
107. Afifi M, El-Kereamy A, Legrand V, Chervin C, Monje MC, Nepveu F et al (2003) Control of anthocyanin biosynthesis pathway gene expression by eutypine, a toxin from Eutypa lata, in grape cell tissue cultures. J Plant Physiol 160(8):971–975. PubMed PMID: WOS:000184874200016
108. Camire ME, Chaovanalikit A, Dougherty MP, Briggs J (2002) Blueberry and grape anthocyanins as breakfast cereal colorants. J Food Sci 67(1):438–441. PubMed PMID: WOS:000173893100077
109. Deroles S (2009) Anthocyanin biosynthesis in plant cell cultures: a potential source of natural colourants. In: Winefield C, Davies K, Gould K (eds) The anthocyanins. Springer-Verlag, New York, pp 107–167
110. Cormier F, Crevier HA, Do CB (1990) Effects of sucrose concentration on the accumulation of anthocyanins in grape (*vitis-vinifera*) cell-suspension. Can J Bot-Revue Canadienne De Botanique 68(8):1822–1825. PubMed PMID: WOS:A1990DZ76000029
111. Do CB, Cormier F (1991) Effects of low nitrate and high sugar concentrations on anthocyanin content and composition of grape (*vitis-vinifera* l) cell-suspension. Plant Cell Rep 9(9):500–504. PubMed PMID: WOS:A1991EU30300008
112. Do CB, Cormier F (1991) Effects of high ammonium concentrations on growth and anthocyanin formation in grape (*vitis-vinifera* l) cell-suspension cultured in a production medium. Plant Cell Tissue Organ Cult 27(2):169–174. PubMed PMID: WOS:A1991GR32000008
113. Dedaldechamp F, Uhel C, Macheix JJ (1995) Enhancement of anthocyanin synthesis and dihydroflavonol reductase (dfr) activity in response to phosphate deprivation in grape cell-suspensions. Phytochemistry 40(5):1357–1360. PubMed PMID: WOS:A1995TE27500005
114. Cai Z, Kastell A, Mewis I, Knorr D, Smetanska I (2012) Polysaccharide elicitors enhance anthocyanin and phenolic acid accumulation in cell suspension cultures of *Vitis vinifera*. Plant Cell Tissue Organ Cult 108(3):401–409. PubMed PMID: WOS:000301601600005

Chapter 10
Pigments in Strawberry

Jeremy Pillet and Kevin M. Folta

Strawberry Background

Strawberries are fruits recognized for their sweet, fruity flavor and nutrient content. They are cultivated principally throughout the Northern Hemisphere and typically consumed fresh, with significant acreage dedicated toward processing berries used in jams and jellies or as frozen fruits. The commercial strawberry *(Fragaria × ananassa)* is octoploid, an interspecific hybrid between *Fragaria virginiana* (a forest-floor strawberry from North America) and *F. chiloensis* (a wild strawberry from Chile). Strawberries are rich in many phytonutrients, including vitamin C, vitamin A, and a variety of minerals, while healthful attributes of strawberries are attractive to consumers, who are also highly motivated by their sweet flavors and rich red coloration [1].

The red coloration is a conspicuous feature of just about all cultivated strawberries, ranging from orange leaning cultivars (e.g., "Elsanta") to the extremely dark cultivar (e.g., "Purple Wonder"). White, yellow, peach, or pink-blushed strawberries exist throughout the world, yet are not grown in substantial acreage. The internal coloration also varies among varieties and correlates with cultural preferences. Various breeding programs prefer selection for pigmentation inside the strawberry, leading to an array of cultivars ranging from those with consistent red flesh throughout, to others with an attractive stark white contrast beneath the deep red epidermis. For the purposes of this chapter, the familiar red strawberry is discussed.

The strawberry is not a true fruit. Botanically speaking, it is an accessory fruit. The actual "fruits" are the achenes that dot the outside of the berry and the familiar red flesh is the expanded receptacle. The expansion and maturation are controlled by auxin and other phytohormones, with accumulation of pigment occurring typically once the fruit reaches full expansion.

K. M. Folta (✉) · J. Pillet
Horticultural Sciences, University of Florida, 1251 Fifield Hall,
Gainesville, FL 32611, USA
e-mail: kfolta@ufl.edu

C. Chen (ed.), *Pigments in Fruits and Vegetables,* DOI 10.1007/978-1-4939-2356-4_10

The Molecules Behind the Colors

The principle pigments in commercial strawberries are anthocyanins [2]. In strawberry, anthocyanins (from Greek anthos = flower and kyanos = blue) are responsible for pink, red, and purple pigments in the petals of some cultivars, like "Pink Panda" and "Lipstick." These pigments also provide the characteristic coloration of petioles, leaflets, stolons, and fruits.

Strawberry, like other plants, presents an anthocyanin-based color diversity that can range from blue to red with mauve, pink, or orange derivatives [3]. The pigments represent a wide class including at least 6000 molecules found across the plant kingdom that may be divided into several families, namely the flavonols, proanthocyanidins, and anthocyanins. The majority of these molecules have a C6–C3–C6 structure. Their classifications are based on the oxidation of central heterocyclic ring, the hydroxyl and methyl substitutions of the two other ring structures, glycosylation by glucose, galactose, arabinose, or rhamnose (more rarely by disaccharides), and flavonoid polymerization [4, 5]. Almost all flavonoids have associations with various sugars. Glycosylation, and thereafter acylation of sugars (association with one or more acids, in particular p-coumaric and cafeic acids), as well as methylation, modifies the biological and physicochemical properties of flavonoids. These modifications can affect the color of anthocyanins, their absorption spectrum, and increase their stability and solubility, which allows for vacuolar accumulation [6–8].

The strawberry pigments share basic structural similarities with other plants, namely the central unit of flavylium cation or 2-phenyl-1-benzopyrilium. They differ from each other by the number of acylations, hydroxylations, methylations, by the nature and the number of sugars for the molecule, as well as by the nature and number of aliphatic or aromatic acids related to sugar. The basic structures without sugar attached to the aromatic ring (aglycone) are called anthocyanidins. The most common forms of anthocyanidins found in plants are cyanidin, delphinidin, malvidin, pelargonidin, peonidin, and petunidin [9]. These molecules are generally present in plant tissues in a heterosid form. Each anthocyanidin type may be glycosylated and acylated at different sites and with different groups and acyl sugars. Once glycosylated, anthocyanidins become anthocyanins. The most common sugars associated with anthocyanidins are glucose, galactose, rhamnose, and arabinose [10]. These sugars can then be esterified with aliphatic or aromatic acids. The glycosylation and methylation processes are involved in the stabilization of anthocyanins protecting them from oxidation. The acylation process promotes intermolecular interactions with other molecules like flavonoids or carotenoids and these interactions are called co-pigmentation. This phenomenon can increase the stability of anthocyanins and their solubility in water [11].

Like in all plants, the accumulation and retention of these compounds in strawberry is regulated by modification. Sugar bonding capacity plays a critical role in the transport and storage of anthocyanins in the vacuole. All of these modifications (glycosylation, acylation, and methylation) not only increase the number of anthocyanins (approximately 400 identified to date) but also diversify their biological activities [11]. In an acidic aqueous medium, anthocyanidin color is determined by the degree of hydroxylation of the B core, producing a bathochromic shift. On the contrary, methylation and glycosylation reactions produce a hypsochromic shift [12].

Types of Anthocyanins

In strawberry fruit, the main anthocyanins are derived from pelargonidin and cyanidin aglycones [13]. The major pigment in cultivated strawberries has been identified as pelargonidin-3-glucoside which confers a bright red color to the receptacle. The presence of cyanidin-3-glucoside, in smaller proportions, has also been quantified in a large number of varieties [2, 14]. This minor pigment is responsible for the dark red colors. Pelargonidin-3-rutinoside can also be commonly found in a much smaller proportion in cultivars like "Camarosa" or "Parker," but this molecule is absent from native species like the diploid *F. vesca* or octoploid *F. chiloensis* [2, 15]. Other anthocyanins have been identified in other cultivars. For example, cyanidin-3-rutinoside can be found in "Camarosa," "Carisma," "Eris," and "Oso Grande" but not in "Tudnew" [2]. As another example, Tamura, Takada, and Yoshida (1995) have identified pelargonidin-3-(6-malonylglucoside) in several Japanese cultivars. In these cultivars, it is one of the major pigments comprising 5–30 % of total anthocyanin content [16, 17]. Co-pigmentation is common in strawberry fruits; Lopes da Silva et al. (2007) found four condensed products containing C–C linked anthocyanin and flavanol residues [2].

The developmental stage plays a key role in flavonoid metabolism in strawberry fruits and seems to be predominant over genotypes and environmental factors [18]. Anthocyanins are not present in all green stages. Their synthesis starts at approximately 25 days after anthesis (turning stage) and reaches a peak at full ripening [14, 18]. Anthocyanins mostly accumulate in epidermis and core fruit tissues from white to red stages [19].

Flavonoid and Anthocyanin Biosynthesis in Strawberry

Flavonoid biosynthesis in plants has been the subject of many studies over the past three decades [20, 21]. The basic substrate for flavonoid synthesis is phenylalanine. Phenylalanine ammonia-lyase (PAL), cinnamate 4-hydroxylase (C4H), and 4-coumaroyl CoA ligase (4CL) catalyzed the first steps of flavonoid synthesis and are leading to the production of 4-coumaroyl CoA. A chalcone is synthetized by the action of a chalcone synthase (CHS) on the 4-coumaroyl CoA. The gene coding for this enzyme is the first gene identified and described in this metabolic pathway [22]. This chalcone is then transformed into naringenin under the action of a chalcone isomerase. Hydroxylation of the C core of this molecule occurs then in position 3 by a flavanone 3-b hydroxylase (F3H) to synthesize dihydrokampferol.

This molecule can be hydroxylated on B core either in position 3′ or into 3′ and 5′ by a flavonoid 3′ or 3′-5′ hydroxylase (F3′H or F3′5′H). This operation generates, respectively, dihydroquercetin or dihydromyricetin. From those dihydroflavonols, dihydroflavonol 4-reductase (DFR) catalyzes the formation of anthocyanin precursors: leucocyanidin, leucodelphinidin, and leucopelargonidin. Leucoanthocyanidins oxidation catalyzed by the leucoanthocyanidin dioxygenase/anthocyanidin synthase (LDOX/ANS) leads to the formation of anthocyanidins: cyanidin, delphinidin, and perlargonidin. It is important to note that the flavonoid biosynthesis pathway is a complex and ramified network and there are numerous other enzymes involved and linked to those above that can lead to other types of flavonoids such as flavonols or flavanols. For example, flavonol synthase (FLS) can synthetize flavonol from dihydroflavonol precursors and so this enzyme is in competition with DFR enzyme.

Strawberry Genes for Anthocyanin Biosynthesis

In strawberry, different enzymes involved in the flavonoid biosynthetic pathway have been cloned and characterized at the genetic, biochemical, and molecular levels [18, 19, 23, 24]. The CHS gene family includes four isoforms, and while all *FaCHS* genes share a similar developmental expression pattern in fruits, a significant spatial regulation of the *FaCHS* family was observed. *FaCHS2* and *FaCHS5* are more expressed in petals, while *FaCHS1* and *FaCHS3* transcripts are more abundant in red fruits. *FaDFR* is encoded by a small multigene family but one of the isoforms, cloned from late ripening fruit complementary DNA (cDNA), seems to be associated to anthocyanin biosynthesis [19, 25]. Recently, Griesser et al. (2008) have shown that *FaGT1,* a putative glycosyltransferase, is fruit-associated, and *FaGT1* transcripts increase dramatically during turning and red stages [24]. Anthocyanidin-3-*O*-glucosyltransferase (GT) can catalyze the formation of the first stable intermediate in the anthocyanin pathway.

As noted previously, critical stages for anthocyanin biosynthesis start at the "turning" stage, a time when the fruit approaches full expansion and tracks toward maturation. At this time, there is an upregulation of the anthocyanin-related genes and a downregulation of the other flavonoid pathways genes [19]. These gene expression shifts determine the massive synthesis of pelargonidin and cyanidin derivatives against flavonols and flavan-3-ols.

Regulation of Anthocyanin Synthesis in Strawberry

The strawberry fruit is an emerging model for functional study of the genes contributing to pigmentation. Flavonoid biosynthesis has been measured using RNA interference (RNAi) silencing approaches in attached fruits. Strawberry lends itself well to this approach. Briefly, developing fruits may be injected with Agrobacterium containing hairpin constructs for a given gene [26]. After typically 14 days, the fruit expands and gene expression, physiology, and development may be assessed. This simple assay and the conspicuous effects metabolites of altering fruit color have enabled functional evaluation of strawberry's pigment-producing genes.

Anthocyanin regulation in strawberry parallels that in other organisms with some interesting deviations. Analysis of gene expression profiles related to the biosynthesis of flavonoids indicates a temporal and developmental regulation. Gene expression is regulated transcriptionally by a set of MYB/bHLH complexes paralleling what is seen in maize, petunia, and snapdragon [27]. Studies in *Arabidopsis,* petunia, and *Antirrhinum majus* have shown that each complex activates a specific pool of genes [28–30]. A similar situation has been observed in strawberry. In 2001, Aharoni et al. characterized *FaMYB1,* a ripening regulated gene and first member of the strawberry MYB family implicated in flavonoid synthesis [31]. Using transgenic expression in tobacco, the authors showed the potential role of this gene as a repressor of the anthocyanin and flavonol synthesis in the cultivated strawberry. Recently, Salvatierra et al. (2013) showed that *FaMYB1* ortholog in *F. chiloensis FcMYB1* acts also as a repressor of anthocyanin accumulation and is more active in the white-fruited strawberries than in the red ones [32]. Another MYB member, FaMYB10 is strongly associated with anthocyanin production in fruit [33]. Furthermore, the transcription activating capacity of FaMYB10 can be enhanced by the presence of a heterologous bHLH supporting the MYB–bHLH regulation hypothesis in strawberry. Finally, Schaart et al. (2013) report the regulation of proanthocyanidin biosynthesis by a MYB–bHLH–WD40 complex in strawberry fruits in the same way than the AtTT2–AtTT8–AtTTG1 complex in *Arabidopsis* [34].

Regulation in strawberry is proposed to follow a two-unit regulation model, a group upstream in the pathway (CHS, CHI, F3H) referred to as "early biosynthetic genes" (EBGs) and a set downstream called "late biosynthetic genes" (LBGs). The flavone-3-hydrolase (*F3H*) gene was functionally analyzed using the agro-infiltration method (35). *F3H* is highly expressed in fruits, and the protein hydroxylates the C-ring of flavanone, converting it into dihydroflavanol as an early step in anthocyanin synthesis. Successful suppression of this transcript significantly reduced pigmentation in injected fruits.

The function of CHS has been demonstrated using RNAi silencing in stable transgenic lines. Lunkenbein et al. (2006) used the antisense to suppress CHS levels, and saw an effect on fruits that had less than 25 % of the native transcript level [36]. Strong effects on pigmentation were only observed in strongly suppressing lines, giving rise to pink fruits. Strongly suppressing lines showed predicted decreases in flavonoids, anthocyanins, and proanthocyanidins, as well as the accumulation of p-coumaryl alcohol and p-coumaryl acetate, two unexpected metabolites lignin biosynthesis. These later compounds are thought to arise from redirection of substrates to the phenylpropanoid pathway. Similar effects have also been observed after agro-infiltration [37].

Similarly, DFR has been suppressed using the agro-infiltration approach. The treated fruits exhibited lower levels of DFR transcript and were markedly pale compared to controls [38]. Anthocyanin levels were suppressed, with pelargonidin, cyanidin, and kaempferol glycosides decreased significantly, while quercetin glycosides increased. Agro-infiltration methods also examined the function of anthocyanidin synthase, the step that oxidizes leucoanthocyanidins to anthocyanidins, providing color to the fruit. The cloning and in vitro characterization of the gene [19] was followed by functional tests in developing fruits, showing that the enzyme positively regulates pelargonidins and cyanidins [37].

Another functional study examined the association between the lack of an allergen and the lack of coloration in strawberry fruits. White fruits are tolerated well by those with sensitivity to red strawberry fruits. A proteomic study revealed that the fruits lacking pigmentation also were lacking *Fra a,* a protein related to the major birch allergen Bet v1 superfamily [39]. The same study noted that the fruits showed low levels of CHS, DFR, and F3H, leading to the compelling hypothesis that the allergen was related to pigment synthesis. To test the direct association, Munoz et al. (2011) used the agro-infiltration protocol to suppress the allergen in developing fruits [15]. The results show that *Fra a* suppression leads to decreases in pigmentation, and leads to the speculation that the protein may be required in the transportation of substrates between flavonoid pathway enzymes.

In other plants, such as in petunia and *Arabidopsis,* the separation between the two early and late steps occurs after F3H while in *Antirrhinum majus,* F3H is part of LBGs [40, 41]. In 2004, Hernandez et al. proposed a model for the discrimination of the different promoters of the early and late pathway genes [42]. In fact, MYB factors may act on the DNA physically or independently depending of bHLH cofactors. These cofactors act as activators by increasing the affinity of MYB for the DNA sequence and/or cancelling a specific inhibitor site located on the MYB sequence. In *Arabidopsis* and petunia, WD-repeat proteins (WDR) play a stabilizing role for the bHLH–MYB complex [43–45].

The temporal accumulation of these compounds is carefully regulated by these transcription factors. Throughout berry development, several families of compounds accumulate simultaneously. As the dihydroflavonols and leucoanthocyanidins are the same precursors of flavonols, anthocyanins, and flavanols, a competition between specific enzymes of the three pathways can be predicted. To avoid this competition, multienzyme complexes (also called metabolons) drive the synthesis flux toward one or more of the three final products. These complexes are associated with the endoplasmic reticulum (ER) cytosolic membranes or the vacuole by the P450 cytochrome enzyme family (F3′H, F3′5′H, C4H). Two hypotheses about the nature of these metabolons are proposed. In the model proposed by Ralston and Yu (2006), the enzymes are in contact with more than one partner, allowing coexistence of several biosynthetic pathways simultaneously in a same complex (flavonols, anthocyanins, etc.) [46]. Such complexes not only are useful to sequester toxic or unstable intermediates but also they can drive the synthesis to one or more of the many products of this pathway (anthocyanins, proanthocyanidins, flavonols, etc.). In contrast, in the model of Jørgensen et al. (2005) and Lepiniec et al. (2006), the multienzyme complex is linear, each enzyme having for partners the previous and next enzymes, promoting only one biosynthetic pathway [21, 47].

Other Flavonoid Pigments in Strawberry

It is important to note that despite the importance of anthocyanins in red fruits, phenylpropanoid metabolite synthesis is not completely directed into anthocyanin synthesis. Among these molecules, flavonols and flavanols influence strawberry pigmentation. Flavonols are yellow, white, and ivory pigments found in most plants. They are involved in ultraviolet (UV) protection and berry color by co-pigmentation with anthocyanins. They contribute to fruit coloration by stabilizing the unstable anthocyanin chromophore [11]. In most cases, the most common flavonols found in strawberry fruit are kaempferol-3-*O*-glucoside and quercetin-3-*O*-glucoside with a higher amount of kaempferol glucoside especially in *Fragaria* × *ananassa* cv "Camarosa" and "Parker" [15]. Glucuronide conjugates can also be found bound with kaempferol and quercetin aglycones. For example, in cultivar Queen Elisa, quercetin-3-*O*-glucuronide is the major flavonol found in the red fruit [19]. It seems that glucoside conjugates mostly accumulate in achenes during red stages, while glucuronide conjugates are most present in the receptacle during all developmental stages [48, 49]. More studies are needed to confirm these findings.

Flavanols, also called proanthocyanidins, are slightly brown or colorless polymerized compounds. They are known to be an essential factor affecting astringency and bitterness tastes of many fruits like apple or grape. The proanthocyanidin basic structures are the flavan-3-ols monomeric stereoisomers, (+)-catechin, and (−)-epicatechin [50]. It is important to note that they share the same substrates with anthocyanins, and they exhibit curious developmental and spatial accumulation. Indeed, during green and white stages PA accumulates in the receptacle and the achenes but a radical shift occurs during turning. As anthocyanin biosynthesis starts in the receptacle, PA accumulation tends to be restricted to vascular tissues and achenes throughout ripening [19]. Co-pigmentation of anthocyanin with flavanol can also occur in strawberry fruit. Fossen et al. (2004) report the identification of four dimeric flavonoids, which are pelargonidin 3-*O*-β-glucopyranoside units connected to four different flavan-3-ols with a covalent binding [51].

Anthocyanins in Postharvest Conditions

One of the areas of interest in strawberry is retention of pigmentation post harvest, both in fresh fruits and in processed materials. The main goal is to retain attractive coloration in the fruits to inspire purchase, as well as provide health benefits of related compounds.

The use of 1-methylcyclopropene is used on fruits to block ethylene effects, but the use of this compound on strawberry decreases anthocyanin levels due to effects on PAL [35]. Strawberry anthocyanin levels are also sensitive to carbon dioxide. Controlled environment experiments showed that while carbon dioxide slowed pathogen growth on fruits, it decreased anthocyanin accumulation in internal tissues while the external colors did not change [52]. It was later shown that the higher carbon dioxide levels affected PAL and GT activity, primarily in the internal tissues [53]. On the other hand, UV-C irradiation and heat have been shown to increase anthocyanin accumulation and have antimicrobial activities [54]. Other methods have been used to manipulate pigment accumulation, including the use of growth regulators and calcium.

In processed materials like jams and spreads, anthocyanin retention been improved by addition of pectins [55], sucrose [56], and lower storage temperatures [57]. Light has also been shown to decrease pigment stability, suggesting that processed materials need to be stored in darkness [55]. Anthocyanins have been stabilized in strawberry products using polyphenol oxidase inhibitors, and different compounds reflected in variable retention of monomeric and polymeric pigments [58]. Colorants are often added to such preparations to compensate for anthocyanin loss. These may include extracts from black carrots or elderberries [59].

Environmental Factors

Environmental factors can also affect the flavonoid pathway. In other species, anthocyanins respond to abiotic factors such as light and temperature. For example, high temperatures cause a decrease in the synthesis of anthocyanins in the reproductive organs of red apples [60], chrysanthemums [61], and asters [62] but in grapevine berry high temperatures induce production of p-coumaroyl derivatives [63]. Only malvidin-3-glucoside seems not affected by high temperatures but it nevertheless causes an increase of p-coumaroyl and acylated forms of anthocyanidin [64]. Light can also impact flavonoid composition. Shadowed grapevine berries show a decreased in anthocyanin and flavonol content, an effect also witnessed in strawberry [65]. In other plants, the anthocyanin composition changes and presents a higher level of trihydroxylated anthocyanins [63, 66].

Specific to strawberry, environmental factors have an influence on pigment accumulation. Dark and light mulch color has a significant effect, with more anthocyanins produced on brown mulch [65]. The same study examined planting date, forcing, and fruit order as presented on the plant, all of which had significant effects. The work by Carbone et al. (2009) examined gene expression, protein activity, and metabolites and found that location has a significant effect on proanthocyanidin levels [18]. The report notes that while environmental effects are observed, they have a higher amplitude in specific tissues and developmental states, and that genetic factors tend to override environmental variables. The authors do conclude that manipulation of the environment could be a means to increase the pigment content of strawberries with a focus on healthful compounds.

Conclusion

Strawberries are coveted for their sensory attributes, and while flavors and aromas are important, they are not the first sensory experience a consumer has with the fruit. The rich, red coloration is the first attraction, and an important quality in the purchase decision. This economic reality, tied to the potential health benefits of red fruits, has led scientists to understand the mechanisms controlling color in strawberry. Over the past decade, significant progress has been made in translating what is known from model organism pigment production to strawberry. This process has been facilitated by the sequenced reference genome and a groundswell of new genomics tools and data. The ability to make transgenic plants is a great asset to functional characterization, and the ability to obtain clear results from simple infiltration of fruits with agrobacterium-based silencing constructs has accelerated study in this area. There is a consensus in the strawberry community that the mechanisms governing pigment accumulation still leave many opportunities untouched, and all of these tools will be leveraged to hasten gene discovery.

References

1. Colquhoun T et al. (2012) Framing the perfect strawberry: an exercise in consumer-assisted selection of fruit crops. J Berry Res 2:45–61
2. Lopes da Silva F, Escribano-Bailón MT, Pérez Alonso JJ, Rivas-Gonzalo JC, Santos-Buelga C (2007) Anthocyanin pigments in strawberry. LWT Food Sci Technol 40:374–382
3. Tanaka Y, Sasaki N, Ohmiya A (2008) Biosynthesis of plant pigments: anthocyanins, betalains and carotenoids. Plant J 54:733–749
4. Ferrer JL, Austin MB, Stewart C, Noel JP (2008) Structure and function of enzymes involved in the biosynthesis of phenylpropanoids. Plant Physiol Biochem 46:356–370
5. Waterhouse AL (2002) Wine phenolics. Ann NY Acad Sci 957:21–36
6. Aron PM, Kennedy J (2008) Flavan-3-ols: nature, occurrence and biological activity. Mol Nutr Food Res 52:79–104
7. Macheix JJ, Fleuriet A, Jay-Allemand C (2005) Les composés phénoliques des végétaux. Presses polytechniques et universitaires romandes, Lausanne, p 131
8. Kong JM, Chia LS, Goh NK, Chia TF, Brouillard R (2003) Analysis and biological activities of anthocyanins. Phytochemistry 64:923–933
9. Harborne JB, Williams CA (2001) Anthocyanins and other flavonoids. Nat Prod Rep 18:310–333
10. Brouillard R, Dangles O (1993) Flavonoids and flower colour. In: Harborne JB (ed) The flavonoids. Advances in research since 1986. Chapman & Hall, London, pp 565–588
11. Grotewold E (2006) The genetics and biochemistry of floral pigments. Ann Rev Plant Biol 57:761–780
12. Brouillard R, Dubois JE (1977) Mechanism of the structural transformations of anthocyanins in acidic media. J Am Chem Soc 99:1359–1364
13. Mazza G, Miniati E (1993) Anthocyanins in fruits, vegetables and grains. CRC, Boca Raton
14. Kosar M, Kafkas E, Paydas S, Baser K (2004) Phenolic composition of strawberry genotypes at different maturation stages. J Agric Food Chem 52:1586–1589

15. Muñoz C, Sánchez-Sevilla JF, Botella MA, Hoffmann T, Schwab W, Valpuesta V (2011) Polyphenol composition in the ripe fruits of *Fragaria* species and transcriptional analyses of key genes in the pathway. J Agric Food Chem 59(23):12598–12604
16. Tamura H, Takada M, Yoshida Y (1995) Pelargonidin 3-O-(6-O-malonyl-β-D-glucoside) in *Fragaria* × *ananassa* Duch. cv. Nyoho. Biosci Biotechnol Biochem 59(6):1157–1158
17. Yoshida Y, Koyama N, Tamura H (2002) Color and anthocyanin composition of strawberry fruit: changes during fruit development and differences among cultivars, with special reference to the occurrence of pelargonidin 3-malonylglucoside. J Jpn Soc Horticult Sci 71(3):355–361
18. Carbone F, Preuss A, de Vos RCH, D'Amico E, Perrotta G, Bovy AG, Martens S, Rosati C (2009) Developmental, genetic and environmental factors affect the expression of flavonoid genes, enzymes and metabolites in strawberry fruits. Plant Cell Environ 32(8):1117–1131
19. Almeida JR, D'Amico E, Preuss A, Carbone F, de Vos RCH, Deiml B, Mourgues F, Perrotta G, Fischer TC, Bovy AG, Martens S, Rosati C (2007) Characterization of major enzymes and genes involved in flavonoid and proanthocyanidin biosynthesis during fruit development in strawberry (*Fragaria* × *ananassa*). Arch Biochem Biophys 465(1):61–71
20. Winkel-Shirley B (2006) The biosynthesis of flavonoids. In: E. Grotewold (Ed.) The science of flavonoids. Springer, New York, pp 71–95
21. Lepiniec L, Debeaujon I, Routaboul JM, Baudry A, Pourcel L, Nesi N, Caboche M (2006) Genetics and biochemistry of seed flavonoids. Ann Rev Plant Biol 57:405–430
22. Reimold U, Kroger M, Kreuzaler F, Hahlbrock K (1983) Coding and 3' non-coding nucleotide sequence of chalcone synthase mRNA and assignment of amino acid sequence of the enzyme. EMBO J 2:1801–1805
23. Muñoz C, Hoffmann T, Escobar NM, Ludemann F, Botella MA, Valpuesta V, Schwab W (2010) The strawberry fruit Fra a allergen functions in flavonoid biosynthesis. Mol Plant 3:113–124
24. Griesser M, Hoffmann T, Bellido ML, Rosati C, Fink B, Kurtzer R, Aharoni A, Muñoz-Blanco J, Schwab W (2008) Redirection of flavonoid biosynthesis through the down-regulation of an anthocyanidin glucosyltransferase in ripening strawberry fruit. Plant Physiol 146(4):1528–1539
25. Moyano E, Portero-Robles I, Medina-Escobar N, Valpuesta V, Muñoz-Blanco J, Caballero JL (1998) A fruit-specific putative dihydroflavonol 4-reductase gene is differentially expressed in strawberry during the ripening process. Plant Physiol 117(2):711–716
26. Hoffmann T, Kalinowski G, Schwab W (2006) RNAi-induced silencing of gene expression in strawberry fruit (*Fragaria* × *ananassa*) by agroinfiltration: a rapid assay for gene function analysis. Plant J 48:818–826
27. Vom Endt D, Kijne JW, Memelink J (2002) Transcription factors controlling plant secondary metabolism: what regulates the regulators? Phytochemistry 61(2):107–114
28. Mol J, Grotewold E, Koes R (1998) How genes paint flowers and seeds. Trends Plant Sci 3(6):212–217
29. Quattrocchio F, Wing JF, Va K, Mol JN, Koes R (1998) Analysis of bHLH and MYB domain proteins: species-specific regulatory differences are caused by divergent evolution of target anthocyanin genes. Plant J 13(4):475–488
30. Pelletier MK, Winkel-Shirley B (1996) Analysis of flavanone 3-hydroxylase in Arabidopsis seedlings (Coordinate regulation with chalcone synthase and chalcone isomerase). Plant Physiol 111(1):339–345
31. Aharoni A, de Vos RCH, Wein M, Sun Z, Greco R, Kroon A, Mol J, O'Connell AP (2001) The strawberry FaMYB1 transcription factor suppresses anthocyanin and flavonol accumulation in transgenic tobacco. Plant J 28(3):319–332
32. Salvatierra A, Pimentel P, Moya-León MA, Herrera R (2013) Increased accumulation of anthocyanins in *Fragaria chiloensis* fruits by transient suppression of FcMYB1 gene. Phytochemistry 90:25–36

33. Lin-Wang K, Bolitho K, Grafton K, Kortstee A, Karunairetnam S, McGhie TK, Espley RV, Hellens RP, Allan AC (2010) An R2R3 MYB transcription factor associated with regulation of the anthocyanin biosynthetic pathway in Rosaceae. BMC Plant Biol 10(1):50
34. Schaart JG, Dubos C, Romero De La Fuente I, van Houwelingen AM, de Vos RCH, Jonker HH, Xu W, Routaboul JM, Lepiniec L, Bovy AG (2013) Identification and characterization of MYB-bHLH-WD40 regulatory complexes controlling proanthocyanidin biosynthesis in strawberry (*Fragaria* × *ananassa*) fruits. New Phytol 197(2):454–467
35. Jiang Y, Joyce, DC, Terry LA (2001) 1-Methylcyclopropene treatment affects strawberry fruit decay. Postharvest Biol Technol 23:227–232
36. Lunkenbein S, Coiner H, deVos CHR, Schaart J, Boone MJ, Krens FA, Schwab W, Salentjin EMJ (2006) Molecular characterization of a stable antisense chalcone synthase phenotype in strawberry. J Agric Food Chem 54:2145–2153
37. Schwab W, Hoffmann T, Kalinowski G, Preuss A (2011) Functional genomics in strawberry fruit through RNAi-mediated silencing. Genes Genom Genomics 5:91–101
38. Lin X, Xiao M, Luo Y, Wang J, Wang H (2013) The effect of RNAi-induced silencing of FaDFR on anthocyanin metabolism in strawberry (*Fragaria* × *ananassa*) fruit. Sci Hortic 160:123–128
39. Hjernø K, Alm R, Canbäck B, Matthiesen R, Trajkovski K, Björk L, Roepstorff P, Emanuelsson C (2006) Down-regulation of the strawberry Bet v 1-homologous allergen in concert with the flavonoid biosynthesis pathway in colorless strawberry mutant. Proteomics 6:1574–1587
40. Kubasek WL, Winkel-Shirley B, McKillop A, Goodman HM, Briggs W, Ausubel FM (1992) Regulation of flavonoid biosynthetic genes in germinating *Arabidopsis* seedlings. Plant Cell 4:1229–1236
41. Martin C, Prescott A, Mackay S, Bartlett J, Vrijlandt E (1991) Control of anthocyanin biosynthesis in flowers of *Antirrhinum majus*. Plant J 1(1):37–49
42. Hernandez JM, Heine GF, Irani NG, Feller A, Kim MG, Matulnik T, Chandler VL, Grotewold E (2004) Different mechanisms participate in the R-dependent activity of the R2R3 MYB transcription factor C1. J Biol Chem 279:48205–48213
43. Broun P (2005) Transcriptional control of flavonoid biosynthesis: a complex network of conserved regulators involved in multiple aspects of differentiation in Arabidopsis. Curr Opin Plant Biol 8:272–279
44. Ramsay NA, Glover BJ (2005) MYB-bHLH-WD40 protein complex and the evolution of cellular diversity. Trends Plant Sci 10:63–70
45. deVetten N, Quattrocchio F, Mol J, Koes R (1997) The an11 locus controlling flower pigmentation in petunia encodes a novel WD-repeat protein conserved in yeast, plants, and animals. Genes Dev 11:1422–1434
46. Ralston L, Yu O (2006) Metabolons involving plant cytochrome P450s. Phytochem Rev 5:459–472
47. Jørgensen K, Rasmussen AV, Morant M, Nielsen AH, Bjarnholt N, Zagrobelny M, Bak S, Møller BL (2005) Metabolon formation and metabolic channeling in the biosynthesis of plant natural products. Curr Opin Plant Biol 8:280–291
48. Halbwirth H, Puhl I, Haas U, Jezik K, Treutter D, Stich K (2006) Two-phase flavonoid formation in developing strawberry (*Fragaria* × *ananassa*) fruit. J Agric Food Chem 54(4):1479–1485
49. Fait A, Hanhineva K, Beleggia R, Dai N, Rogachev I, Nikiforova VJ, Fernie AR, Aharoni A (2008) Reconfiguration of the achene and receptacle metabolic networks during strawberry fruit development. Plant Physiol 148:730–750
50. Dixon RA, Xie DY, Sharma SB (2005) Proanthocyanidins–a final frontier in flavonoid research? New Phytol 165:9–28
51. Fossen T, Rayyan S, Andersen ØM (2004) Dimeric anthocyanins from strawberry (*Fragaria ananassa*) consisting of pelargonidin 3-glucoside covalently linked to four flavan-3-ols. Phytochemistry 65(10):1421–1428

52. Gil MI, Holcroft DM, Kader AA (1997) Changes in strawberry anthocyanins and other polyphenols in response to carbon dioxide treatments. J Agric Food Chem 45:1662–1667
53. Holcroft DM, Kader AA (1999) Carbon dioxide-induced changes in color and anthocyanin synthesis in stored strawberry fruit. HortScience 34:1244–1248
54. Pan J, Vicente AR, Martinez GA, Chaves AR, Civello PM (2004) Combined use of UV-C irradiation and heat treatment to improve postharvest life of strawberry fruit. J Sci Food Agric 84:1831–1838
55. Holzwarth M, Korhummel S, Siekmann T, Carle R, Kammerer DR (2013) Influence of different pectins, process and storage conditions on anthocyanin and colour retention in strawberry jams and spreads. Food Sci Technol 52:131–138
56. Watanabe Y, Yoshimoto K, Okada Y, Nomura M (2011) Effect of impregnation using sucrose solution on stability of anthocyanin in strawberry jam. LWT Food Sci Technol 44:891–895
57. García-Viguera P, Zafrilla P, Tomás-Barberán FA (1999) Influence of processing and storage conditions in strawberry jam color. Food Sci Technol Int 5:487–492
58. Holzwarth M, Wittig J, Carle R, Kammerer DR (2013b) Influence of putative polyphenoloxidase (PPO) inhibitors on strawberry PPO, anthocyanin and color stability in stored purees. LWT Food Sci Technol 52:116–122
59. Kammerer DR, Schillmöller S, Maier O, Schieber A, Carle R (2007) Colour stability of canned strawberries using black carrot and elderberry juice concentrates as natural colourants. Eur Food Res Technol 224:667–679
60. Tomana T, Yamada H (1988) Relationship between temperature and fruit quality of apple cultivars grown at different locations. J Jpn Soc Horticult Sci 56:391–397
61. Shibata M, Amano M, Kawata J, Uda M (1988) Breeding process and characteristics of "Summer Queen", a spray-type chrysanthemum. Bull Natl Res Inst Veg Orn Plants Tea. Series A 2:245–255
62. Shaked-Sachray L, Weiss D, Reuveni M, Nissim-Levi A, Oren-Shamir M (2002) Increased anthocyanin accumulation in aster flowers at elevated temperatures due to magnesium treatment. Phys Plant 114:559–565
63. Downey MO, Dokoozlian NK, Krstic MP (2006) Cultural practice and environmental impacts on the flavonoid composition of grapes and wine: a review of recent research. Am J Enol Vitic 57:257–268
64. Tarara JM, Lee J, Spayd SE, Scagel CF (2008) Berry temperature and solar radiation alter acylation, proportion, and concentration of anthocyanin in merlot grapes. Am J Enol Vitic 59:235–247
65. Anttonen MJ, Hoppula KI, Nestby R, Verheul MJ, Karjalainen RO (2006) Influence of fertilization, mulch color, early forcing, fruit order, planting date, shading, growing environment, and genotype on the content of selected phenolics in strawberry (*Fragaria* × *ananassa* Duch.) fruits. J Agric Food Chem 54:2614–2620
66. Spayd SE, Tarara JM, Mee DL, Ferguson JC (2002) Separation of sunlight and temperature effects on the composition of *Vitis vinifera* cv. Merlot berries. Am J Enol Vitic 53:171–182

Chapter 11
Carotenoids in Carrot

Claudia Stange Klein and Manuel Rodriguez-Concepcion

Introduction

Carrot (*Daucus carota* L., $2n=18$) is a biennial plant of the Umbelliferae (or Apiaceae) family, which also includes parsley (*Petroselinum hortense*), celery (*Apium graveolens*), anise (*Pimpinella anisum*), and dill (*Anethum graveolens)* among other thousands of species and hundreds of genera. The main particularity of carrot is the massive accumulation of carotenoids that takes place in the root, which causes its characteristic color in most cultivars. Carrots are actually the origin of the name for carotenoids. In other plants, these plastidial isoprenoid pigments provide yellow to red colors to flowers, fruits, and seeds [1–4]. For example, carotenoids provide the yellow color of corn and bananas, the orange color of pumpkin and oranges, and the red color of tomato and watermelon. However, it is very uncommon that these pigments are produced at levels high enough to color the root, as it is the case of some carrot varieties.

Young carrot roots are pale but they acquire a characteristic color by accumulating carotenoids after the first month of growth, reaching a maximum shortly before secondary growth is completed in about 3 months [5–7]. Yellow, orange, or red carrots (i.e., mature *D. carota* roots) result from differences in the levels and types of carotenoids they produce [8–13]. Mature roots of orange varieties (the most popular carrots, at least in Western countries) accumulate mainly β-carotene and lower level of α-carotene (Fig. 11.1). In red carrots, lycopene is accumulated even at higher levels than those in tomato (*Solanum lycopersicum*) fruits. Yellow carrots get their color from preferentially accumulating lutein, although the level of total

M. Rodriguez-Concepcion (✉)
Centre for Research in Agriculture Genomics (CRAG) CSIC-IRTA-UAB-UB, Campus Uab Bellaterra, Barcelona, Spain
e-mail: manuel.rodriguez@cragenomica.es

C. S. Klein
Department of Biology, University of Chile, Santiago, Chile

C. Chen (ed.), *Pigments in Fruits and Vegetables,* DOI 10.1007/978-1-4939-2356-4_11

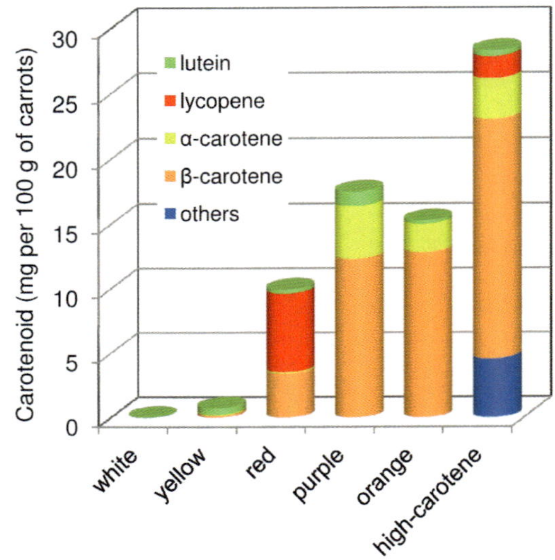

Fig. 11.1 Carotenoid profile in carrots of different colors. (Based on data, Ref. [8])

root carotenoids is much lower than that in orange or red varieties (Fig. 11.1). The presence of only trace levels of lutein and other carotenoids in white carrots results in the characteristic lack of pigmentation of these cultivars [8, 9, 11–14]. By contrast, the color of purple carrots is not caused by carotenoids but by a hydrophilic class of pigments, anthocyanins [11, 12, 15, 16].

The range of carrot root colors resulted from breeding activities during cultivation of this plant. Wild carrots are likely to be native from central Asia (Iran and Afghanistan), where domestication probably took place around the tenth century [17]. Initially, carrot plants were likely cultivated because of their aromatic leaves and seeds, since their uncolored roots were bitter and had a woody core. The first Eastern carrots were purple or yellow, and they were brought to Spain in the twelfth century via North Africa. In Europe, selective breeding over the centuries led to the development of new cultivars of white and orange carrots in the seventeenth century. Red carrots appeared in India and China around the eighteenth century [17]. Despite intensive breeding exchanges since the nineteenth century, the background structure coming from demographic and early cultivation history is still conserved in currently cultivated carrot germplasm [18].

At present, orange carrots are consumed worldwide and are popular in a variety of foods because of their pleasant flavor mainly from volatile isoprenoids and sugars [12, 19]. Although pigments (carotenoids and anthocyanins) do not appear to impact carrot flavor [8], they contribute to make the product more colorful and appealing to consumers. Most importantly, both carotenoids and anthocyanins are powerful antioxidants and phytonutrients that contribute to human health [3, 20]. In particular, orange varieties rich in β-carotene, including Nantaise, Bolero, Chantenay, and genetically selected strains like the high-carotene mass carrot [21] shown in Fig. 11.1, represent a major source for provitamin A in the human diet.

Additionally, carrots contain other vitamins (C, D, E, K, B1, and B6), minerals, and nutrients (see http://ndb.nal.usda.gov/ndb/foods/list).

Developmental and Environmental Cues Influence Carotenoid Accumulation in Carrot Roots

Carotenoid content in carrots varies according to genotypes, growing conditions, season, maturity, and storage [11, 13, 22, 23]. Two of the best characterized factors influencing carotenoid accumulation in carrots of the same cultivar growing under controlled conditions are the root developmental program and the presence of light. The effects of these two factors on final root carotenoid profiles are highly intertwined (Fig. 11.2).

When grown underground (in the absence of light), the roots of colored carrot cultivars like the orange carrot Nantaise are thin and colorless at early stages of development (4 weeks) but thicken and become colored by carotenoid accumulation at 8 weeks (Fig. 11.2a). From 8 to 12 weeks, secondary growth of these dark-grown roots (referred to as R/D) takes place, causing a substantial enlargement and a dramatic accumulation of carotenoids in chromoplasts (Fig. 11.2a). Chromoplasts are plastids specialized in the production and accumulation of carotenoids at massive levels [24]. The carotenoid profile of chromoplasts varies widely in different species and organs [25]. For example, the chromoplasts of red ripe tomato fruit and red carrots accumulate high levels of lycopene, whereas those in orange and yellow carrots mainly accumulate β-carotene or lutein, respectively (Fig. 11.1). Chromoplasts sequester large amounts of carotenoids in plastoglobules or/and carotenoid storage structures of globular, tubular, fibrillar, membranous, or crystalline shapes composed of lipids and proteins [1, 24, 26–28]. In orange carrots, root chromoplasts accumulate carotenoids as large crystals [6, 14, 29]. Plastoglobuli were also found in these chromoplasts, but they were small in size and number and only carried minute amounts of carotene *cis* isomers unlike the *trans* isomers stored as crystals [29, 30]. In R/D carrot segments, carotenoid-containing chromoplasts are likely derived from starch-containing amyloplasts and are particularly abundant in the secondary phloem of the root [10, 29, 31]. Chromoplasts are much smaller in size and number in white carrots, which instead contain abundant amyloplasts with large starch grains [29].

When the upper section of carrot roots is left aboveground, i.e., is exposed to light, chloroplasts differentiate instead of chromoplasts [6, 7] (Fig. 11.2b). Unlike chromoplasts, chloroplasts show a very similar carotenoid profile in different species and plant tissues, with high levels of carotenoids (mainly lutein and β-carotene) associated with photosynthetic systems, where they contribute to light harvesting and photoprotection [32–34]. Therefore, it is not surprising that the light-exposed carrot root sections (referred to as R/L) show a carotenoid profile similar to that of other chloroplast-containing organs such as leaves (Fig. 11.2c). For example, lutein accumulates similarly in R/L root segments and leaf samples, whereas it is absent

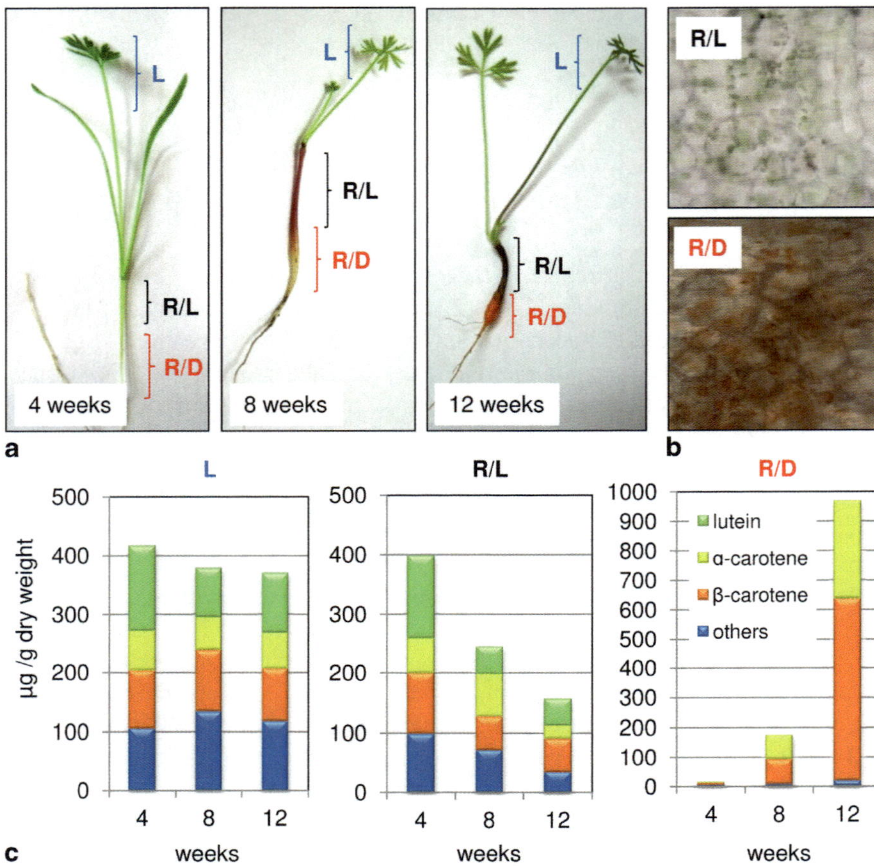

Fig. 11.2 Root development and carotenoid accumulation in Nantaise carrots. **a** Carrot plants grown for 4, 8, and 12 weeks. *L* leaf, *R/L* root section continuously exposed to light during development, *R/D* root section grown in darkness. **b** Light microscopy images of the secondary phloem zone of 12-week-old R/L and R/D sections. Chloroplasts in the R/L panel are *green*, and chromoplasts in the R/D panel are *orange*. **c** Carotenoid content and composition of samples like those shown in **a.** (Adapted from Fuentes et al. [6]. With permission from Springer Verlag)

from R/D segments (Fig. 11.2c). Most strikingly, total carotenoid levels decrease during root development in R/L sections but dramatically increase in R/D segments (Fig. 11.2c). Another important difference between R/L and R/D sections is that R/L segments remain thin and stem-like (Fig. 11.2a), illustrating a negative effect of light on the normal development of carrot storage roots. However, R/L sections showed a cellular organization similar to that in R/D root sections, with a central secondary xylem, a vascular cambium, and an extensive secondary phloem [6]. This observation confirms that light does not change the organ identity of the carrot root but it influences plastid development, promoting the differentiation of chloroplasts and preventing the differentiation of chromoplasts (Fig. 11.2b). Consistent with the conclusion that light treatment does not transform the carrot root into a stem, R/L sections expand and become orange (R/D-like) when transferred back to the dark [7].

Genes Encoding Most of the Enzymes Involved in Carotenoid Biosynthesis Have Been Identified in Carrots

The characteristic color of carotenoids in the yellow to red range is due to a polyene chain with a number of conjugated double bonds that functions as a chromophore. The core pathway for the formation of this polyene chain and the biosynthesis of major carotenoids is well established [35]. Plant carotenoids are synthesized in plastids and derive from isoprenoid precursors synthesized by the 2-C-methyl-D-erythritol 4-phosphate (MEP) pathway [3, 36]. MEP-derived units of five carbons (5-C) are used to produce the 20-C molecule geranylgeranyl diphosphate (GGPP), the precursor not only for 40-C carotenoids (Fig. 11.3) but also for other plastidial isoprenoids involved in photosynthesis (side chain of chlorophylls, tocopherols, plastoquinones, and phylloquinones) and regulation of plant growth and development (gibberellins).

The first committed step in carotenoid biosynthesis is the condensation of two GGPP molecules to synthesize 15-*cis*-phytoene catalyzed by the enzyme phytoene synthase (PSY). The rest of the pathway has been best characterized at the genetic level in *Arabidopsis thaliana* [35]. The colorless 15-*cis*-phytoene molecule is first desaturated and isomerized to form all-*trans* lycopene, a reddish carotenoid. As shown in Fig. 11.3, the enzymes involved in these reactions are, sequentially, phytoene desaturase (PDS), 15-*cis*-ζ–carotene isomerase (ZISO), ζ–carotene desaturase (ZDS), and carotenoid (prolycopene) isomerase (CISO). Next, the pathway branches with the cyclization of the two ends of the lycopene molecule for the production of either β-carotene (containing two β rings) or α-carotene (with one β ring and one ε ring). The cyclization reactions are catalyzed by lycopene cyclases of the β (LCYB) or/and ε (LCYE) type (Fig. 11.3). Hydroxylation of orange carotenes leads to the synthesis of oxygenated carotenoids of a yellowish color collectively known as xanthophylls. Hydroxylation of β-carotene carried out preferentially by carotenoid β-hydroxylase (CHYB) enzymes of the nonheme diiron type (BCH) leads to the production of zeaxanthin, violaxanthin, and neoxanthin, whereas hydroxylation of α-carotene carried out by β- and ε-hydroxylase (CHYB and CHYE) enzymes, mainly of the cytochrome P450 (CYP97) type, results in the production of lutein [35]. Violaxanthin is produced from zeaxanthin by the activity of zeaxanthin epoxidase (ZEP) and then transformed into neoxanthin by a neoxanthin synthase (NSY) enzyme (Fig. 11.3). Violaxanthin deepoxidase (VDE) can transform violaxanthin back into zeaxanthin in the so-called xanthophyll cycle, which is very important for the photoprotection of photosynthetic tissues against excess light [37]. Along with the pathway, carotenoids can be cleaved by carotenoid cleavage dioxygenase (CCD) and 9-*cis*-epoxycarotenoid dioxygenase (NCED) enzymes into apocarotenoids [1, 38–40], including the hormones abscisic acid (ABA) and strigolactones (Fig. 11.3).

Carrot sequences showing homology to most of the identified genes involved in carotenoid biosynthesis have been isolated [41] and are summarized in Table 11.1. Similar to that reported in most plants [35], single-copy genes are likely to encode the pathway enzymes PDS, CISO, LCYE, ZEP, and VDE. The same is probably true for ZISO and NSY, although sequences encoding these enzymes have not been reported yet in carrot (Table 11.1). By contrast, several genes encode BCH iso-

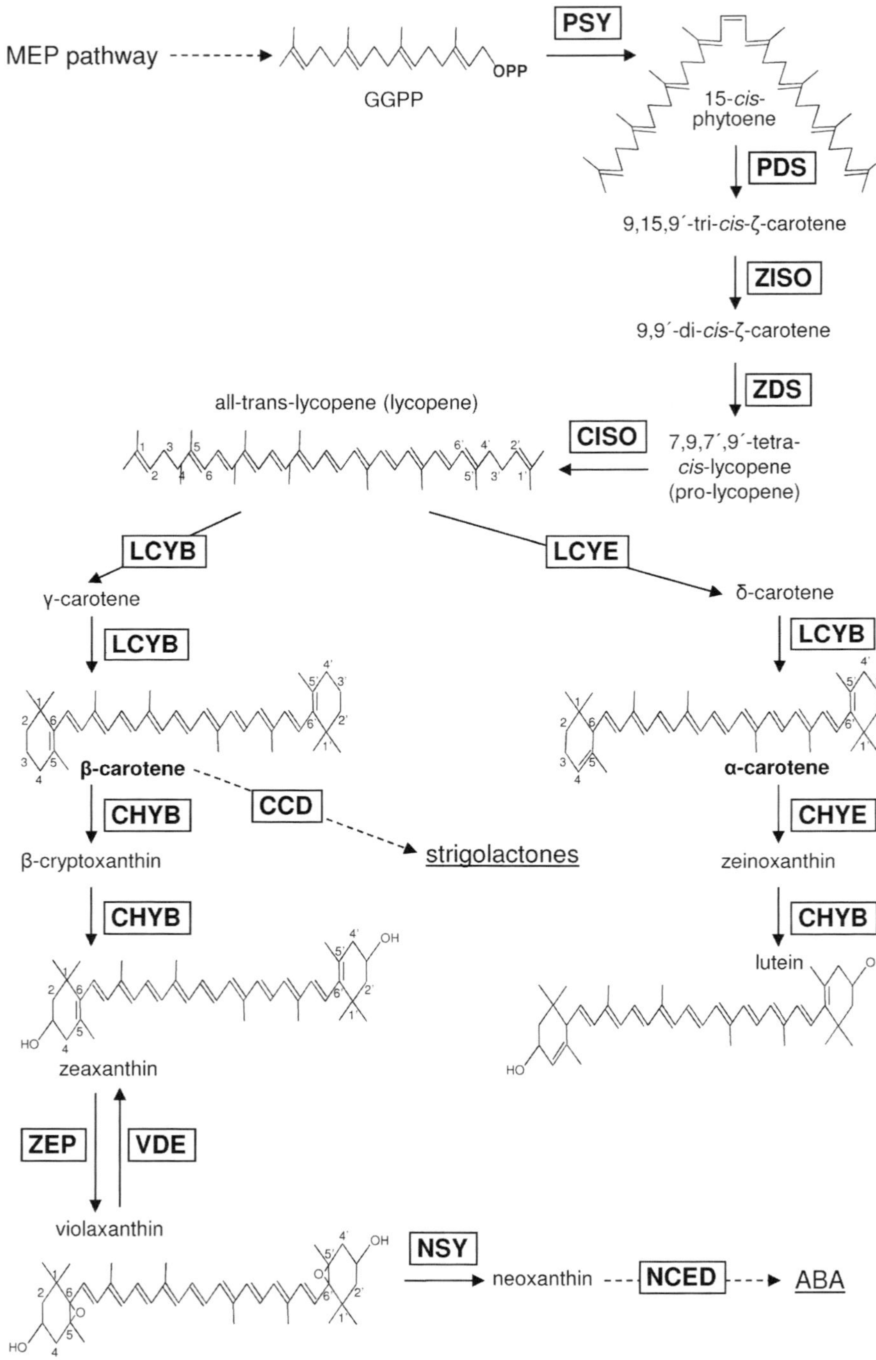

Fig. 11.3 Pathway for carotenoid biosynthesis in plants. *Dashed arrows* represent multiple steps. Enzymes are *boxed*. *PSY* phytoene synthase, *PDS* phytoene desaturase, *ZISO* 15-*cis*-ζ–carotene isomerase, *ZDS* ζ–carotene desaturase, *CISO* carotenoid (prolycopene) isomerase, *LCYB*

Table 11.1 Genes and proteins involved in carotenoid biosynthesis in *Arabidopsis* and carrot. (See text and Fig. 11.3 for enzyme acronyms.)

	Arabidopsis	*Daucus carota* var. *sativus*			
Enzyme	Gene	cDNA	Protein	Length	Name
PSY	At5g17230	DQ192186	ABB52067	398	PSY1
		DQ192187	ABB52068	438	PSY2
PDS	At4g14210	DQ222429	ABB52082	573	PDS
ZISO	At1g10830	NI	NI		
ZDS	At3g04870	DQ222430	ABB52083	573	ZDS1
		DQ192189	ABB52070	575	ZDS2
CISO	At1g06820	DQ192188	ABB52069	615	CISO
	At1g57770				
LCYB	At3g10230	DQ192190	ABB52071	508	LCYB1
		DQ192191	ABB52072	492	LCYB2
LCYE	At5g57030	DQ192192	ABB52073	530	LCYE
BCH	At4g25700	DQ192193	ABB52074	309	CHYB1
	At5g52570	DQ192194	ABB52075	303	CHYB2
		DQ192195	P.S.		CHYB3
CYP97	At1g31800	DQ192196	ABB52076	548	CHYE
	At4g15110				
	At3g53130				
ZEP	At5g67030	DQ192197	ABB52077	668	ZEP
VDE	At1g08550	DQ192198	PS		VDE
NSY	At1g67080	NI	NI		NSY

NI not identified, *PS* partial sequence

forms in carrot (Table 11.1). Although only one CYP97-encoding gene has been reported in carrot, it might be expected that more isoforms are present in the genome of this plant. Unlike that observed in *Arabidopsis,* at least two isoforms for PSY, ZDS, and LCYB are found in carrot [35, 41]. Carrot appears to be one of the few plants containing more than one ZDS isoform, whereas the presence of more than one PSY or LCYB enzymes is common in the plant kingdom [2, 3, 35, 42]. The existence of several enzyme isoforms often implies that particular isozymes are involved in the production of carotenoids in specific plastid types. For example, the tomato PSY1 isoform is required for the production of carotenoids in the chromoplasts of ripening fruit, whereas PSY2 is involved in the biosynthesis of carotenoids in chloroplasts of photosynthetic tissues [43]. In carrot, the analysis of gene expression suggests a preferential involvement of the PSY2 isoform in root

lycopene β-cyclase, *LCYE* lycopene ε-cyclase, *CHYB* carotenoid β-hydroxylase, *CHYE* carotenoid ε-hydroxylase, *ZEP* zeaxanthin epoxidase, *VDE* violaxanthin deepoxidase, *NSY* neoxanthin synthase, *CCD* carotenoid cleavage dioxygenase, *NCED* 9-*cis*-epoxycarotenoid dioxygenase carotenoid biosynthesis and a primary role for PSY1 in producing carotenoids in mature leaves [5–7]. The ZDS2 and LCYB1 isoforms in carrot are expressed at higher levels in chromoplast-containing R/D sections of the root compared to illuminated, chloroplast-containing R/L segments, whereas the ZDS1 and LCYB2 isoforms show the opposite expression pattern [6]. It is therefore conceivable that isoforms PSY1, ZDS1, and LCYB2 are mainly responsible for the production of carotenoids for photosynthesis and photoprotection (chloroplasts) whereas PSY2, ZDS2, and LCYB1 would be most involved in the biosynthesis of carotenoid in dark-grown carrot roots (chromoplasts). Recent work confirmed the involvement of LCYB1 in carotenoid biosynthesis in storage roots but also in leaves [44]. Further work will be required to define the specific role of all enzyme isoforms in carrot.

Carotenoid Production and Accumulation in Carrot Can Be Regulated at Multiple Levels

Carrots are one of the richest provitamin A sources in the human diet, but our knowledge of how carotenoid biosynthesis and accumulation are regulated in carrot roots is still scarce. Phylogenetic analyses have shown greater constraints on genes encoding upstream enzymes compared to those encoding downstream enzymes, suggesting a preferential selection during carrot domestication for mutations affecting upstream genes of the pathway that might have a greater impact on flux during carotenoid biosynthesis [45]. Although major quantitative trait loci (QTLs) for carrot color (i.e., carotenoid content) have been positionally associated with genes encoding enzymes involved in carotenoid biosynthesis (ZEP, CHYE/CYP97) and degradation (NCED), the genetic determinism of carotenoid composition in carrots is still far from clear [18, 41, 46]. Gene expression analyses have shown that carrot genes involved in carotenoid biosynthesis are upregulated when carotenoid levels increase during root development [5, 6]. However, the same genes are also expressed and even upregulated in white varieties that do not accumulate carotenoids. These results, together with the fact that transcript accumulation patterns hardly correlate with carotenoid profiles in orange roots, suggest that other mechanisms besides the transcriptional control of carotenoid pathway genes regulate the production of carotenoids in carrots [41].

The differentiation of chromoplasts in roots grown normally underground (i.e., in the absence of light) is a major determinant for the accumulation of carotenoids [6, 29, 30]. Plastid ultrastructure has actually been shown to be relevant for the activity of key enzymes required for carotenoid biosynthesis [35]. For example, PSY activity in the etioplasts of dark-grown *Arabidopsis* seedlings is very low, but it increases upon association to thylakoid membranes in the chloroplasts that develop after illumination [47]. It is possible that the differentiation of chromoplasts from amyloplasts during carrot root development also involves an increase in PSY activity. Additionally, higher PSY levels and activity have been shown to promote the accumulation of carotenoids in crystals and the differentiation of chromoplasts not only in the root of white carrot cultivars but also in other plant systems, therefore improving the capacity of plastids to store carotenoids [14, 48–50]. Phytoene synthesis and desaturation appear to be limiting for the production of carotenoids in roots of white carrot, but not in chromoplast-containing orange carrots [14, 51]. The activity of multienzyme complexes involving carotenoid biosynthetic enzymes like those found for PDS in carrot [52] might also change during chromoplast differentiation, contributing to a dramatic increase in carotenoid levels observed in mature roots.

Future Perspectives

Carrot plants can be easily transformed, and this property has been already exploited to modify the carotenoid profile of storage roots [14, 53]. Future approaches using particular carotenoid biosynthetic genes should help to establish the role of the encoded enzymes in the control of carotenoid production in this unique system. Additionally, transgene-mediated modification of the levels of regulatory proteins such as *Arabidopsis* phytochrome-interacting factors shown to regulate the expression of biosynthetic genes [54] or the cauliflower (*Brassica oleracea*) *or* gene involved in chromoplast development [55] should contribute not only to understanding of the mechanisms of carotenoid biosynthesis and accumulation in carrot roots but also to modifying their carotenoid content for an even higher nutritional and economic value.

Acknowledgments We acknowledge the financial support from the Catalan AGAUR (2009SGR-26, 2014SGR-1434 and XRB), Spanish DGI (BIO2011-23680 and PIM2010IPO-00660), and European Union FP7 (TiMet, contract 245143) to MRC, and Chilean Fondef (D10I1022) and Fondecyt (1130245) to CS. Our collaborative work was funded by grants CSIC-2010CL0039 and CSIC/ESPAÑA 10/11-2, and it is currently ongoing in the frame of the IBERCAROT network funded by Programa Iberoamericano de Ciencia y Tecnologia para el Desarrollo (CYTED, code 112RT0445).

References

1. Walter MH, Strack D (2011) Carotenoids and their cleavage products: biosynthesis and functions. Nat Prod Rep 28:663–692
2. Hirschberg J (2001) Carotenoid biosynthesis in flowering plants. Curr Opin Plant Biol 4:210–218
3. Fraser PD, Bramley PM (2004) The biosynthesis and nutritional uses of carotenoids. Prog Lipid Res 43:228–265
4. Tanaka Y, Sasaki N, Ohmiya A (2008) Biosynthesis of plant pigments: anthocyanins, betalains and carotenoids. Plant J 54:733–749
5. Clotault J, Peltier D, Berruyer R, Thomas M, Briard M, Geoffriau E (2008) Expression of carotenoid biosynthesis genes during carrot root development. J Exp Bot 59:3563–3573
6. Fuentes P, Pizarro L, Moreno JC, Handford M, Rodriguez-Concepcion M, Stange C (2012) Light-dependent changes in plastid differentiation influence carotenoid gene expression and accumulation in carrot roots. Plant Mol Biol 79:47–59

7. Stange C, Fuentes P, Handford M, Pizarro L (2008) *Daucus carota* as a novel model to evaluate the effect of light on carotenogenic gene expression. Biol Res 41:289–301
8. Surles RL, Weng N, Simon PW, Tanumihardjo SA (2004) Carotenoid profiles and consumer sensory evaluation of specialty carrots (*Daucus carota*, L.) of various colors. J Agric Food Chem 52:3417–3421
9. Mech-Nowak A, Swiderski A, Kruczek M, Luczak I, Kostecka-Gugala A (2012) Content of carotenoids in roots of seventeen cultivars of *Daucus carota* L. Acta Biochim Pol 59:139–141
10. Baranska M, Baranski R, Schulz H, Nothnagel T (2006) Tissue-specific accumulation of carotenoids in carrot roots. Planta 224:1028–1037
11. Grassmann J, Schnitzler WH, Habegger R (2007) Evaluation of different coloured carrot cultivars on antioxidative capacity based on their carotenoid and phenolic contents. Int J Food Sci Nutr 58:603–611
12. Alasalvar C, Grigor JM, Zhang D, Quantick PC, Shahidi F (2001) Comparison of volatiles, phenolics, sugars, antioxidant vitamins, and sensory quality of different colored carrot varieties. J Agric Food Chem 49:1410–1416
13. Nicolle C, Simon G, Rock E, Amouroux P, Remesy C (2004) Genetic variability influences carotenoid, vitamin, phenolic, and mineral content in white, yellow, purple, orange, and dark-orange carrot cultivars. J Am Soc Hortic Sci 129:523–529
14. Maass D, Arango J, Wust F, Beyer P, Welsch R (2009) Carotenoid crystal formation in *Arabidopsis* and carrot roots caused by increased phytoene synthase protein levels. PLoS ONE 4:e6373
15. Montilla EC, Arzaba MR, Hillebrand S, Winterhalter P (2011) Anthocyanin composition of black carrot (*Daucus carota* ssp. sativus var. atrorubens Alef.) cultivars antonina, beta sweet, deep purple, and purple haze. J Agric Food Chem 59:3385–3390
16. Glassgen WE, Wray V, Strack D, Metzger JW, Seitz HU (1992) Anthocyanins from cell suspension cultures of *Daucus carota*. Phytochemistry 31:1593–1601
17. Rubatzky VE, Quiros CF, Simon PW (1999) Carrots and related vegetable Umbelliferae. CABI, New York, pp 1–21
18. Clotault J, Geoffriau E, Lionneton E, Briard M, Peltier D (2010) Carotenoid biosynthesis genes provide evidence of geographical subdivision and extensive linkage disequilibrium in the carrot. Theor Appl Genet 121:659–672
19. Simon PW, Lindsay RC, Peterson CE (1980) Analysis of carrot volatiles collected on porous polymer traps. J Agric Food Chem 28:549–552
20. He J, Giusti MM (2010) Anthocyanins: natural colorants with health-promoting properties. Annu Rev Food Sci Technol 1:163–187
21. Simon PW, Wolff XY, Peterson CE, Kammerlohr DS, Rubatzky VW, Strandberg JO et al (1989) High carotene mass carrot population. Hort Sci 24:174–175
22. Heinonen M (1990) Carotenoids and provitamin A activity of carrot (*Daucus carota* L.) cultivars. J Agric Food Chem 38:609–612
23. Holden JM, Eldridge AL, Beecher GR, Buzzard IM, Bhagwat S, Davis CS et al (1999) Carotenoid content of U.S. foods: an update of the database. J Food Compos Anal 12:169–196
24. Egea I, Barsan C, Bian W, Purgatto E, Latche A, Chervin C et al (2010) Chromoplast differentiation: current status and perspectives. Plant Cell Physiol 51:1601–1611
25. Howitt CA, Pogson BJ (2006) Carotenoid accumulation and function in seeds and non-green tissues. Plant Cell Environ 29:435–445
26. Deruere J, Romer S, d'Harlingue A, Backhaus RA, Kuntz M, Camara B (1994) Fibril assembly and carotenoid overaccumulation in chromoplasts: a model for supramolecular lipoprotein structures. Plant Cell 6:119–133
27. Vishnevetsky M, Ovadis M, Vainstein A (1999) Carotenoid sequestration in plants: the role of carotenoid-associated proteins. Trends Plant Sci 4:232–235
28. Simkin AJ, Gaffe J, Alcaraz JP, Carde JP, Bramley PM, Fraser PD et al (2007) Fibrillin influence on plastid ultrastructure and pigment content in tomato fruit. Phytochemistry 68:1545–1556

29. Kim JE, Rensing KH, Douglas CJ, Cheng KM (2010) Chromoplasts ultrastructure and estimated carotene content in root secondary phloem of different carrot varieties. Planta 231:549–558
30. Vasquez-Caicedo AL, Heller A, Neidhart S, Carle R (2006) Chromoplast morphology and beta-carotene accumulation during postharvest ripening of Mango Cv. 'Tommy Atkins'. J Agric Food Chem 54:5769–5776
31. Koch TC, Goldman IL (2005) Relationship of carotenoids and tocopherols in a sample of carrot root-color accessions and carrot germplasm carrying Rp and rp alleles. J Agric Food Chem 53:325–331
32. Demmig-Adams B, Adams WW (2002) Antioxidants in photosynthesis and human nutrition. Science 298:2149–2153
33. Xiao FG, Shen L, Ji HF (2011) On photoprotective mechanisms of carotenoids in light harvesting complex. Biochem Biophys Res Commun 414:1–4
34. Stange C, Flores C (2012) Carotenoids and photosynthesis—regulation of carotenoid biosyntesis by photoreceptors. In: Advances in photosynthesis—fundamental aspects. Najafpour M (Ed.). InTech, Rijeka, Croatia, pp 77-96. ISBN: 978-953-307-928-8
35. Ruiz-Sola MA, Rodriguez-Concepcion M (2012) Carotenoid biosynthesis in *Arabidopsis*: a colorful pathway. Arabidopsis Book 10:e0158
36. Rodriguez-Concepcion M (2010) Supply of precursors for carotenoid biosynthesis in plants. Arch Biochem Biophys 504:118–122
37. Niyogi KK (1999) Photoprotection revisited: genetic and molecular approaches. Annu Rev Plant Physiol Plant Mol Biol 50:333–359
38. Giuliano G, Al-Babili S, von Lintig J (2003) Carotenoid oxygenases: cleave it or leave it. Trends Plant Sci 8:145–149
39. Bouvier F, Isner JC, Dogbo O, Camara B (2005) Oxidative tailoring of carotenoids: a prospect towards novel functions in plants. Trends Plant Sci 10:187–194
40. Auldridge ME, McCarty DR, Klee HJ (2006) Plant carotenoid cleavage oxygenases and their apocarotenoid products. Curr Opin Plant Biol 9:315–321
41. Rodriguez-Concepcion M, Stange C (2013) Biosynthesis of carotenoids in carrot: an underground story comes to light. Arch Biochem Biophys 539:110-116
42. Cunningham FX, Gantt E (1998) Genes and enzymes of carotenoid biosynthesis in plants. Ann Rev Plant Physiol Plant Mol Biol 49:557–583
43. Fraser PD, Kiano JW, Truesdale MR, Schuch W, Bramley PM (1999) Phytoene synthase-2 enzyme activity in tomato does not contribute to carotenoid synthesis in ripening fruit. Plant Mol Biol 40:687–698
44. Moreno JC, Pizarro L, Fuentes P, Handford M, Cifuentes V, Stange C (2013) Levels of lycopene β-cyclase 1 modulate carotenoid gene expression and accumulation in Daucus carota. PLoS One 8:e58144.
45. Clotault J, Peltier D, Soufflet-Freslon V, Briard M, Geoffriau E (2012) Differential selection on carotenoid biosynthesis genes as a function of gene position in the metabolic pathway: a study on the carrot and dicots. PLoS ONE 7:e38724
46. Just BJ, Santos CA, Yandell BS, Simon PW (2009) Major QTL for carrot color are positionally associated with carotenoid biosynthetic genes and interact epistatically in a domesticated x wild carrot cross. Theor Appl Genet 119:1155–1169
47. Welsch R, Beyer P, Hugueney P, Kleinig H, von Lintig J (2000) Regulation and activation of phytoene synthase, a key enzyme in carotenoid biosynthesis, during photomorphogenesis. Planta 211:846–854
48. Shewmaker CK, Sheehy JA, Daley M, Colburn S, Ke DY (1999) Seed-specific overexpression of phytoene synthase: increase in carotenoids and other metabolic effects. Plant J 20:401–412
49. Ronen G, Cohen M, Zamir D, Hirschberg J (1999) Regulation of carotenoid biosynthesis during tomato fruit development: expression of the gene for lycopene epsilon-cyclase is downregulated during ripening and is elevated in the mutant delta. Plant J 17:341–351

50. Bramley PM (2002) Regulation of carotenoid formation during tomato fruit ripening and development. J Exp Bot 53:2107–2113
51. Santos CAF, Senalik D, Simon PW (2005) Path analysis suggests phytoene accumulation is the key step limiting the carotenoid pathway in white carrot roots. Genet Mol Biol 28:287–293
52. Lopez AB, Yang Y, Thannhauser TW, Li L (2008) Phytoene desaturase is present in a large protein complex in the plastid membrane. Physiol Plant 133:190–198
53. Jayaraj J, Devlin R, Punja Z (2008) Metabolic engineering of novel ketocarotenoid production in carrot plants. Transgenic Res 17:489–501
54. Toledo-Ortiz G, Huq E, Rodriguez-Concepcion M (2010) Direct regulation of phytoene synthase gene expression and carotenoid biosynthesis by phytochrome-interacting factors. Proc Natl Acad Sci U S A 107:11626–11631
55. Lu S, Van Eck J, Zhou X, Lopez AB, O'Halloran DM, Cosman KM et al (2006) The cauliflower Or gene encodes a DnaJ cysteine-rich domain-containing protein that mediates high levels of beta-carotene accumulation. Plant Cell 18:3594–3605

Chapter 12
Carotenoids in Green Vegetables and Health Aspects

Julie Garden-Robinson

"Try some spinach, please." "Finish your broccoli. It's good for you." Anecdotal evidence suggests that many children have heard, and sometimes obeyed, these dinnertime admonishments. Scientific evidence is catching up with long-time adult encouragement to eat more green vegetables, and adults as well as children worldwide can reap benefits from the advice.

The natural components of fruits and vegetables, including carotenoid pigments found in gold/orange and green types of produce, may help protect people from chronic diseases including cancer and heart disease [1, 2, 3]. Although bright orange carrots often are considered the gold standard for promoting healthy eyes, researchers have studied leafy greens for their potential role in reducing the risk for potentially blinding eye diseases such as macular degeneration [4, 5]. Along with gold/orange vegetables and cruciferous vegetables such as broccoli and cabbage, leafy greens rank among the "powerhouse" vegetables and fruits that could play a major role in improving public health [6]. Despite these potential health benefits, many children and adults do not eat the recommended amount of fruits and vegetables, especially dark green vegetables, to meet their nutritional needs and to promote long-term health [7–9].

The goals of this chapter are: (1) to present current vegetable recommendations based on the US Dietary Guidelines for Americans; (2) to review literature related to sources and nutritional and disease-fighting substances in dark green vegetables, especially leafy greens; (3) to provide an overview of relevant studies of the health benefits associated with green vegetables, particularly eye health; and (4) to review strategies to encourage vegetable consumption among children and adults.

J. Garden-Robinson (✉)
Health, Nutrition, and Exercise Sciences, North Dakota State University, 1400 Centennial Blvd. Dept. 7270, Box 6050, Fargo, ND 58105, USA

C. Chen (ed.), *Pigments in Fruits and Vegetables*, DOI 10.1007/978-1-4939-2356-4_12

US Dietary Guidelines Promote Fruit and Vegetable Intake

The US Dietary Guidelines for Americans recommend increased fruit and vegetable intake [8, 9]. In particular, a key recommendation is to "eat a variety of vegetables, especially dark green and red and orange vegetables and beans and peas" [9]. Dark green vegetables, which include broccoli, spinach, romaine, collard, turnip, and mustard greens, comprise one of the five vegetable subgroups. Subgroups are classified based on color and/or similar nutritional composition. The other four subgroups are red and orange vegetables, such as tomatoes, red peppers, carrots, and sweet potatoes; beans and peas (legumes), including kidney beans, lentils, and chickpeas; starchy vegetables, such as white potatoes, corn, and green peas; and other vegetables, including iceberg lettuce, green beans, and onions [9].

Table 12.1 shows the current recommendations for overall and subgroup vegetable consumption based on various calorie levels established by the US Dietary Guidelines for Americans [9]. At least 2 cups of vegetables per day and at least 1.5 cups of dark green vegetables per week are recommended based on the 2000-cal-per-day dietary pattern. A 2000-cal-per-day diet is used as a reference on the Nutrition Facts labels on US food products [10]. As shown in Fig. 12.1, the MyPlate icon serves as a visual reminder for the US Dietary Guidelines for Americans, and further information and educational tools are available at the website, www.choosemyplate.gov [11]. A key MyPlate message is that half of one's plate should consist of fruits and vegetables.

Table 12.1 Vegetable intake recommendations based on calorie level. (Based on data from [9])

	Calorie level of dietary pattern										
	1000	1200	1400	1600	1800	2000	2200	2400	2600	2800	3000
Total vegetables (in cups per day)	1	1½	1½	2	2½	2½	3	3	3½	3½	4
Dark green vegetables (cups/week)	½	1	1	1½	1½	1½	2	2	2½	2½	2½
Red and orange vegetables (cups/week)	2½	3	3	4	5½	5½	6	6	7	7	7½
Beans and peas (legumes) (cups/week)	½	½	½	1	1½	1½	2	2	2½	2½	3
Starchy vegetables (cups/week)	2	3½	3½	4	5	5	6	6	7	7	8
Other vegetables (cups/week)	1½	2½	2½	3½	4	4	5	5	5½	5½	7

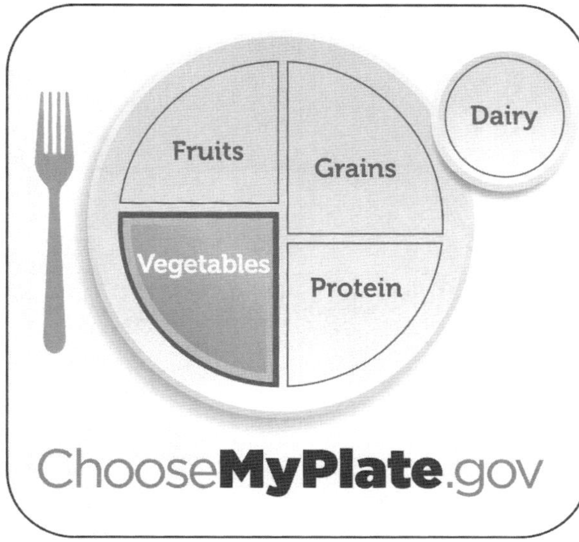

Fig. 12.1 MyPlate, the icon for the US Dietary Guidelines for Americans. (Reprinted from [11])

Nutritional and Phytochemical Composition of Green Leafy Vegetables

Green leafy vegetables (GLVs) are considered nutrient-dense food sources with relatively low calories [9]. GLVs are good sources of vitamins K, A (as precursor carotenoids such as β-carotene), C, and the B vitamin folate, and they provide minerals including potassium, magnesium, iron, and calcium [12]. As shown in Table 12.2, green vegetables such as broccoli, kale, romaine lettuce, and spinach contain a high percentage of water and are fairly low in calories. Oxalates found in abundance in some GLVs, however, may reduce the availability of minerals, such as calcium, for use by the body. For example, among 11 Indian GLV samples studied, spinach was particularly high in oxalates [13].

GLVs are considered the best food source of vitamin K [14]. This fat-soluble nutrient plays a role in blood clotting, cardiovascular health, and bone health. People on anticoagulant drugs such as warfarin (Coumadin) are advised to exercise caution in their consumption of vitamin K sources, especially GLVs, as a result of potential nutrient-drug interactions [14].

Vitamin K levels in vegetables such as leaf lettuce, spinach, and kale are influenced by the growing location, soil, climate, and other factors. For example, comparing green vegetables grown in Montreal, Canada, and Boston, MA, USA, researchers reported higher vitamin K levels among the Canadian-grown vegetables [15]. Using gas chromatography, researchers quantified vitamin K levels, showing spinach to be the best vitamin K source among samples of lettuce, cabbage, and spinach [16].

Table 12.2 Nutrition analysis of selected green vegetables. (Based on data from [11])

	Units	Broccoli, raw, chopped, per cup (91 g)	Kale, raw, per cup (67 g)	Romaine lettuce, raw, shredded, per cup (47 g)	Spinach, raw, per cup (30 g)	Daily reference values[a]
Water	g	81.26	56.31	44.47	27.42	Na[b]
Energy	kcal	31	33	8	7	2000
Protein	g	2.57	2.87	0.58	0.86	50
Total lipid (fat)	g	0.34	0.62	0.14	0.12	65
Carbohydrate (by difference)	g	6.04	5.86	1.55	1.09	300
Fiber (dietary)	g	2.4	Na	1.0	0.7	25
Sugars, total	g	1.55	Na	0.56	0.13	Na
Calcium	mg	43	100	16	30	1000
Iron	mg	0.66	0.98	0.46	0.81	18
Magnesium	mg	19	31	7	24	400
Phosphorus	mg	60	62	14	15	1000
Potassium	mg	288	329	116	167	3500
Sodium	mg	30	25	4	24	2400
Zinc	mg	0.37	0.38	0.11	0.16	15
Vitamin C, total ascorbic acid	mg	81.2	80.4	1.9	8.4	60
Thiamin	mg	0.065	0.074	0.034	0.023	1.5
Riboflavin	mg	0.106	0.087	0.031	0.057	1.7
Niacin	mg	0.581	0.670	0.147	0.217	20
Folate/folic acid	Dietary Folate Equivalents (DFE) or micrograms (μg)	57 DFE	21 DFE	64 DFE	58 DFE	400 μg
Vitamin A	International Units (IU)	567	6693	4094	2813	5000
Vitamin D	IU	0	0	0	0	400
Vitamin E (alpha tocopherol)	mg	0.71	Na	0.06	0.61	30
Vitamin K	μg	92.5	472.2	48.2	144.9	80
Lutein and zeaxanthin[c]	μg	1488	23,720	1295	3659	Na

[a] Daily reference values are based on a 2000-cal diet and used on US food labels [10]

[b] *Na* not available/applicable

[c] Values for lutein and zeaxanthin are per-cup raw vegetables, with the exception of kale, where the value was available from cooked kale only

Nutrition experts promote eating more dark green and orange vegetables [9]. Carotenoid pigments are responsible for the dark gold, orange, and some of the red pigmentation of fruits and vegetables, but these natural compounds are also found in abundance in dark green leafy vegetables [12]. The levels of the carotenoid pigments lutein and zeaxanthin per cup of raw spinach, broccoli, and romaine lettuces, and cooked kale are shown in Table 12.2 [12].

About 600 carotenoid pigments have been identified in nature; of these, 40 are found in the human diet, and these "phytochemicals" (plant chemicals) may play a role in promoting health [17]. For example, although apricots and carrots are leading sources of β-carotene, this carotenoid is also found in spinach, collard greens, beet greens, and broccoli. Lutein and zeaxanthin, which are carotenoid pigments classified as xanthophylls, are found in kale, spinach, collards, turnip greens, romaine lettuce, beet greens, broccoli, and green peas [18]. Among 18 vegetables most commonly consumed in Spain, spinach had the highest quantity of lutein and zeaxanthin. Further, lutein was found in higher quantity than β-carotene in green vegetables, but the red-orange vegetables analyzed contained a wider variety of carotenoids [19].

Some compounds within GLVs affect their palatability among consumers. Kale, for example, provides an abundance of lutein and β-carotene; however, this GLV contains glucosinolate and S-methylcysteine sulfoxide, sulfur-containing compounds which may impart a bitter flavor that affects the palatability and consumption of this vegetable [20]. Researchers studied three kale cultivars grown in a greenhouse in nutrient solution cultures with five different sulfur levels. The amount of sulfur-containing compounds and carotenoid pigments in the leaves was determined. Kale cultivars grown in media with lower sulfur content had lower levels of sulfur compounds associated with negative flavors, but the carotenoid pigment levels remained the same [20].

Plant scientists have made efforts to increase the carotenoid concentration in vegetable crops [21]. Carotenoid levels may be influenced by seasonal factors, maturity level, and processing prior to storage. In studies of the carotenoid composition of two kale cultivars grown in Brazil, seasonal differences in carotenoid levels were shown [22]. One kale cultivar grown in the summer showed higher levels of total carotenoids, lutein, and β-carotene, while another cultivar showed higher carotenoid values during the winter season. Both cultivars had significantly higher levels of carotenoids when grown without agrochemicals.

When kale was grown with the addition of agrochemicals, lutein and β-carotene levels were significantly higher among mature leaves of kale compared to immature leaves. No differences in carotenoid levels were reported among immature or mature samples grown without agrochemicals [23]. Higher carotenoid levels, with the exception of β-carotene, were found among kale samples grown in the summer compared to kale grown in the winter. Storage for 5 days at 7–9 °C significantly reduced the amount of carotenoids, including β-carotene and lutein levels, particularly during the first 2 days of storage. The researchers suggested that lower temperature and modified packaging may help retain carotenoid values [23].

Hydroponic farming has been used to grow vegetables year-round in areas deemed unsuitable for traditional farming, to produce multiple crops of vegetables throughout the year, as a means of using less water and fertilizer and for other reasons [24]. GLVs, including several varieties of lettuce, grown in Brazil using hydroponic methods were analyzed for their carotenoid content. The researchers reported lower levels of β-carotene, lutein, and other carotenoids among hydroponically grown lettuce compared with conventionally grown lettuce [24].

The antioxidant capacity and content of phenolic compounds were analyzed in fresh and cooked tree spinach, which is grown in dry areas of the tropics [25]. Phenolic compounds are considered powerful antioxidants with the potential to act as free radical scavengers and the capacity for health benefits. Using the oxygen radical absorbance capacity (ORAC) assay, high-performance liquid chromatography, and gas chromatography for assessing, identifying, and quantifying compounds, the researchers determined tree spinach was a good source of natural antioxidants, but cooking decreased the quantity of natural antioxidants [25].

Researchers continue to identify and quantify nutrients, carotenoids, and other phytochemicals in various GLVs found throughout the world [26–31]. Wild GLVs found in South Africa are a potential source of nutritious food with negligible cost [30]. Researchers determined the protein, fiber, mineral, and antioxidant content of wild GLVs. The vegetables contained similar or greater mineral levels including calcium, iron, magnesium, and manganese, and lower levels of antinutritional factors (phytates, saponins), compared with cultivated spinach or cabbage.

In their studies of curly kale grown in Norway, "one of the nutritionally richest vegetables," at least 32 phenolic compounds were identified, which was more than previously known [32]. Other research teams have identified phenolic compounds in collard greens, kale, Chinese broccoli, mustard greens, and other vegetables in the *Brassica* family. The researchers identified 45 flavonoids in broccoli, kale, and collard greens [33, 34].

Wild greens used in traditional diets in the Mediterranean region have been determined to be significant contributors of dietary carotenoids [35]. In response to published studies showing a reduced risk of heart disease among Greek immigrants to Australia who have maintained a traditional diet, researchers determined the carotenoid composition of wild green vegetables common in traditional diets. The wild greens, including sow thistle, amaranth, purslane, and dandelion greens, contained high levels of lutein and β-carotene.

In an analysis of the antioxidant content of wild greens in the Mediterranean, researchers gathered vegetation in three different growing areas (mountains, country, seaside) [36]. The greens were boiled in a minimal amount of water using traditional recipes. The samples were analyzed for antioxidant capacity. Researchers reported high levels of antioxidants, including carotenoids, ascorbic acid, and polyphenols, in the plants, at levels comparable to those of spinach. The researchers promoted regular consumption of the wild greens [36].

Traditional GLVs grown in Kenya have been shown to be contributors of calcium, iron, zinc, and many other nutrients. As a nutritious food source, the wild vegetables could aid in food security and potentially provide additional income for families [37]. Using atomic absorption spectroscopy techniques, researchers studied the mineral composition of 54 wild and cultivated leafy vegetables grown in Kenya. They compared the mineral content of the wild plants with spinach, kale, and cabbage. Many of the traditional plants had significantly higher levels of calcium, iron, and zinc, and the researchers urged further study to determine bioavailability and identification of potential antinutritional factors such as oxalates [37].

Global Health Benefits Associated with GLVs

Low intake of fruits and vegetables is considered among the top ten leading causes of mortality worldwide [38]. According to the World Health Organization (WHO), vitamin A deficiency is a common nutritional deficiency among children in developing nations, which is linked to blindness, and child and maternal mortality [39]. Vitamin A can be produced in the body when carotenoid-containing vegetables, including dark green vegetables, are consumed. Among the global nutrition guidelines is the recommendation to increase dietary carotenoids, which act as antioxidants in the body [39]. Carotenoids have been linked to reducing the risk for chronic diseases including cancer, diabetes, heart disease, stroke, HIV, age-related macular degeneration (AMD), and cataracts [17]. Further, some research has linked the consumption of lutein and zeaxanthin to reducing the risk for lung and breast cancer, although results are inconsistent [18, 40]. The relationship between vitamins, carotenoid intake, and oral cancer risk has been studied [41]. Using dietary intake data from 42,340 men in the Health Professionals Follow-up Study, researchers determined the risk for oral premalignant lesions related to intake of vitamins A, C, and E, and carotenoids from food or dietary supplements. Intake of vitamin E and β-carotene was associated with increased oral cancer risk among smokers. The relationship between lutein/zeaxanthin and oral lesions was not clear, but β-carotene intake among smokers was associated with an increased risk [41].

Kale has been studied for its potential role in cancer prevention in vitro [42]. Extracts from fresh and processed kale were used to determine the antiproliferative effects of kale on colon cancer cells in cell culture. Kale was processed by blanching, freezing, and boiling in a bag. Compared with heat-processed kale extracts, extracts of fresh kale had significantly greater antioxidant capacity and levels of vitamin C, flavonols, phenolic acids, and glucosinolates, but similar levels of isothiocyanates. Fresh kale extracts showed a greater ability to inhibit cancer cell growth compared with processed kale extracts.

Eating more GLVs may lower the risk for type 2 diabetes, a chronic disease that is increasing worldwide [43]. In a meta-analysis of six studies with a combined population of 223,512 participants, researchers showed that a greater intake of *green* vegetables, including spinach, Brussels sprouts, lettuce, and parsley, was associated with a 14% lower risk of diabetes; however, eating more vegetables and/or fruits of any color did not have the same protective effect. Data analysis revealed that increased intake of fruits only, vegetables only, and fruits and vegetables combined was not linked with a reduction in diabetes risk [43].

Some interventions to improve vitamin A and overall health status have identified indigenous sources of β-carotene-containing plants (including GLVs), to provide supplemental vegetables [31, 44–46]. Using gerbils as an animal model, researchers studied the ability of GLVs indigenous to Africa to maintain vitamin A status. Gerbils were fed one of six diets formulated to be equal in calorie and protein content. The animals were depleted of vitamin A and then fed a treatment diet including powders made from various leafy greens. The gerbils fed indigenous plants maintained their vitamin A status, as measured by conversion of β-carotene to retinol [44].

To help prevent vitamin A deficiencies among people in Jordan, researchers determined carotenoid levels in commonly consumed fruits and vegetables native to their region [31]. Using a Food Frequency Questionnaire (FFQ), researchers determined the characteristic foods in the diets of Jordanians and the best sources of carotenoids. Rice, olive oil, carrots, tomatoes, mint, chickpeas, and parsley were among the foods consumed in the highest quantity. As determined by ultraviolet (UV) spectrophotometry, leafy greens including mint leaves and parsley were among the foods with highest amount of dietary carotenoids. The nutritional composition information is encouraged to use in meal planning to help prevent vitamin A deficiency.

Following a food consumption survey among households in an urban area of Madagascar with children under age 5, a laboratory-based nutritional analysis of 50 common recipes was conducted to determine their protein, fat, fiber, iron, zinc, and vitamin A content [45]. The primary concern was the retinol activity equivalent (RAE) because of the local vitamin A deficiency issue. The recipes were high in β-carotene and RAE because of the high content of GLVs, which comprised 81 % on a wet basis.

Indigenous GLVs in the Limpopo and KwaZulu-Natal provinces of South Africa were analyzed to determine their β-carotene content and potential use as food [46]. Using field walks, structured interviews with key informants, and focus groups, researchers determined that the use of the indigenous plants not only improved food security but also provided a readily available, economical source of β-carotene-rich food. The leaves of the spider plant and amaranth were the most popular leafy plants consumed by the participants in the study, who claimed the spider plant "has vitamins" and "prevents us from going hungry" [46].

Supplementation with GLVs may improve plasma β-carotene and iron status, and the amount of oil used in cooking affects the absorption of the carotenoid and other nutrients [47]. Following a 3-week supplementation study, researchers determined the plasma levels of β-carotene, vitamin C, iron, and zinc among 50 healthy adult volunteers divided into five groups consuming meals containing: (1) 100 g of GLV cooked with 5 g of oil; (2) GLV cooked with 10 g of oil; (3) 100 g GLV cooked with 5 g of oil plus added vitamin C; (4) 100 g GLV cooked with 5 g of oil plus vitamin E; or (5) a cereal-based diet with a β-carotene supplement but no GLV. The researchers concluded that consuming 100 g of dark green leafy vegetables daily with 10 g of oil resulted in a significant increase in plasma iron, β-carotene, vitamin C, and zinc levels [47].

The effects of adding GLVs or pumpkin to the diets of children were studied in an area of Bangladesh known for its high incidence of vitamin A deficiency [48]. In the 6-week study, 110 children aged 8–12 years, were treated to control *Ascaris lumbricoides* (worms) and then fed one supplemental meal per day for 35 days. The children were randomly assigned to receive a diet with β-carotene from GLVs (4.4 mg), β-carotene from sweet pumpkin (1.4 mg), or a control diet without a β-carotene source. The serum β-carotene levels of the participating children successfully treated for worms increased statistically; however, a greater increase was seen among those children who received supplemental GLVs [48].

A home gardening intervention in South Africa targeted 2- to 5-year-olds living in economically disadvantaged rural households during a 20-month period [49]. Prior to the intervention, the growth of children was assessed, and the serum retinol concentration was determined as a measure of vitamin A status. In addition, a survey of the primary caregiver determined demographics, health status, and knowledge, and consumption practices related to vitamin A-rich foods. Garden production teams with nutrition monitors helped implement the gardening project and provided education and support to the experimental villages. Post-surveys indicated significant improvements in knowledge related to vitamin A sources and roles. Children in the intervention villages consumed more GLVs and dark gold vegetables, which resulted in significant increases in their serum retinol concentrations compared with those of children in the nearby control villages [49].

The plasma levels of carotenoids can be increased in a fairly short amount of time with diets high in fruits and vegetables [50]. During two 15-day residency periods, 36 healthy men and women consumed a controlled, high-carotenoid diet that was moderate in fat and comprised typical American foods. During one of the residency periods, the diets were supplemented with 205 g of microwaved broccoli daily (102.4 g at lunch and dinner) for 5 days. The short-term high-carotenoid diets significantly increased plasma carotenoid levels [50].

Carotenoids and Eye Health

Many studies have indicated the relation of carotenoid consumption to the risk of AMD, cataracts, and glaucoma [51, 52]. AMD is a leading cause of blindness among the elderly population [53]. The macula lutea (yellow spot) of the human retina is composed of the xanthophyll pigments lutein and zeaxanthin, and these pigments are primarily obtained from the diet, especially from GLVs. An adequate intake of these pigments from food may reduce the risk for AMD by protecting the retina from oxidation [4]. However, some studies have not shown a protective role of lutein and zeaxanthin in reducing the risk of AMD [54, 55].

Researchers conducted a meta-analysis of the effect of lutein and zeaxanthin intake on early and late AMD. Six studies, primarily with Western populations, were selected for inclusion from more than 73 articles screened [56]. Intake of lutein and zeaxanthin was not associated with a lower risk of early-stage AMD; however, intake of these xanthophyll pigments showed protective effects among those with late-stage AMD.

The Eye Disease Case-Control Study in five ophthalmology centers in the USA included 356 subjects with advanced AMD ranging in age from 55 to 80 years, and 520 control subjects without AMD but from similar geographic areas [5]. The case subjects showed visual acuity less than 20/20, drusen (yellow or white spots that can form in the retina) in either eye, and at least one clinical sign of exudative AMD. Data were gathered through a physical examination, interview, and blood specimens, and the subjects completed a 60-item FFQ to determine their intake of carotenoids including β-carotene, lutein, zeaxanthin, and lycopene. Consuming more lutein and zeaxanthin from spinach or collard greens was associated with a significantly reduced risk for AMD; however, intake of vitamins A, C, and E was not.

The Age-Related Eye Disease Study (AREDS) included 4519 participants ranging in age from 60 to 80 years who completed eye examinations and questionnaires to determine smoking, sun exposure, and medication and dietary supplement use [57]. Participants completed a 90-item FFQ to determine dietary intake of carotenoids, vitamins, and minerals. Consuming more lutein and zeaxanthin was associated with a significant decrease in AMD; however, vitamins A, C, and E, and other carotenoids were not associated with a lower risk of AMD.

In the population-based, longitudinal Blue Mountains Eye Study conducted in Australia, baseline eye examinations were conducted with 3654 participants ages 49 and older, and the subjects were assessed at 5 and 10 years. Participant numbers in the 5-year and 10-year follow-up studies were 2335 (75.1 % of survivors) and 1952 (76.6% of survivors), respectively [58, 59]. At baseline, a majority (85%) of participants completed a 145-item FFQ, and the food components of interest included β-carotene, lutein, zeaxanthin, lycopene, retinol, vitamins A and C, and zinc. Within 5 years, 192 participants had developed early AMD. Neither zinc nor dietary carotenoids were associated with a protective effect against early AMD [58]. However, at the 10-year follow-up, the incidence of AMD was affected by diet and supplements. Participants with higher intake of total zinc (from food and supplements) and of dietary lutein and zeaxanthin were significantly less likely to have AMD. In addition, participants with a higher intake of all vegetables were less likely to have AMD [59].

In a study of 71,494 female nurses and 41,564 male health professionals at least 50 years of age, 673 cases of early AMD and 442 cases of neovascular AMD were documented during the 18 years of follow-up. Neovascular AMD is the "wet" form of advanced AMD, which is associated with vision loss [54]. Dietary intake of lutein- and zeaxanthin-containing foods and dietary supplements, including vitamins C and E, were assessed using a 130-item FFQ. Lutein and zeaxanthin intake was not associated with early AMD risk, nor were vitamins C and E, or smoking status; however, among those who never smoked, an inverse, nonlinear association was shown between lutein/zeaxanthin intake and neovascular AMD [54].

The Carotenoids in Age-Related Eye Disease Study (CAREDS) examined the relationship between either a high or a low lutein and zeaxanthin intake and AMD [60]. Among the overall CAREDS population, an inverse relationship was not found between the high intake of these xanthophylls and AMD. The researchers reported an inverse relationship between lutein and zeaxanthin intake and intermediate AMD among healthy women under age 75 without a history of heart disease or diabetes [60].

In an animal study, researchers showed a relationship between the type of oil consumed and the absorption of the carotenoids lutein and zeaxanthin in blood plasma or eye samples [61]. A powder made from leafy greens with a high amount of lutein and zeaxanthin was fed to rats depleted of lutein. In addition to a standard diet supplemented with the GLV powder, the animals consumed olive oil, sunflower oil, or groundnut (peanut) oil. The rats who consumed olive oil showed a significantly greater absorption of lutein and zeaxanthin measured in either their plasma or eye samples.

Researchers studied the risk of cataracts related to the dietary intake of lutein, zeaxanthin, and vitamins C and E among 35,551 participants in the Women's Health Study, who completed a 131-item FFQ [62]. During 10 years of follow-up, 2031 cataract cases were confirmed. In their assessment of cataract risk compared with their dietary intake of lutein and zeaxanthin or vitamin E, researchers reported a 14 % lower risk of cataracts with higher vitamin E intake and an 18 % lower risk associated with a high intake of lutein and zeaxanthin. In addition, a higher intake of GLVs resulted in a borderline significant decreased risk of cataracts.

The relationship between antioxidant intake and glaucoma was studied using data from the Nurses' Health Study, with 76,200 female participants followed up from 1980 to 1996, and the Health Professionals Follow-up Study, with 40,284 participants followed up from 1986 to 1996 [63]. The participants completed FFQs every 2 years. Although the researchers found no association between consuming more carotenoids, vitamins C, E, or A and reduction in glaucoma risk, they noted a potential role of lutein/zeaxanthin in reducing glaucoma risk, a finding for which they encouraged further investigation. [63].

The effectiveness of consuming eye-protecting supplements versus carotenoid-rich foods remains inconclusive. Some studies have shown a role for supplements in addition to foods, while others have not. People at risk for AMD are advised to eat more fruits and vegetables, especially dark green leafy vegetables. They are also advised to wear eyewear that protects against UV light and an eye-shading hat when outdoors [52].

Strategies to Encourage Vegetable Consumption Among Children

The US Dietary Guidelines for Americans promotes increased consumption of vegetables among adults and children, and educational tools to help families consume more vegetables at home are available [9, 11]. The diets of children are influenced by a variety of factors, including the school environment [64]. Some studies have examined the fruit and vegetable consumption among toddlers, while others have focused the effort to reach elementary-school-age children and their parents or caregivers to determine factors that influence both fruit and vegetable intake [65, 66]. Published intervention studies have promoted both fruits and vegetables, and researchers have developed techniques to promote and measure behavior changes regarding fruit and vegetable consumption [66, 67]. Some pilot studies have used a combination of taste testing, modeling, and rewards [68]. Assessing the fruit and vegetable intake among children can be challenging; children's self-reported intake often differs from parents' assessment of their children's intake of fruits and vegetables [69].

Using data from the Feeding Infants and Toddlers Study (FITS), a random sample including 3022 infants and toddlers, researchers examined food consumption among US toddlers from the age of 4–24 months [65]. Among the most striking findings was the low intake of vegetables consumed as separate foods, rather than combined with other foods. Consumption of GLVs was particularly low, with fewer than 10 % of infants and toddlers consuming green vegetables daily. French fries and other fried potatoes were the vegetables consumed in highest amount by 24 months.

In a survey of students in grades 4–6, strategies to increase fruit and vegetable intake among children in the USA were reported [66]. The children were more likely to consume fruits than vegetables, and these parents were more likely to buy fruits than vegetables for either snacks or meals. The strongest correlation with vegetable intake was found among children who asked their parents to purchase vegetables for snacks or meals at home. In addition, children who knew the number of servings recommended were more likely to consume more fruits and vegetables, which indicates that nutrition education (NE) and follow-up assessments are significant factors in promoting behavior change related to fruit and vegetable intake.

In focus groups with children in grades 4–6 and their parents of African American, Euro-American, and Mexican American descent, researchers explored the influences on children's diets, particularly related to fruit, juice, and vegetable intake [70]. Among the influences on children's food choices were availability and accessibility of food, parents, peers, television viewing behavior, and eating out. Few ethnic differences were noted, and parents cited concerns with consumption of "junk food" and too little fruit, juice, and vegetables. Peers were cited as a negative influence on consumption of vegetables. Further, participants in the focus groups considered "salad" different from vegetables, which could influence future research regarding consumption of leafy greens [70].

In a survey of 130 school food authorities combined with 24-h dietary recalls with 2314 children in 398 US schools, researchers determined the types of foods offered by schools and consumed by children during school breakfast and lunch [67]. In terms of leafy greens, lettuce salad was available in about one third of the elementary school menus and one half of the high school menus. Less than one third of schools provided dark green and orange vegetables such as broccoli and carrots. The researchers reported that the majority of the children did not consume dark green vegetables at breakfast or lunch and encouraged improvements in preparation methods or presentation, NE, and tasting opportunities to encourage consumption of produce during school lunch [67].

In another school-based study, US researchers used 24-h recall food intake data from 414 third-grade children (86 % European American) participating in the High 5 project of the National Cancer Institute's 5 a Day for Better Health initiative [71]. The researchers developed a theory-based model of the determinants of fruit and vegetable consumption, including motivation and self-efficacy, among the children. Children, for example, may improve their self-efficacy from practice in selecting vegetables or in developing skills to ask their parents to purchase more vegetables or fruits.

Environmental changes have been shown to support greater fruit and vegetable intake. The 2-year Cafeteria Power Plus project included 1668 children in grades 1 and 3 in year 1, and follow-up with 1168 children in grades 2 and 4 in year 2, in 26 US schools randomly assigned to control or intervention groups [72]. Students in the 13 intervention schools had increased opportunities to eat a variety of fruits and vegetables (through greater availability of fruits and vegetables, taste testing), who were exposed to healthful role models (posters of life-sized fruit and vegetable characters known as the High 5 Flyers), and had access to social support, including encouragement from the food-service staff. The intervention group participated in challenges/competitions, samplings, and a finale meal. The fruit and vegetable consumption among randomly selected students was determined by trained observers in the cafeteria. Students in the intervention schools had a significantly greater consumption of overall fruits and vegetables, but the increases were attributed to fruit consumption. Intake of vegetables and juice did not change. Encouragement from food service staff improved the students' consumption of produce. The researchers promoted a multifaceted approach to include not only cafeteria modifications but also classroom instruction and parent involvement [72].

School gardens combined with NE may pave the way for increasing consumption of leafy greens and other vegetables among children. In particular, researchers have shown that gardening combined with NE can increase children's preferences for spinach, broccoli, and other vegetables [73]. In a study of six second-grade classrooms with a total of 115 students, surveys and lunchroom observation were used to evaluate whether the experiential addition of gardening influenced knowledge gain and consumption of produce. Two of the classes participated in NE, two of the classes participated in nutrition education and gardening (NE + G), and two served as the control group. The children in the gardening group worked with seeds and plants, and they maintained the garden through watering and pest control. The researchers reported statistically significant gains among the children in the NE + G group in their ability to identify spinach, zucchini, and cabbage. The children in both the NE and the NE + G groups significantly increased their liking of broccoli, cabbage, and zucchini, with the NE + G group showing the greater change. In addition, the NE + G group increased their preference scores for spinach [73].

Local foods initiatives have promoted the consumption of food from community gardens. In one pilot project, with poster promotions and student-developed public service announcements, 115 pounds of kale were grown and served to 2235 students in a US high school. Of those who tried the kale, 66 % of students liked it [74].

Key Messages to Support Increased Dark Green Vegetable Intake

To support the consumer behaviors to increase vegetable intake, especially consumption of dark green vegetables, the US Dietary Guidelines for Americans provided several strategies [9]. Consumers should include vegetables in meals and snacks, keeping in mind that fresh, frozen, canned, and dehydrated forms of vegetables count toward the daily recommendation. Dark green vegetables can be added to soups, casseroles, stews, and stir fries. To enhance nutritional value, consumers are advised to include dark green leafy vegetables, such as romaine lettuce and spinach, in salads. Raw vegetables should be available for quick snacks, and if served with a dip, calcium-rich yogurt-based dressings, hummus, or other nutrient-dense dips should be used instead of full-fat sour cream or cream-cheese-based dips. When eating out, vegetables can be ordered as side dishes, and the dressings or sauces can be served on the side or in smaller amounts [9].

Conclusion

Dark green vegetables, particularly leafy greens, are nutrient-dense foods that may have a range of health benefits when included in a varied, balanced diet. Scientists worldwide continue to identify and characterize substances in wild and cultivated green vegetables that may be responsible for associations of plant components with decreased risk for chronic disease, including cancer, heart disease, diabetes, and eye diseases. More research is needed to elucidate the components of dark green vegetables that are responsible for their potential health benefits. Continued promotion of vegetable consumption through effective consumer messaging and research-tested multifaceted educational strategies that reach adults and children may help reach the goals set forth for vegetable consumption.

Acknowledgments The author expresses appreciation to graduate students Rebecca West, for conducting a thorough literature search, and Kimberly Beauchamp, for her thoughtful review.

References

1. Zhang S, Hunter DJ, Forman MR, Rosner BA, Speizer FE, Colditz GA et al (1999) Dietary carotenoids and vitamins A, C, and E and risk of breast cancer. J Natl Cancer Inst 91(6):547–556
2. Liu S, Lee IM, Ajani U, Cole SR, Buring JE, Manson JE (2001) Intake of vegetables rich in carotenoids and risk of coronary heart disease in men: the physician's health study. Int J Epidemiol 30:130–135
3. Joshipura KJ, Hu FB, Manson JE, Stampfer MJ, Rimm EB, Speizer FE, et al (2001) The effect of fruit and vegetable intake on risk for coronary heart disease. Ann Intern Med 134:1106–1114

4. Whitehead JA, Mares JA, Danis RP (2006) Macular pigment: a review of current knowledge. Arch Ophthalmol 124:1038–1045
5. Seddon JM, Umed AA, Sperduto RD, Hiller R, Blair N, Burton TC, et al (1994) Eye disease case-control study group. Dietary carotenoids, vitamins A, C, and E, and advanced age-related macular degeneration. JAMA 272:1413–1420
6. Nanney MS, Haire-Joshu D, Hessler K, Brownson RC (2004) Rationale for a consistent "powerhouse" approach to vegetable and fruit messages. J Am Diet Assoc 104(3):352–356
7. Johnston CS, Taylor AS, Hampl JS (2000) More Americans are eating "5 a day" but intakes of dark green and cruciferous vegetables remain low. J Nutr 130:3063–3067
8. U.S. Department of Health and Human Services and U.S. Department of Agriculture (2005) Dietary Guidelines for Americans, 6th edn. U.S. Government Printing Office, Washington, DC. January 2005. http://www.health.gov/dietaryguidelines/dga2005/document/pdf/DGA2005.pdf. Accessed 20 Dec 2012
9. U.S. Department of Health and Human Services and U.S. Department of Agriculture (2010) Dietary Guidelines for Americans, 7th edn. U.S. Government Printing Office, Washington, DC. January 2010. http://www.cnpp.usda.gov/Publications/DietaryGuidelines/2010/PolicyDoc/PolicyDoc.pdf. Accessed 20 Dec 2012
10. U.S. Food and Drug Administration (2011) Guidance for industry: a food labeling guide. Appendix F: calculate the percent daily value for the appropriate nutrients, 2011. http://www.fda.gov/Food/GuidanceComplianceRegulatoryInformation/GuidanceDocuments/FoodLabelingNutrition/FoodLabelingGuide/ucm064928.htm. Accessed 20 Dec 2012
11. ChooseMyPlate.gov [2012]. United States Department of Agriculture. http://www.choosemyplate.gov. Accessed 20 Dec 2012
12. U.S. Department of Agriculture, Agricultural Research Service (2012) USDA National Nutrient Database for Standard Reference, Release 25, 2012. http://www.ars.usda.gov/ba/bhnrc/ndl. Accessed 20 Dec 2012
13. Radek M, Savage GP (2008) Oxalates in some Indian green leafy vegetables. Int J Food Sci Nutr 59(3):246–260
14. Vitamin K2. Altern Med Rev (2009);14(3):284–293
15. Ferland G, Sadowski JA (1992) Vitamin K1(phylloquinone) content of green vegetables: effects of plant maturation and geographical growth location. J Agric Food Chem 40:1874–1877
16. Seifert RM (1979) Analysis of vitamin K1 in some green leafy vegetables by gas chromatography. J Agric Food Chem 27(6):1301–1304
17. Rao AV, Rao LG (2007) Carotenoids and human health. Pharmacol Res 55:207–216
18. Ribaya-Mercado JD, Blumberg JB (2004) Lutein and zeaxanthin and their potential roles in disease prevention. J Am Coll Nutr 23(6):567S–587S
19. Granado F, Olmedilla B, Blanco I, Rojas-Hidalago E (1992) Carotenoid composition in raw and cooked Spanish vegetables. J Agric Food Chem 40:2135–2140
20. Kopsell DE, Kopsell DA, Randle WM, Coolong TW, Sams CE, Curran-Celentano J (2003) Kale carotenoids remain stable while flavor compounds respond to changes in sulfur fertility. J Agric Food Chem 51:5319–5325
21. Kopsell DA, Kopsell DE (2006) Accumulation and bioavailability of dietary carotenoids in vegetable crops. Trends Plant Sci 11(10):499–507
22. Mercadante AZ, Rodriquez-Amaya DB (1991) Carotenoid composition of a leafy vegetable in relation to some agricultural variables. J Agric Food Chem 39:1094–1097
23. de Azevedo CH (2005) Carotenoid composition of kale as influenced by maturity, season and minimal processing. J Sci Food Agric 85:591–597
24. Kimura M, Rodriguez-Amaya DB (2003) Carotenoid composition of hydroponic leafy vegetables. J Agric Food Chem 51:2603–2607
25. Kuti JO, Konuru HB (2004) Antioxidant capacity and phenolic content in leaf extracts of tree spinach (*Cnidoscolus* spp.). J Agric Food Chem 52:117–121
26. Olaiya C, Adebisi J (2009) Phytoevaluation of the nutritional values of ten green leafy vegetables in South-Western Nigeria. Internet J Nutr Wellness 9(2). doi:10.5580/1150

27. Takeiti CY, Antonio GC, Motta EMP, Collares-Queiroz FP, Park KJ (2009) Nutritive evaluation of a non-conventional leafy vegetable (*Pereskia aculeate* Miller). Int J Food Sci Nutr 60(S1):148–160
28. Gupta S, Prakash J (2009) Studies on Indian green leafy vegetables for their antioxidant activity. Plant Foods Hum Nutr 64:39–45
29. Niizu PY, Rodriguez-Amaya DB (2005) New data on the carotenoid composition of raw salad vegetables. J Food Compos Anal 18:739–749
30. Afolayan AJ, Jimoh FO (2009) Nutritional quality of some wild leafy vegetables in South Africa. Int J Food Sci Nutr 60(5):424–431
31. El-Qudah JM (2008) Dietary intake of selected common vegetable foods and their total carotenoids determination. Am J Agric Biol Sci 3(4):729–733
32. Olsen H, Aaby K, Borge GIA (2009) Characterization and quantification of flavonoids and hydroxycinnamic acids in curly kale *(Brassica oleracea* L. Convar. *acephala* Var. *sabellica*) by HPLC-DAD-ESI-MSn. J Agric Food Chem 57:2816–2825
33. Lin L-Z, Harnly JM (2009) Identification of the phenolic components of collard greens, kale, and Chinese broccoli. J Agric Food Chem 57:7401–7408
34. Lin L-Z, Harnly JM (2010) Phenolic component profiles of mustard greens, yu choy, and 15 other *Brassica* vegetables. J Agric Food Chem 58:6850–6857
35. Su Q, Rowley KG, Itsiopoulos C, O'Dea K (2002) Identification and quantitation of major carotenoids in selected components of the Mediterranean diet: green leafy vegetables, figs and olive oil. Eur J Clin Nutr 56:1149–1154
36. Salvatore S, Pellegrini N, Brenna OV, Del Rio D, Frasca G, Brighenti F et al (2005) Antioxidant characterization of some Sicilian edible wild greens. J Agric Food Chem 53:9465–9471
37. Orech FO, Christensen DL, Larsen T, Friis H, Aagaard-Hansen J, Estambale BA (2007) Mineral content of traditional leafy vegetables from western Kenya. Int J Food Sci Nutr 58(8):595–602
38. Ezzati M, Lopez AD, Rodgers A, Vander Hoorn S, Murray CJL (2002) Selected major risk factors and global and regional burden of disease. Lancet 360:1347–1360
39. World Health Organization (2009) Global prevalence of vitamin A deficiency in populations at risk 1995–2005. WHO global database on vitamin A deficiency. World Health Organization, Geneva
40. Astorg P (1997) Food carotenoids and cancer prevention: an overview of current research. Trends Food Sci Technol 8:406–413
41. Maserejian NN, Giovannucci E, Rosner B, Joshipura K (2006) Prospective study of vitamins C, E, and A and carotenoids and risk of oral premalignant lesions in men. Int J Cancer 120:970–977
42. Olsen H, Grimmer S, Aaby K, Saha S, Borge GIA (2012) Antiproliferative effects of fresh and thermal processed green and red cultivars of curly kale *(Brassica oleracea* L. convar. *acephala* var. *sabellica*). J Agric Food Chem 60:7375–7383
43. Esposito K, Giugliano D (2011) Increased consumption of green leafy vegetables, but not fruit, vegetables or fruit and vegetables combined, is associated with reduced incidence of type 2 diabetes. Evid Based Med 16(1):27–28
44. Ejoh RA, Dever JT, Mills JP, Tanumihardjo SA (2010) Small quantities of carotenoid-rich tropical green leafy vegetables indigenous to Africa maintain vitamin A status in Mongolian gerbils (*Meriones unguiculatus*). Br J Nutr 103(11):1594–1601
45. Randrianatoandro VA, Avallone S, Picq C, Ralison C, Trèche S (2010) Recipes and nutritional value of dishes prepared from green-leafy vegetables in an urban district of Antananarivo (Madagascar). Int J Food Sci Nutr 61(4):404–416
46. Faber M, Oelofse A, Van Jaarsveld PJ, Wenhold FAM, Jansen van Rensburg WS (2010) African leafy vegetables consumed by households in the Limpopo and KwaZulu-Natal provinces in South Africa. S Afr J Clin Nutr 23(1):30–38
47. Agte V, Jahagirdar M, Chiplonkar S (2006) GLV supplements increased plasma β-carotene, vitamin C, zinc and hemoglobin in young healthy adults. Eur J Nutr 45:29–36

48. Persson V, Ahmed F, Gebre-Medhin M, Greiner T (2001) Increase in serum beta-carotene following dark green leafy vegetable supplementation in Mebendazole-treated school children in Bangladesh. Eur J Clin Nutr 55:1–9
49. Faber M, Phungula MAS, Venter SL, Dhansay MA, Spinnler Benadé AJ (2002) Home gardens focusing on the production of yellow and dark-green leafy vegetables increase the serum retinol concentrations of 2–5-y-old children in South Africa. Am J Clin Nutr 76:1048–1054
50. Yeum K-J, Booth SL, Sadowski JA, Liu C, Tang G, Krinsky NI et al (1996) Human plasma carotenoid response to the ingestion of controlled diets high in fruits and vegetables. Am J Clin Nutr 64:594–602
51. Moeller SM, Jacques PF, Blumberg, JB (2000) The potential role of dietary xanthophylls in cataract and age-related macular degeneration. J Am Coll Nutr 19:522S–527S
52. Mozaffarieh M, Sacu S, Wedrich A (2003) The role of the carotenoids, lutein and zeaxanthin, in protecting against age-related macular degeneration: a review based on controversial evidence. Nutr J 2(20). http://www.nutritionj.com/content/2/1/20. Accessed 20 Dec 2012
53. Lee PP, Feldman ZW, Ostermann J, Brown DS, Sloan FA et al (2003) Longitudinal prevalence of major eye diseases. Arch Ophthalmol 121:1303–1310
54. Cho E, Hankinson SE, Rosner B, Willett WC, Colditz GA (2008) Prospective study of lutein/zeaxanthin intake and risk of age-related macular degeneration. Am J Clin Nutr 87(6):1837–1843
55. LaRowe TL, Mares JA, Snodderly DM, Klein ML, Wooten BR, Chappell R (2008) Ancillary study women's health initiative. Macular pigment density and age-related maculopathy in the carotenoids in age-related eye disease study. Ophthalmology 115:876–883
56. Ma L, Dou H-L, Wu Y-Q, Huang Y-M, Huang Y-B, Xu X-R et al (2011) Lutein and zeaxanthin intake and the risk of age-related macular degeneration: a systematic review and meta-analysis. Br J Nutr 107(3):350–359
57. SanGiovanni JP, Chew EY, Clemons TE, Ferris III FL, Gensler G, Lindblad AS (2007) Age-related eye disease study research group. The relationship of dietary carotenoid and vitamin A, E, and C intake with age-related macular degeneration in a case-control study [AREDS report no. 22]. Arch Ophthalmol 125(9):1225–1232
58. Flood V, Smith W, Wang JJ, Manzi F, Webb K, Mitchell P (2002) Blue mountains eye study. Dietary antioxidant intake and incidence of early age-related maculopathy. Ophthalmology 109:2272–2278
59. Tan JSL, Wang JJ, Flood VF, Rochtchina E, Smith W, Mitchell P (2008) Blue mountains eye study. Dietary antioxidants and the long-term incidence of age-related macular degeneration. Ophthalmology 115:334–341
60. Moeller SM, Parekh N, Tinker L, Ritenbaugh C, Blodi B, Wallace RB et al (2006) Ancillary study women's health initiative. Associations between intermediate age-related macular degeneration and lutein and zeaxanthin in the carotenoids in age-related eye disease study (CAREDS). Arch Ophthalmol 124:1151–1162
61. Lakshminarayana R, Raju M, Krishnakantha TP, Baskaran V (2007) Lutein and zeaxanthin in leafy greens and their bioavailability: olive oil influences the absorption of dietary lutein and its accumulation in adult rats. J Agric Food Chem 55:6395–6400
62. Christen WG, Liu S, Glynn RJ, Gaziano JM, Buring JE (2008) A prospective study of dietary carotenoids, vitamins C and E, and risk of cataract in women. Arch Ophthalmol 126(1):102–109
63. Kang JH, Pasquale LR, Willett W, Rosner B, Egan KM, Faberowski N et al (2003) Antioxidant intake and primary open-angle glaucoma: a prospective study. Am J Epidemiol 158:337–346
64. French SA, Wechsler, H (2004) School-based research and initiatives: fruit and vegetable environment, policy, and pricing workshop. J Prev Med 39:S101–S107.
65. Fox MK, Pac S, Devaney B, Jankowski L (2004) Feeding infants and toddlers study: what foods are infants and toddlers eating? J Am Diet Assoc 104(Suppl):S22–S30
66. Sandeno C, Wolf G, Drake T, Reicks M (2000) Behavioral strategies to increase fruit and vegetable intake by fourth- through sixth-grade students. J Am Diet Assoc 100(7):828–830

67. Condon EM, Crepinsek MK, Fox MK (2009) School meals: types of foods offered to and consumed by children at lunch and breakfast. J Am Diet Assoc 109(Suppl):S67–S78
68. Wengreen HJ, Madden GJ, Aguilar SS, Smits RR, Jones BA (2012) Incentivizing children's fruit and vegetable consumption: results of a United States pilot study of the Food Dudes program. J Nutr Educ Behav 45(1):54-59 (Epub 2012 Nov 20. http://dx.doi.org/10.1016/j.jneb.2012.06.001)
69. Tak NI, te Velde SJ, de Vries JHM, Brug J (2006) Parent and child reports of fruit and vegetable intakes and related family environmental factors show low levels of agreement. Brit Diet Assoc 19:275–285
70. Cullen KW, Baranowski T, Rittenberry L, Olvera N (2000) Social-environmental influences on children's diets: results from focus groups with African-, Euro- and Mexican-American children and their parents. Health Educ Res 15(5):581–590
71. Reynolds KD, Hinton AW, Shewchuk RM, Hickey CA (1999) Social cognitive model of fruit and vegetable consumption in elementary school children. J Nutr Educ 31:23–30
72. Perry CL, Bishop DB, Taylor GL, Davis M, Story M, Gray C et al (2004) A randomized school trial of environmental strategies to encourage fruit and vegetable consumption among children. Health Educ Behav 31(1):65–76
73. Parmer SM, Salisbury-Glennon J, Shannon D, Struempler B (2009) School gardens: an experiential learning approach for a nutrition education program to increase fruit and vegetable knowledge, preference, and consumption among second-grade students. J Nutr Educ Behav 41(3):212–217
74. Geering D (2012) Food fight. What's keeping local produce out of school cafeterias? Atlanta 2012:65–67, 84

Chapter 13
Anthocyanins in Staple Crops

Katia Petroni, Roberto Pilu, and Chiara Tonelli

Abbreviations

PAL	Phenylalanine ammonia lyase
C4H	Cinnamic acid 4-hydroxylase
4CL	4 coumarate CoA ligase
CHS	Chalcone synthase
CHI	Chalcone isomerase
F3H	Flavanone 3-hydroxylase
F3′H	Flavanone 3′-hydroxylase
DFR	Dihydroflavonol reductase
FLS	Flavonol synthase
ANS	Anthocyanidin synthase
UFGT, UDP	Flavonoid glucosyl transferase
GST	Glutathione S-transferase
ANR	Anthocyanidin reductase
LAR	Leucoanthocyanidin reductase
DHK	Dihydrokaempferol
DHQ	Dihydroquercetin
bHLH-LZ	Basic helix-loop-helix-leucine zipper
RIF1	R-interacting factor 1
^{ha}PBS	High-affinity P1-binding sites
^{la}PBS	Low-affinity PBS
Pp	*Purple pericarp*
VIGS	Virus-induced gene silencing
RNAi	RNA interference

K. Petroni (✉) · C. Tonelli
Dipartimento di Bioscienze, Università degli Studi di Milano, Via Celoria 26,
20133 Milano, Italy
e-mail: katia.petroni@unimi.it

R. Pilu
Di.S.A.A.—Produzione, Territorio, Agroenergia, Università degli Studi di Miano, Via Celoria 26,
20133 Milano, Italy

C. Chen (ed.), *Pigments in Fruits and Vegetables,* DOI 10.1007/978-1-4939-2356-4_13

r1	*Colored 1*
b1	*Booster 1*
pl1	*Purple plant 1*
p1	*Pericarp color 1*
lpa1	*Low phytic acid 1*
sulf	*Sulfurea*
R-st	*R-stippled*
B-I	*Booster-Intense*
Pl-Rh	*Pl-Rhoades*
sRNAs	Small RNAs
HPT	Hygromycin phosphotransferase
mop1	*Mediator of paramutation1*
mop2	*Mediator of paramutation2*
rmr1	*Required to maintain repression1*
rmr2	*Required to maintain repression2*
rmr6	*Required to maintain repression6*
Ufo1	*Unstable factor for orange 1*
siRNA	Small interfering RNA
dsRNA	Double strand RNA
miRNA	microRNAs
CBBP	CXC domain *b1*-repeat binding protein
RISC	RNA-induced silencing complex
RITS	RNA-induced transcriptional silencing
EPA	Eicosapentanoic acid
DHA	Docosohexanoic acid
PUFA	Polyunsaturated fatty acid
SD	Standard diet
PCC	Purple corn color
HFD	High fat diet
SREBP-1	Sterol-binding protein-1
FAS	Fatty acid synthase
ACS1	Acyl-CoA synthase 1
AMPK	AMP-activated protein kinase
apoE	Apoliprotein E
HDL	High-density lipoprotein
AST	Aspartate transaminase
ALT	Alanine transaminase
GGT	γ glutamyl transferase

Introduction

Anthocyanin synthesis is one of the most studied secondary metabolic pathways in plants. Most knowledge about transcriptional regulation in plants has initially come from studies of the regulation of anthocyanin biosynthesis in maize. Furthermore, genetic and molecular studies aimed at understanding its regulation in maize have

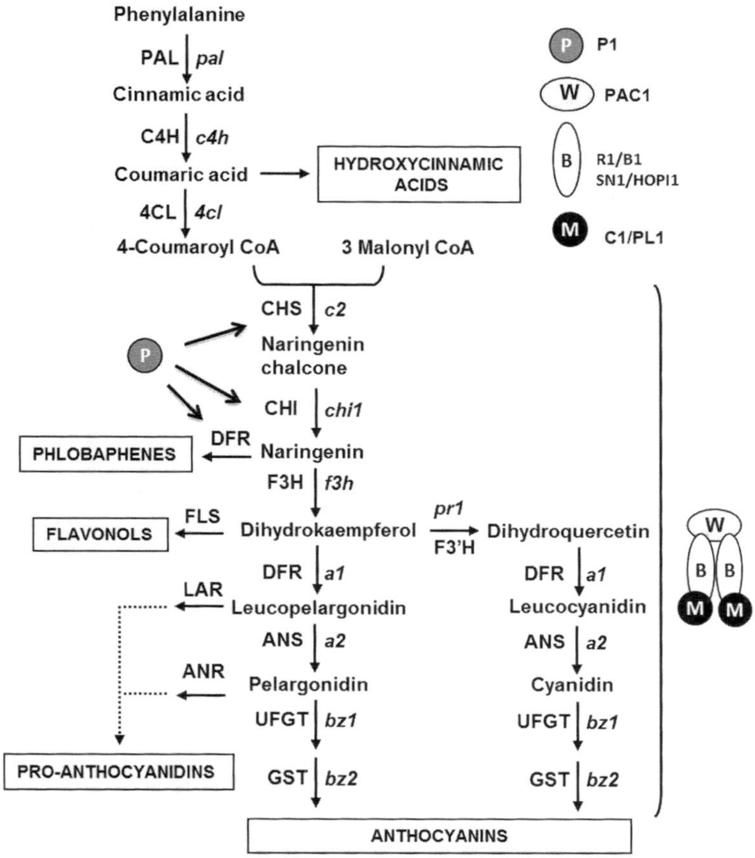

Fig. 13.1 Simplified scheme of the flavonoid pathway, comprising the general phenylpropanoid pathway (*PAL, C4H, 4CL*), the anthocyanin branch and other subgroups of flavonoids. MYB (*M*), bHLH (*B*), WD40 (*W*), and P1 (*P*) transcription factors controlling the anthocyanin pathway in maize and putative MBW complexes are indicated. Genes (enzyme) names are indicated as follows: *pal* (*PAL*) phenylalanine ammonia lyase; *c4h* (*C4H*) cinnamic acid 4-hydroxylase; *4cl* (*4CL*) 4 coumarate CoA ligase, *c2* (*CHS*) chalcone synthase, *chi1* (*CHI*) chalcone isomerase, *f3h* (*F3H*) flavanone 3-hydroxylase, *pr1* (*F3'H*) flavanone 3′-hydroxylase, *a1* (*DFR*) dihydroflavonol reductase, FLS flavonol synthase, *a2* (*ANS*) anthocyanidin synthase, *bz1* (*UFGT*) UDP-flavonoid glucosyl transferase, *bz2* (*GST*) glutathione S-transferase, ANR anthocyanidin reductase, LAR leucoanthocyanidin reductase. *Brackets* indicate the anthocyanin biosynthesis genes activated coordinately by WD40/bHLH/MYB proteins

led to the discovery of transposable elements [1], of the epigenetic phenomena of paramutation and silencing of duplicated genes [2, 3], and of the first plant transcription factor [4]. Anthocyanins are the final products from a specific branch of the flavonoid pathway. Flavonoid biosynthesis genes are highly conserved at structural and functional level among plant species and are organized to produce different subclasses of flavonoids in different branches of the pathway, including hydroxycinnamic acids, flavonols, phlobaphenes, pro-anthocyanidins, and anthocyanins (Fig. 13.1). Some flavonoids are species specific, such as phlobaphenes in maize, whereas others are common in more than one species (e.g., pro-anthocyanidins in *Arabidopsis,* petunia, and rice seeds) or almost ubiquitous, such as anthocyanins and flavonols. Anthocyanin biosynthesis and its regulation have been extensively studied in different plants. Several enzymes involved in the production of different subclasses of flavonoids have been characterized and the corresponding biosynthesis genes isolated [5, 6]. In maize, the regulatory genes of anthocyanin biosynthesis are divided in two families, encoding MYB and bHLH transcription factors respectively. Such factors interact to form an MYB–bHLH heterodimer able to activate all anthocyanin biosynthesis genes in a coordinate manner [7]. Anthocyanin biosynthesis genes have also been isolated in rice and wheat, but knowledge about regulatory loci is still rudimentary.

Anthocyanins are health-protecting components in the human diet as a result of their capacity to activate endogenous antioxidant defense systems and signaling pathways [8, 9]. A number of studies suggest that they have a protective effect against cardiovascular disease, obesity, cancer, and neurodegenerative diseases [10–13]. Here, we report recent advances in understanding the regulation of anthocyanin biosynthesis in staple crops, updates on epigenetic regulatory mechanisms in maize, and recent findings demonstrating important health benefits of anthocyanin-rich cereals against chronic diseases.

Genomics

The Anthocyanin Biosynthetic Pathway in Staple Crops

In maize, the anthocyanin biosynthesis (Fig. 13.1) begins from 4-coumaroyl-CoA and malonyl-CoA, which are converted into naringenin chalcone by the chalcone synthase (CHS) enzyme. Naringenin chalcone is, then, the substrate for chalcone isomerase (CHI) which catalyzes the intramolecular cyclization of bicyclic chalcones into tricyclic (S)-flavanones, thus producing naringenin. The subsequent step is the hydroxylation of naringenin into dihydrokaempferol (DHK), which is the substrate for three different enzymes and towards three biosynthesis directions: (i) dihydroflavonol reductase (DFR) that leads to anthocyanin synthesis (i.e., pelargonidin); (ii) flavanone 3′-hydroxylase (F3′H) that converts DHK to dihydroquercetin (DHQ); and (iii) flavonol synthase (FLS) that catalyzes the formation of flavonols (i.e., kaempferol). Similarly, DHQ is directed to the anthocyanin biosynthesis (i.e., cyanidin) by DFR or to flavonol synthesis (i.e., quercetin) by FLS. Pelargonidin and cyanidin are then glycosylated and acylated by UDP-glucose flavonoid 3-*O*-glucosyltransferase (UFGT) and glutathione S-transferase (GST) respectively and then transferred to the vacuole.

In the past years, the availability of maize mutants affecting each enzymatic step of the anthocyanin pathway has led to the isolation of most anthocyanin biosynthesis genes in maize, except *pr1,* encoding F3′H, which was recently cloned [6, 14]. Anthocyanin biosynthesis genes are highly conserved among maize, rice, and wheat. Rice orthologous genes encoding CHS, CHI, F3′H, DFR, and ANS with high amino acid identities ranging from 66 to 92% to the corresponding maize genes have been isolated [15]. Both CHS and ANS are encoded by duplicated genes located on different chromosomes with high amino acid sequence identities (80% for *OsCHS1* and *OsCHS2*; 93% for *OsANS1* and *OsANS2*). However, only *OsCHS1* and *OsANS1* were found to be expressed in rice. The function of the expressed anthocyanin biosynthetic genes was confirmed by complementation of the corresponding *Arabidopsis transparent testa* flavonoid mutants. In addition, yeast two-hybrid assay suggested the formation of a macromolecular complex composed by an OsCHS1 homodimer interacting with OsF3H, OsF3′H, OsDFR, OsANS1, similarly to what previously reported in *Arabidopsis* [15]. Anthocyanin genes encoding CHS, CHI, F3H, DFR, and ANS have been also cloned and mapped in wheat and duplicated genes were found for *CHS, CHI,* and *DFR*. High expression levels were found for *CHS, CHI, F3H,* and *DFR* in the grain coat tissue and in coleoptiles of red wheat lines [16, 17].

The Regulation of Anthocyanin Biosynthesis in Staple Crops

In maize, one MYB-related protein and one bHLH-containing protein interact and activate the anthocyanin biosynthesis genes (encoding CHS, CHI, F3H, DFR, ANS, UFGT, GST) as a single unit (Fig. 13.1). MYB and bHLH proteins are encoded by two families of regulatory genes (*c1/pl1* and *r1/b1*, respectively), whose members encode functionally equivalent proteins. Each member has a tissue or development-specific expression, so that the pigmentation pattern of a maize plant depends on its allelic combination at the *r1/b1* and *c1/pl1* loci. For example, the *r1/b1* genes (*R1, B1, Sn1, Lc1, Hopi1*) determine the tissue distribution of anthocyanins [18–21], whereas *c1* in the seed [22] or *pl1* in vegetative tissues [23] contributes to the developmental and light-dependent regulation of anthocyanin synthesis. In maize seed, the dominant *R1 C1* genes activate anthocyanin pigmentation of the aleurone layer in blue corn, whereas the dominant *B1 Pl1* genes induce anthocyanin synthesis in the pericarp of purple corn (Fig. 13.2) [18]. By contrast, MYB P1, which confers phlobaphene synthesis in kernels and cobs, activates a subset of anthocyanin genes

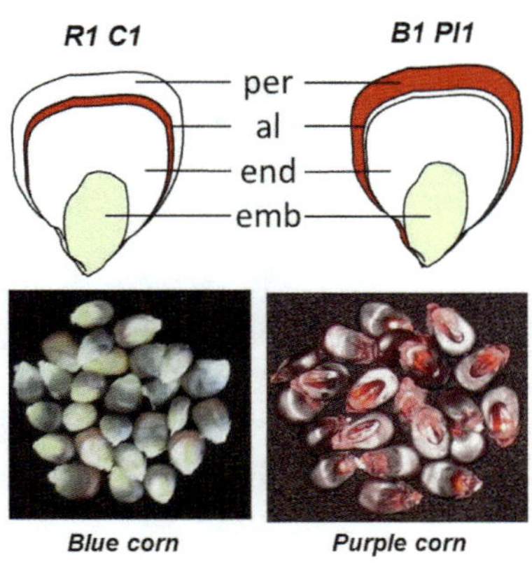

Fig. 13.2 Scheme depicting the tissue-specific anthocyanin pigmentation determined by *R1 C1* in aleurone of blue corn seeds (*left panel*) and by *B1 Pl1* in pericarp of purple corn seeds (*right panel*). *per* pericarp, *al* aleurone, *end* endosperm, *emb* embryo

(i.e., *CHS, CHI, DFR,* and *FLS*) without bHLH interactors [24, 25]. Recent studies have demonstrated that the newly isolated *pr1* gene, encoding the F3′H enzyme responsible for the conversion of DHK to DHQ (Fig. 13.1), is also activated by R1/C1 in aleurone of maize seeds [14] and by P1 in pericarp of seeds and in cob glumes, where it contributes to the conversion of apiferol to luteoforol, both precursors of phlobaphenes [26]. In addition to the MYB and bHLH regulators, the *PAC1* gene encoding a WD40 protein is required by either bHLH B1 or R1 proteins for full activation of anthocyanin biosynthesis genes in seeds [27, 28]. Although the stoichiometry of the R1/C1 complex is currently under investigation (see below), it is likely that the bHLH proteins function as homodimers and as docking proteins between MYB and WD40 proteins (Fig. 13.1).

Despite being activated in a coordinate manner by R1/C1, anthocyanin biosynthesis genes lack obviously conserved *cis*-regulatory elements. Recent studies suggested that this coordinate regulation may be achieved by two different R1/C1 complexes, each able to bind specific *cis*-elements and thus to activate a subset of anthocyanin biosynthesis genes [29]. These alternative R1/C1 complexes depend on two different dimerization domains found in bHLH proteins. Three interaction domains were identified in bHLH proteins (Fig. 13.3a): (i) a N-terminal acidic region for interaction with MYB proteins [19]; (ii) an ACT-like dimerization domain located C-terminal of the bHLH motif, already known as required for homodimer formation [30]; (iii) a newly discovered homodimerization domain consisting of the bHLH domain and a nearly located short leucine zipper (LZ) motif. Interestingly, the bHLH–LZ domain can form a homodimer *in vitro* only when the ACT-like dimerization domain is mutated or deleted [29]. In the proposed model, when R1 homodimerize via the ACT-like domain (Fig. 13.3b), the bHLH region of R1 interacts with R-interacting factor 1 (RIF1), an EMSY-like maize nuclear factor, necessary to the R1/C1 complex for transcriptional activation of anthocyanin biosynthesis genes through increased histone acetylation of the promoter region [31], whereas the N-terminal acidic region interacts with the MYB C1, necessary to bind the high-affinity P1-binding sites (^{ha}PBS) and low-affinity PBS (^{la}PBS) found in the promoters of some anthocyanin biosynthesis genes, such as *a1* [24, 32]. As an alternative (Fig. 13.3c), when R1 homodimerizes via the bHLH-LZ domain, it directly recognizes and binds E-box elements, such as those present in the *bz1* promoter, and interacts with the C1 protein, necessary to provide a strong transcriptional activation domain [29]. Which factor(s) is responsible for the switch between the two alternative R1/C1 complexes remains to be determined.

Most rice varieties have white seeds with small quantities of flavones and flavonols, but some varieties with dark purple (black) seeds containing high levels of anthocyanins (mainly cyanidin 3-glucoside) or red seeds contanining pro-anthocyanidins are known. Three rice loci (*A, B,* and *C*) control anthocyanin synthesis in seeds of purple rice varieties [33], but none of the genes has yet been isolated. Only one *bHLH* gene, *Rc,* controlling pro-anthocyanidin synthesis in pericarp of red rice varieties has been identified and shown to carry a 14-bp deletion in all white rice varieties [34], suggesting that the wild progenitors of rice were colored, but this trait was lost during domestication. Many other putative anthocyanin regulators have

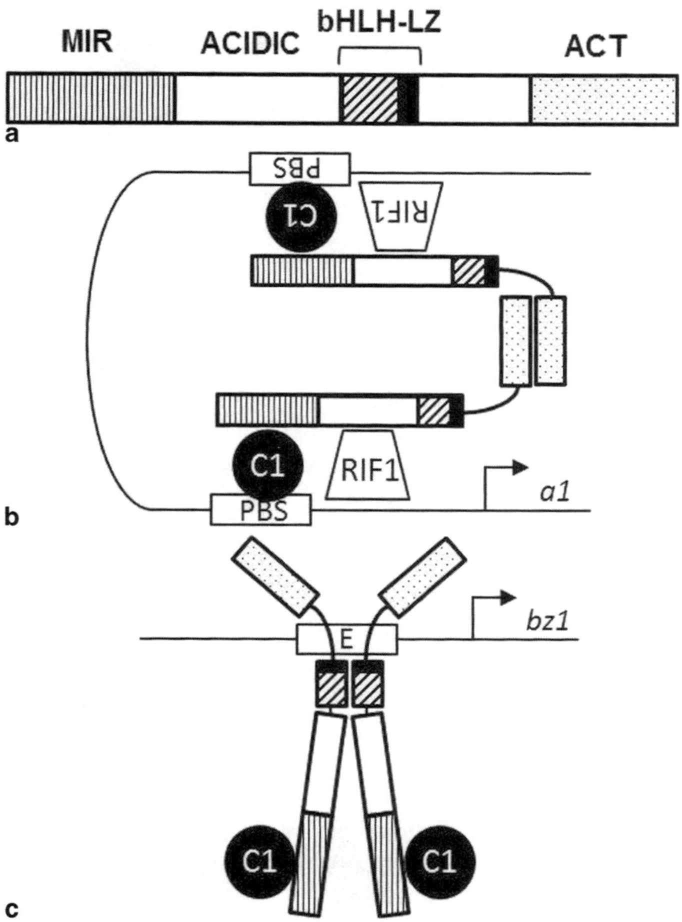

Fig. 13.3 Alternative R1/C1 complexes activating anthocyanin biosynthesis genes in maize. **a** Schematic representation of the R1 functional domains. *MIR* MYB interacting region, *ACIDIC* acidic region, *bHLH* basic helix-loop-helix domain, *LZ* leucine zipper, *ACT* ACT-like domain. **b** In the first R1/C1 complex, R1 homodimerize via the ACT-like domain, the bHLH region of R1 interacts with R-interacting factor 1 (RIF1) and the N-terminal acidic region interacts with C1, which binds the P1-binding sites (PBS) found in promoters of some anthocyanin biosynthesis genes, such as *a1*. **c** In the second R1/C1 complex, R1 homodimerize via the bHLH-LZ domain, which binds E-box elements, such as those present in the *bz1* promoter, and interacts with the C1 protein, providing a strong transcriptional activation domain

been recently identified by comparing white and black rice cultivars using a micro-array approach, but still lack functional characterizations [35]. Some rice varieties with brown grains are also known, carrying a mutation in the *DFR* gene which determines accumulation of the intermediates dihydroquercetin and dihydrorhamnetin [34, 36]. Finally, it has been demonstrated that *ANS* overexpression in *Rc* red rice genotypes redirects metabolic flux towards anthocyanin synthesis with a concomitant reduction of pro-anthocyanidins [37].

Similarly to rice, most wheat varieties have no pigmentation in grains, but purple, red, and blue wheat grain varieties are known. Pigmentation in purple wheat is mainly due to cyanidin 3-glucoside and its acylated derivatives [38], red wheat grains contain pro-anthocyanidins [39], whereas the main anthocyanin in blue wheat is delphinidin 3-glucoside [38]. The synthesis of delphinidin in blue wheat suggests that this variety carries an active gene encoding flavanone 3′5′-hydroxylase converting dihydroquercetin to dihydromyricetin, in turn converted to delphinidin by DFR and ANS. Pigmentation in red grain wheat is controlled by the *R* gene located on chromosome 3 [40], whereas red wheat coleoptile is determined by the *Rc* gene located on chromosome 7 [17]. Comparing the expression pattern of anthocyanin biosynthesis genes in red and white lines, it appears that *R* activates *CHS, CHI, F3H,* and *DFR* in coat tissue of red grain wheat, whereas *Rc* is responsible for the activation of only *DFR,* therefore suggesting that *R* and *Rc* encode transcription factors regulating different steps of the flavonoid biosynthesis pathway in a tissue-specific manner [17]. Another regulatory gene of anthocyanin biosynthesis in wheat seeds is *Pp* (*Purple pericarp*), but no molecular characterization is currently available [40].

Updates on Epigenetic Regulatory Mechanisms

All the cells of a multicellular organism carry the same DNA information but only a small subset of the genes must be active at a certain point of development and growth [41]. The fine tuning of genome expression make possible the cell specialization in complex organisms and this "molecular memory" regarding the level of expression of every gene must be faithfully transmitted through cell division while also allowing the adaptation of the organism to the environmental stimuli during its life. In this context, one of the first exceptions to the Mendel's laws was found in the 1940s by Barbara McClintock while working on anthocyanin pigments in maize; this work led to the discovery of transposons and to a Nobel prize in 1983 [42, 43].

Later on, the epigenetic phenomena defined by Riggs and colleagues as "*the study of mitotically and/or meiotically heritable changes in gene function that cannot be explained by changes in DNA sequence*" have disclosed a new level of gene regulation [44]. These phenomena seem as if they may exist in all phyla, and control a number of gene regulation processes ranging from embryo development to human diseases by DNA methylation, chromatin modification (histone methylation and nucleosome position) and noncoding RNA [45–47].

Among the gene silencing phenomena, paramutation occurs when an epigenetic state (of the paramutagenic allele) is transferred in *trans* to another allele (the paramutable allele), resulting in a heritable modification of its gene expression (the frequency of the change can reach as high as 100 %), the paramutable allele acquires the paramutagenic capability in future generations, while alleles that do not take part in paramutation are nominated as neutral (Fig. 13.4). In a typical mutation, the change of the DNA sequence usually causes a switch off of the gene activity, while paramutation due to modulation of gene expression generates different epialleles silenced with variable phenotypes [48].

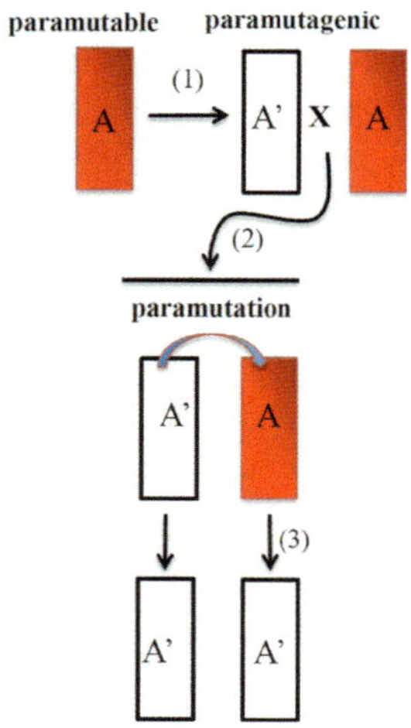

Fig. 13.4 General scheme of paramutation phenomenon. *Color* (*red*=high expression, *white*=low expression) represents phenotypic expression of the A haplotype. A paramutable A allele undergoes spontaneous silencing also inducing paramutagenic activity (*1*). In the A′/A heterozygous, obtained by crossing individual carrying paramutagenic A′ allele with the paramutable A allele (*2*), the haplotypes segregating in the offspring are both A′, because A′ has paramuted A (*3*)

In the past, several gene silencing/paramutation-like phenomena have been discovered in all the kingdoms of eukaryotes, leading to the adoption of different names such as: transvection in Drosophila [49], co-suppression and "virus-induced gene silencing" (VIGS) in the gene silencing phenomena described in transgenic plants [50–52], quelling in the fungus *Neurospora crassa* [51] and RNA interference (RNAi) in the nematode *C. elegans* [52] although classical paramutation was well defined by the maize model.

Five loci of maize: *colored 1* (*r1*), *booster 1* (*b1*), *purple plant 1* (*pl1*), *pericarp color 1* (*p1*), and *low phytic acid 1* (*lpa1*) [53] and the *sulfurea* (*sulf*) locus of tomato [54] have shown classical paramutation phenomena. In maize, the *r1*, *b1*, *pl1,* and *p1* genes encode all for transcription factors involved in the regulation of accumulation of flavonoids and anthocyanins in several plant tissues [55] while *lpa1* locus designated *ZmMRP4,* coding a multidrug resistance-associated protein, is involved in phytic acid transport and storage in the seed [56, 57]. In 1956, a paramutation phenomenon in a regulatory gene (encoding bHLH transcription factors) named *colored1* (*r1*) was first discovered in maize by Alexander Brink that worked on anthocyanin biosynthesis [58]. After crossing the paramutagenic *R-stippled* (*R-st*), conferring tiny spotted aleurone color of the seed, with the paramutable allele *R-r,* conferring full pigmentation, he observed that the progeny carrying *R-r* allele showed a variably reduced pigmentation. The silenced allele (named *R-r′*) was heritable and capable of weak paramutagenic activity (like *R-st*) for some generations, furthermore if *R-r′* was not further exposed to *R-st,* it reverted to *R-r* normal phenotype in few generations [59].

In the case of *b1,* with a frequency ranging from 0.1 to 10%, the paramutable *B-I* (*Booster-Intense*) allele spontaneously becomes partially silent (*B′*). *B′* has paramutagenic activity, in fact crossing *B′* with *B-I* the progeny obtained is 100% *B′* [60, 61], and in contrast with *r1* paramutation *B′* is permanent [61]. In the 1990s, paramutation was discovered at (*pl1*) locus, also in this case, the exposure in *trans* of paramutable allele *Pl-Rhoades* (*Pl-Rh*) to its spontaneously derived silenced paramutagenic *Pl′* allele causes a silencing of *Pl-Rh* (Fig. 13.5) [62].

Fig. 13.5 Spontaneous silencing occurring at the *pl1* locus in maize. *Weakly color seeds on the ears* are shown in a *B-I/B-I Pl-Rh/Pl-Rh* plant (genotype conferring a strong anthocyanin accumulation on the whole plant)

The spontaneously silenced epiallele (*P-rr'*) at the *p1* locus arises by transgene-induced silencing and showed a moderate stability and weak paramutagenic capacity on the original *P-rr* allele [63, 64]. Interestingly, some differences among these paramutation systems can be noted: *p1* and *r1* epigenetic states are stable while *pl1* and *b1* loci are unstable, spontaneously changing to the silenced state with high frequency [55, 64].

A new locus undergoing a paramutation phenomenon which does not involve the anthocyanin pathway has been recently discovered in maize: the *lpa1-241* allele at the *lpa1* locus, that in fact seems somewhat similar to *r1* locus paramutation [57]. The *lpa1-241* mutant reduces phytic acid content, and proportionally increases the level of free phosphate, thus it does not modify the total amount of seed phosphorous but because it is associated to severe negative pleiotropic effects, this mutation is propagated as heterozygous [65, 66]. Also, in this case, the *lpa1-241* paramutagenic allele is able to partially silence the paramutable *Lpa1* allele when exposed in *trans* and this effect is strengthened by the progressive exposure to the paramutagenic allele in the next generations [57].

Another interesting case of paramutation in plants is the recessive *sulfurea* mutant of tomato [67] showing a chlorophyll-deficient phenotype (sulfurous color). This gene was not isolated so far but it seems likely that this phenotype is caused by an auxin deficiency [68]. Because *sulf* homozygous plants do not survive beyond the seedling stage, the paramutation at the tomato sulfurea pigment deficiency appeared at high frequency as somatic sectors in *Sulf/sulf* heterozygous plants. The seeds obtained from sulf sector (where the *sulf* allele has paramutated the + *Sulf* allele) are all *sulf/sulf* while the seeds obtained from the green sectors produce again plants with *sulf* somatic sectors [67, 68]. The level of paramutagenicity of different paramutagenic alleles is different, in fact in the case of *B′* and *Pl′* alleles it is high [62], while for *R-st* [69], *P-rr′* [64, 70] and *lpa1-241* [57] alleles it is variable. In all these cases of paramutation, except the *sulf* locus where so far the corresponding gene has not been isolated, it has been demonstrated that paramutated alleles correlate with a reduction of mRNA levels [48, 55, 57, 63, 64, 71–74].

Two different classes of signal are involved in this self-propagating memory: the *cis*-acting signals physically associated with the gene that they regulate and the *trans*-acting signals. To the first class belong DNA methylation and histones modification associated with a change in the chromatin structure and the consequent accessibility on the gene promoter, although nonhistone proteins also tightly associated with chromatin could be involved [75]. In contrast, *trans* epigenetic signals are maintained by soluble molecules such as transcription factors or small RNAs (sRNAs) acting in feedback loops of self-regulation of own expression level [76, 77]. Recently, it has been argued that prions could also represent a kind of epigenetic inheritance/paramutation-like phenomenon not based on nucleic acid but on the protein folding, resulting in different activity [78]. Three models have been proposed so far to explain the specific molecular mechanism involved in the basis of paramutation: a direct physical interaction between the paramutagenic and the paramutable alleles (pairing model), a gene inactivation mechanism mediated by RNA (small RNA model) and lastly a mix of both [48, 72, 73, 79].

Usually, the DNA of paramutagenic alleles is hypermethylated compared to their paramutable *alter-ego* alleles [57, 64, 80–82] although in some cases the association of paramutation/changes in DNA methylation is not clear or does not appear at all. Also, repeated sequences, in direct as well as inverted orientation, which seem to be associated to the silenced chromatin [83], are present in most plant paramutation systems such as *r1* [69], *pl* [64], and *b1* loci [84]. The repeated sequences can be in the coding regions, such as the *r1* locus, or may be located at the upstream, such as the *b1* and *pl1* genes. In the case of *b1,* seven copies of an 853 base pairs sequence are located about 100 kb from the coding region and they are associated to the paramutation onset (from *B-I* to *B′*) and paramutagenicity. In fact a neutral allele carrying a single copy, furthermore decreasing the number of repeats in *r1* and *b1,* causes a decrease in paramutagenicity [29, 45]. The transition from *B-I* to *B′* correlates with a hypermethylation of tandem repeats and a differential sensitivity to DNaseI suggesting a different chromatin structure [84].

In the case of the *pl1* gene, it is known the presence of repeated sequences although the complex allele named *pl-bol3* containing multiple *pl1* gene copies showing paramutation-like activity has been isolated from a Bolivian maize

Table 13.1 *Trans*-acting genes required in the phenomenon of paramutation. (Modified from Hollick [100]. With permission from Elsevier)

Locus	Molecular identity	Affected examples			
		r1	*b1*	*pl1*	*p1*
mop1/rdr2	RDR2	Yes	Yes	Yes	Yes
rmr1	Rad54-like ATPase	Unreported	No	No	Untested
rmr2	Novel protein	No	Untested	Partially	Untested
rmr6/rpd1	RPD1	Yes	Yes	Yes	Untested
rmr7/mop2	RPD2a	Yes	Yes	Yes	Yes
cbbp	CBBP	Untested	Partially	Untested	Untested
ufo1	un-cloned	Untested	Yes	Untested	Yes

population [85]. Also, dosage effects caused by ploidy changes seem to influence the paramutation as has been demonstrated in the tomato *sulf* locus [86] and in *Arabidopsis* active *hygromycin phosphotransferase* (*HPT*) transgene locus [82].

The isolation and characterization of mutants that perturb the paramutation process can help to dissect the paramutation phenomenon; these mutations can be subdivided roughly into two classes: (i) mutations modifying the establishment of paramutation and (ii) mutations modifying the epigenetic memory [79]. At least ten loci belonging to the first class (named "*mediator of paramutation*") and to the second class (named "*required to maintain repression*") have been isolated by genetic screenings of mutagenized maize populations (carrying *B'* or *Pl'* systems). All the genes cloned so far are involved in the RNA-directed transcriptional silencing: *mediator of paramutation1* (*mop1*) encoding for RNA-dependent RNA polymerase [74, 87], *mediator of paramutation2* (*mop2*) gene encoding for a second-largest subunit of plant-specific RNA polymerases IV and V [88], *required to maintain repression1* (*rmr1*) gene encoding for an SNF2-like ATPase, a chromatin-remodeling enzyme [89], *required to maintain repression2* (*rmr2*) gene encoding for the founding member of a plant-specific clade of predicted proteins [90, 91] and *required to maintain repression6* (*rmr6*) encoding the largest subunit of the plant specific DNA-dependent RNA polymerase [92]. *mop1* in particular is involved in the biogenesis of 24 nt siRNA and synthesis of dsRNA [93] as previously reported for the homologous orthologous *RDR2* gene of *Arabidopsis* [94].

Another important uncloned gene is *unstable factor for orange 1* (*Ufo1*), that is required for maintaining silencing associated with paramutation at the *pericarp color1* and *booster1* loci [95].

Taken together (Table 13.1), the recent findings demonstrate an essential role for RNAi processes in paramutation, the RNAi process includes the gene silencing effects of microRNAs (miRNAs) as well as silencing induced by foreign dsRNA: thus, paramutation and miRNA share in some way the same cellular machinery [96]. It is well known that DNA repeats are able to generate aberrant RNA (such as dsRNA inducing RNAi) even if, in the case of *b1*, it has been reported that the tandem repeats seemed not to be directly involved in the genesis of siRNAs but instead they are required as *cis*-signaling in the paramutation [97].

The yeast one-hybrid technology has been used to identify the proteins able to bind to the repeated sequences present in most paramutagenic alleles. In the case of *b1,* this strategy allowed the identification of a CXC-domain protein CBBP (CXC domain *b1*-repeat binding protein) showing homology with some transposases known to bind *in vivo* and *in vitro* specifically a sequence within the tandem repeats of 853 bp inducing repressive chromatin states [98]. A transgenic maize overexpressing CBBP showed a silent state at the *b1* locus. This change was hereditable and the silent epiallele obtained (in the absence of a transgenic construct) was paramutagenic although with a reduced strength in comparison with *B'*, thus confirming the role played by CBBP. Furthermore, CBBP binds as multimers the *b1* tandem repeats suggesting a correlation between the strength of paramutation and the number of *b1* repeats and a possible *trans* interaction between chromosomes as observed in *Drosophila* in the case of transvection [62]. CBBP mRNA levels are the same in the *B-I* and *B'* while the CBBP protein is only detectable in the *B',* thus it can be hypothesized as a posttranscriptional control of CBBP is involved in the establishment of the *B'* state [98].

So far, the relationship between RNAi machinery and CBBP is not clear, but CBBP might be involved in same way in the chromatin modification complex as hypothesized for Drosophila CXC domain proteins [99]. Hence CBBP defines a new class of protein involved in gene silencing, not sharing similarity to the *Arabidopsis* RNAi silencing pathway [98]. Taking together all the data obtained so far using the best studied *b1* locus, it is possible to speculate regarding a paramutation model: an increase of CBBP protein level (probably due to a stochastic posttranscriptional control) causes the onset of *B'* from *BI,* this state is maintained in the next generation by RNA and/or proteins signals associated with the *b1* repeats during mitosis and meiosis; in some way a pair *trans* interaction between *B'* and *B-I* repeats establishes the paramutation (Fig. 13.6) [100, 101]. Interestingly, another phenomenon involving RNAi-mediated heterochromatin in yeast and *Arabidopsis* does not show paramutation capacity [102, 103] strengthening the idea that although RNA-induced silencing complex (RISC) and RNA-induced transcriptional silencing (RITS) are involved in the paramutation phenomenon, this last could represent a new system to propagate epigenetic information.

Gene silencing phenomena could be considered as rare deregulation mechanisms involved in the establishment and in the maintenance of chromatin state in a particular genome region defining the epigenetic state, but the biological systems where gene silencing has been discovered share two common characteristics: first the genes involved determined a phenotype easy to score such as pigment accumulation [57, 68, 97, 104]; second, all these traits are monogenic characters. These considerations lead us to suppose that gene silencing/paramutation phenomena could be more common than previously thought. In fact any paramutation phenomenon involving QTL would be hard to find due to the small amount of phenotypic modification caused by a change in a single or a few genes expression level involved in the phenotypic complex trait.

Gene silencing and paramutation phenomena are associated in some way to siRNA biogenesis and in most cases to repeated sequences closely linked to the

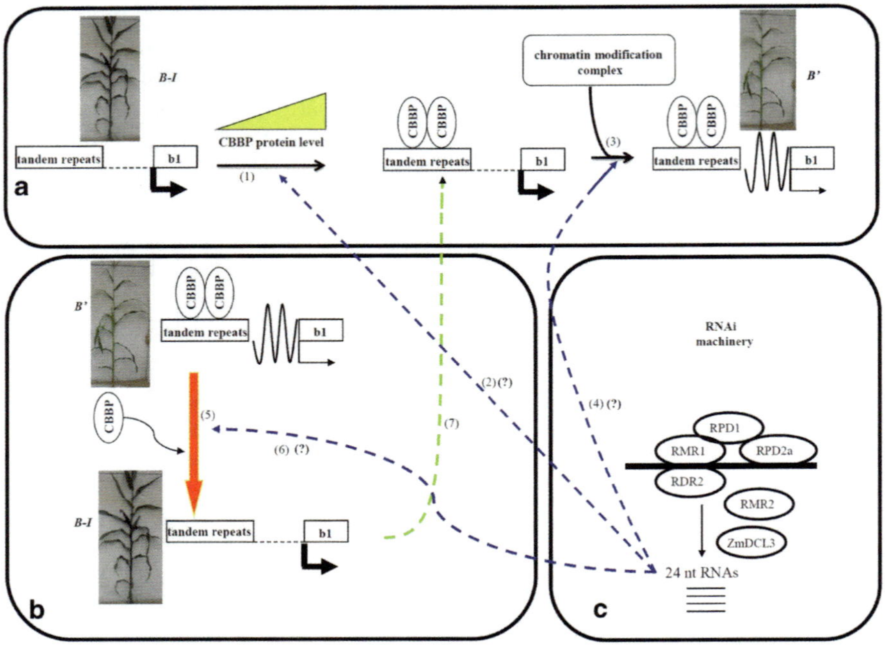

Fig. 13.6 Paramutation models in the *b1* locus **a** the spontaneous appearance of paramutagenic *B′* from *B-I*, **b** the *B′* paramutation activity *versus B-I*, **c** the genes so far discovered involved in the RNAi machinery. In **a** the *B-I* allele (*red pigmented plant*) is depicted by two boxes representing the seven tandem repeats and the *b1* gene; the two boxes are united by hyphens indicating an active conformation of chromatin in this DNA region. *Marked black arrows* starting from *b1* box represents the high transcription levels of *B-I* allele. An increase of CBBP protein level determines the binding of these proteins to the tandem repeats (1), in which step the RNAi machinery could also be involved (2). The CBBP proteins bonded to the tandem repeats in some way trigger the recruitment of the chromatin modification complex (3) that determines an hereditable silent conformation of chromatin structure (depicted by the *sinusoid line* between the two boxes) causing a strong decrease in *b1* transcription levels (depicted by the *thin black arrow* starting from *b1* box) and this new *b1* epiallele named *B′* (*green pigmented plant*) acquires paramutagenic capacity. In **b** when a *B-I* allele is exposed in *trans* with a *B′* paramutagenic allele (*by crossing*), an interaction (5) involving CBBP protein which binds the tandem repeats of *B-I* allele and perhaps a physical interaction between the paired genomic regions on two chromosomes (indicated by the *yellow arrow*) and participation of RNAi machinery (4) cause the paramutation of *B-I* allele as described in A (7). In **c** the proteins so far found involved in the maize paramutation, with the exception of CBBP, are all implicated in the 24 nt RNA biogenesis in maize (RNAi system). On the basis of mutant analyses and presumed orthologies with *Arabidopsis* proteins, a complex containing maize Pol IV (RPD1 and RPD2a), RMR1 and RDR2 produces a double-stranded RNA (dsRNA) that is presumably cleaved by a maize Dicer-like3 (ZmDCL3) into 24 nt RNAs (RMR2 is also required but its action in the pathway remains unknown). Schemes were modified from [100, 101]

gene undergoing paramutation. Although it has been hypothesized for many years that repeated sequences were involved in the transcription of the aberrant RNA triggering an RNA-directed transcriptional silencing, a recent paper regarding the *B1* paramutation phenomenon [98] suggests that these repeated sequences contain target sequences recognized by DNA binding proteins involved in the onset of silencing and correlated with paramutation capacity. So far, the relationship between the siRNA pathway and the regulation of these proteins that are probably involved in the chromatin modification complex is not clear.

In recent years, the increasing interest in epigenetic and paramutation-like phenomenon allowed the release of a huge amount of genomics and transcriptomics data that are shading light on the spread and the mechanism of transmission of epigenetic information.

Dietetics

The importance of plant-based foods in the human diet was first evidenced by the pioneer Lyon-Diet study [105], which led to the conclusion that, despite consuming as much dairy fats as other Northern European Countries, France had a lower mortality due to coronary heart disease, because of a regular moderate consumption of red wine. Since then, many epidemiological studies revealed that consumption of fruit and vegetables, rich in phytochemicals, particularly flavonoids, is associated with a significant reduction of cardiovascular diseases, obesity, and type 2 diabetes [106–109]. However, epidemiology can only show that a preventive factor is associated with a lower incidence of disease in the population exposed to that preventive factor, but it cannot prove a cause–effect relationship. This can be demonstrated using a nutrigenomic approach in preclinical and clinical studies. Aim of nutrigenomics is to identify genes and proteins, whose expression is influenced by nutrients, by using "omic" tools (transcriptomics, proteomics, metabolomics) in animal models. This is normally done with a twofold purpose: (i) the identification of target genes and of the molecular mechanisms underlying nutrient sensing and (ii) the identification of molecular biomarkers, useful to precociously identify a preclinical onset of a diet-related disease and design an appropriate nutritional intervention able to reverse it [110].

A crucial approach in nutrigenomics is to study the effect of specific plant-derived nutrients/bioactives in their whole-food context, that is together with the other metabolites, enzymes, fibre, *etc.,* in which they are normally ingested, since all these external factors may modify the bioavailability and bioactivity of compounds. A careful analytical assessment of the bioactives content of different plant foods is also essential for the reliable evaluation of components in recommended health-promoting diets. Nonetheless, to demonstrate the beneficial activity of specific classes of bioactives, it is also crucial to compare near-isogenic plant-based foods that vary only in the quantity of the bioactives under analysis. This strategy has not been always achieved, but in principle it would allow to reduce some of the complexity of food in the diet–health relationship of foods and provide model foods that can be used for both animal feeding studies and human intervention trials for assessing the role of plant bioactives in the diet.

Cardioprotective Effect of Blue Corn

The extensive knowledge of the regulation of flavonoid pathway in maize can be exploited to obtain "near-isogenic" foods which differ only in the content of specific classes of flavonoids. Taking advantage of proper allelic combination of the *MYB* and *bHLH* regulatory genes controlling activation of the anthocyanin biosynthesis pathway in maize, isogenic lines carrying different levels of anthocyanins in a desired tissue/organ can be obtained. A clear example of such an approach was provided by experiments establishing that anthocyanins in corn could offer cardioprotection against ischemia-reperfusion injury in rats [10]. In this case, the use of maize lines carrying *R1* and *C1* alleles that promote anthocyanin accumulation in kernels [18], allowed the production of isogenic material either containing or not containing anthocyanins in the kernels. Such blue and yellow corn seeds were incorporated as 20% supplements in the standard rat diet and used to feed two groups of rats for 8 weeks on diets with anthocyanins (supplied at 12 mg/kg body weight/day) or without anthocyanins. These studies demonstrated that in rats fed anthocyanin-rich maize the amount of cardiac tissue that was damaged following ischemic conditions was reduced by approximately 30% compared to rats fed anthocyanin-free maize [10]. Cardioprotection was associated with increased myocardial glutathione levels and increased marine omega-3 levels in blood, suggesting that dietary anthocyanins modulate cardiac antioxidant defences and the conversion of plant α-linolenic acid into omega-3 fatty acids [111]. No differences were found in total levels of saturated and monounsaturated fatty acids nor in total n-6 PUFAs between the two dietary groups, but there were significantly higher levels of marine n-3 PUFAs (41% EPA and 16% DHA) and consequently enhanced n-3:n-6 PUFA ratios in the plasma of animals fed the high anthocyanin diet. EPA and DHA synthesis in vertebrates involves as first step Δ6 desaturase, a rate-limiting enzyme in humans which can convert α-linolenic acid into stearidonic acid [112]. These preclinical studies suggest that a combination of plant n-3 PUFA (α-linolenic acid) with polyphenols may promote the enhancement of marine n-3 PUFAs possibly by stimulating Δ6 desaturase. Recent epidemiological studies have revealed that EPA and DHA levels were increased in blood and cells of red wine drinkers and that components of red wine, other than alcohol, probably flavonoids, were responsible for the fish-like effect observed [113, 114].

Considering that a high n-3:n-6 PUFA ratio is known to prevent not only cardiovascular diseases but also other chronic diseases, such as cancer, diabetes and neurodegenerative diseases, it is possible that the diverse effects of anthocyanins on different chronic diseases [10–12, 115, 116] may result from their modulation of plasma n-3 PUFA levels as primary mechanism of action. Alternatively, it may be that the impact of anthocyanins on n-3 PUFA metabolism is just a part of their physiological mode of action.

In these experiments, the level of dietary anthocyanins in the rodent feed was 13 times higher than levels present in an average Western diet [10]. Despite this, the anthocyanin dose tested still represents a physiological dose, that in an average human diet would correspond to a daily intake of 156 mg of anthocyanins, thus being more similar to the anthocyanin levels assumed in a typical Mediterranean diet than to an average Western diet, estimated to be 12 mg/day [117].

Anti-obesity and Anti-diabetic Effects of Purple Corn

Besides cardioprotection, other studies demonstrated that anthocyanin-rich extracts from purple corn also prevents weight gain and obesity in mice under high-fat diet [118]. Purple corn color (PCC) extract normally used as food colorant (containing 70 g/kg of cyanidin 3-glucoside) was incorporated at a final concentration of 2 g/kg to a standard rodent diet or to a high-fat diet. Mice were divided in four groups receiving standard diet (SD), purple corn color (PCC), high-fat diet (HFD) and high-fat diet+PCC (HFD+PCC) for 12 weeks. Considering an average food intake of 3 g/day and a 25 g body weight, the average daily intake of cyanidin 3-glucoside in mice receiving PCC or HFD+PCC can be estimated at 240 mg/kg body weight/day. Results revealed that providing PCC reduced significantly body weight gain and fat accumulation, thus impairing the high-fat-induced obesity.

An increase in adipose tissue is normally associated to enhanced levels of *TNF-α* mRNA in white adipose tissue, a pro-inflammatory cytokine that can inhibit pancreatic β-cell function and insulin signaling, resulting in insulin resistance. Dietary anthocyanins from purple corn prevented the high *TNF-α* expression level and as a consequence prevented hyperglycemia, hyperinsulinemia, hyperleptinemia normally associated to insulin resistance, by impairing adipose tissue accumulation [119]. Microarray analysis of human adipocytes treated with cyanidin 3-glucoside confirmed that administration of anthocyanins also down-regulate the expression of other adipocytokines (e.g., *IL-6* and *PAI-1*), normally highly expressed in obesity and type 2 diabetes [120, 121], and up-regulated adiponectin, specifically highly expressed in adipocytes, whose plasma concentration and mRNA level in white adipose tissue are decreased in the obese and insulin resistance state [122–125].

The anti-obesity property of purple corn was likely achieved by reducing the mRNA level of sterol-binding protein-1 (*SREBP-1*), a transcription factor involved in lipogenesis and cholesterol synthesis. SREBP-1 is synthesized in white adipose tissue as inactive precursor associated to the endoplasmic reticulum and then released by a two-step proteolitic cleavage that allows its entering into the nucleus as a mature transcription factor [126]. Consistent with this idea, mice receiving HFD+PCC had lower transcript levels of *SREBP-1* and its target genes encoding fatty acid synthase (FAS) and acyl-CoA synthase (ACS1), enzymes involved in fatty acid and triacylglycerol synthesis [119].

Recent data suggest that SREBP-1 could be downregulated by anthocyanins through activation of AMP-activated protein kinase (AMPK), an evolutionary conserved sensor of cellular energy [127]. It has been reported that metformin, a drug used for type 2 diabetes, downregulated SREBP-1 expression, thereby reducing FAS mRNA level, through AMPK activation [128] and that dietary anthocyanins from bilberry (supplied at 1.8 mg/kg body weight/day) can activate AMPK in white adipose tissue and scheletric muscle tissue and ameliorates hyperglycemia and insulin resistance in diabetic mice [129].

Chronic hyperglycemia is a major contributor to severe diabetic complications. Diabetic nephropathy is one of the most common microvascular complications of diabetes and a leading cause of end-stage renal disease [130, 131]. The first lesion of diabetes observed in glomeruli is mesangial cell expansion and matrix accumulation, leading to glomerulosclerosis. In particular, accumulation of extracellular matrix can be caused by increased secretion of matrix-synthesizing growth factors and decreased formation of matrix-degrading proteases. Recent studies showed that extracts from anthocyanin-rich purple corn can reduce high glucose-promoted mesangial cell proliferation and matrix accumulation *in vitro* and that the most effective anthocyanins in purple corn reducing mesangial fibrosis and inflammation in vitro are cyanidin 3-glucoside, peonidin 3-glucoside and their acylated derivatives [132]. Anthocyanins from purple corn extracts supplied in drinking water at 10 mg/kg body weight/day to diabetic *db* mice for 8 weeks ameliorated hyperglycemia, insulin sensitivity and retarded diabetic nephropathy [133]. Mice receiving purple corn extracts ameliorated glomerulosclerosis and reduced microalbuminuria, typically detected in *db* mice, by maintaining the glomerular filtration barrier and allowing kidneys to properly operate the urine filtration [133]. The typical glomerular macrophage infiltration and monocyte activation of kidneys associated to diabetes was also inhibited in *db* mice treated with purple corn extracts [134].

Anti-atherogenic and Anti-obesity Effect of Black Rice

Acute coronary syndrome, including *angina pectoris,* myocardial infarction, and sudden coronary death, is always correlated to the presence of atherosclerotique plaques. These plaques severely reduce the vascular lumen and are predisposed to rupture, calcification, and sudden thrombosis formation. Therefore, a major task in cardiovascular disease prevention is to increase stability of the plaques and to inhibit thrombosis formation in the case of plaque rupture, by reducing inflammation which is the underlying cause of vulnerable plaque progression. Concerning this, statins are the most effective agents for the treatment of atherosclerosis, since they lower lipids and act as potent anti-inflammatory agents.

In vivo studies on rabbits have revealed that both red and black rice decrease atherosclerotic plaques and increase the endogenous antioxidant status [135, 136]. This anti-atherogenic effect of black rice was confirmed in feeding experiments performed using 30-week-old apoliprotein E (apoE)-deficient mice, a mouse model characterized by vulnerable plaques that show some morphological features of human advanced atherosclerotic lesions. Old apoE-deficient mice were fed anthocyanin-rich extract from black rice (130 mg/kg body weight/day) supplemented to standard diet for 20 weeks and compared to mice receiving standard diet supplemented with simvastatin. Black rice reduced atherosclerotique plaques and inhibited pro-inflammatory factors similarly to simvastatin. Interestingly, black rice was more effective than simvastatin in decreasing the lipid profile (triglyceride, total cholesterol and HDL- and non-HDL-cholesterol) [137]. The serum lipid profile was also improved in rats with dyslipidemia induced by high-fat diet, fructose-rich diet or in rats under hypercholesterolemic diet [138–140]. Similarly to purple corn, anthocyanin extracts from black rice (supplied at 86 mg/kg body weight/ day) prevented weight gain, insulin resistance, hypertriglyceridemia, and platelet hyperactivity, which normally led to formation of atherosclerotique plaques [138, 140]. The beneficial effects of anthocyanins from black rice was definitely demonstrated by feeding experiments comparing rats fed hypercholesterolemic diet supplemented with 20 % black rice or with 20 % unrefined rice, since it was observed that diet containing black rice was more effective in controlling lipidemia (plasma cholesterol, triglycerides and low-density lipoprotein) compared with the whole rice diet [139]. Supplementation of black rice pigment fraction to patients aged 45–75 years with coronary heart disease for 6 months enhanced the plasma total antioxidant capacity and reduced circulating inflammatory factors more effectively than in patients receiving white rice pigment fraction, thus confirming that black rice can exert cardioprotective effects on patients with coronary heart disease [141].

Interestingly, anthocyanins from black rice (supplied at 112 mg/kg body weight/ day) also have a protective action against alcohol-induced toxicity as demonstrated by the decreased levels of serum and liver triglyceride and total cholesterol. Supplementation with black rice also reduced plasma levels of aspartate transaminase (AST), alanine transaminase (ALT), and γ glutamyl transferase (GGT), hepatic enzymes whose leakage into plasma is a sign of hepatic injury and necrosis. This protective action is probably achieved by the antioxidant properties of anthocyanins, scavenging ethanol-induced free radicals, as demonstrated by lower lipid peroxidation and higher levels of antioxidant enzymes (i.e glutathione peroxidase, superoxide dismutase) in liver of mice fed black rice extracts [142]. Finally, anthocyanins from purple rice were recently demonstrated to exert a neuroprotective effect in rat models of Alzheimer's disease, by preventing memory impairment and hippocampal neurodegeneration [143].

Acknowledgments All authors were supported by the EU FP7 ATHENA collaborative project (Grant Agreement 245121) and by Regione Lombardia Fondo per la promozione di Accordi Istituzionali (project BIOGESTECA 15083/RCC). The authors thank Valentina Calvenzani for final editing of the manuscript.

References

1. Fedoroff N (2001) How jumping genes were discovered. Nat Struct Biol 8(4):300–301 (PubMed PMID: 11276245)
2. Ronchi A, Petroni K, Tonelli C (1995) The reduced expression of endogenous duplications (REED) in the maize R gene family is mediated by DNA methylation. EMBO J 14(21):5318–5328 (PubMed PMID: 7489721)

3. Brink RA (1973) Paramutation. Annu Rev Genet 7:129–152 (PubMed PMID: 4361265)
4. Paz-Ares J, Ghosal D, Wienand U, Peterson PA, Saedler H (1987) The regulatory c1 locus of Zea mays encodes a protein with homology to myb proto-oncogene products and with structural similarities to transcriptional activators. EMBO J 6(12):3553–3558
5. Winkel-Shirley B (2001) Flavonoid biosynthesis. A colorful model for genetics, biochemistry, cell biology, and biotechnology. Plant Physiol 126:485–493
6. Mol J, Grotewold E, Koes R (1998) How genes paint flowers and seeds. Trends Plant Sci 3:212–217
7. Hernandez JM, Heine GF, Irani NG, Feller A, Kim MG, Matulnik T et al (2004) Different mechanisms participate in the R-dependent activity of the R2R3 MYB transcription factor C1. J Biological Chem 279(46):48205–48213
8. Meiers S, Kemeny M, Weyand U, Gastpar R, von Angerer E, Marko D (2001) The anthocyanidins cyanidin and delphinidin are potent inhibitors of the epidermal growth-factor receptor. J Agric Food Chem 49(2):958–962 (PubMed PMID: 11262056)
9. Williams RJ, Spencer JP, Rice-Evans C (2004) Flavonoids: antioxidants or signalling molecules? Free Radic Biol Med 36(7):838–49 (PubMed PMID: 15019969)
10. Toufektsian MC, de Lorgeril M, Nagy N, Salen P, Donati MB, Giordano L et al (2008) Chronic dietary intake of plant-derived anthocyanins protects the rat heart against ischemia-reperfusion injury. J Nutr 138(4):747–752 (PubMed PMID: 18356330)
11. Butelli E, Titta L, Giorgio M, Mock HP, Matros A, Peterek S et al (2008) Enrichment of tomato fruit with health-promoting anthocyanins by expression of select transcription factors. Nat Biotechnol 26(11):1301–1308 (PubMed PMID: 18953354)
12. Titta L, Trinei M, Stendardo M, Berniakovich I, Petroni K, Tonelli C et al (2010) Blood orange juice inhibits fat accumulation in mice. Int J Obes 34(3):578–588 (PubMed PMID: 20029381)
13. Joseph JA, Denisova NA, Arendash G, Gordon M, Diamond D, Shukitt-Hale B et al (2003) Blueberry supplementation enhances signaling and prevents behavioral deficits in an Alzheimer disease model. Nutr Neurosci 6(3):153–162 (PubMed PMID: 12793519)
14. Sharma M, Cortes-Cruz M, Ahern KR, McMullen M, Brutnell TP, Chopra S (2011) Identification of the pr1 gene product completes the anthocyanin biosynthesis pathway of maize. Genetics 188(1):69–79 (PubMed PMID: 21385724. Pubmed Central PMCID: 3120156)
15. Shih CH, Chu H, Tang LK, Sakamoto W, Maekawa M, Chu IK et al (2008) Functional characterization of key structural genes in rice flavonoid biosynthesis. Planta 228(6):1043–1054 (PubMed PMID: 18726614)
16. Yang GH, Li B, Gao JW, Liu JZ, Zhao XQ, Zheng Q et al (2004) Cloning and expression of two chalcone synthase and a flavonoid 3′5′-hydroxylase 3′-end cDNAs from developing seeds of blue-grained wheat involved in anthocyanin biosynthetic pathway. Acta Botanica Sinica 46(5):588–594 (PubMed PMID: WOS:000221531900012. Chinese)
17. Himi E, Nisar A, Noda K (2005) Colour genes (R and Rc) for grain and coleoptile upregulate flavonoid biosynthesis genes in wheat. Genome 48(4):747–754 (PubMed PMID: 16094442)
18. Procissi A, Dolfini S, Ronchi A, Tonelli C (1997) Light-dependent spatial and temporal expression of pigment regulatory genes in developing maize seeds. The Plant Cell 9(9):1547–1557
19. Goff SA, Cone KC, Chandler VL (1992) Functional analysis of the transcriptional activator encoded by the maize B gene: evidence for a direct functional interaction between two classes of regulatory proteins. Genes Dev 6(5):864–875
20. Tonelli C, Consonni G, Dolfini SF, Dellaporta SL, Viotti A, Gavazzi G (1991) Genetic and molecular analysis of Sn, a light-inducible, tissue specific regulatory gene in maize. Mol Gen Genet 225(3):401–410 (PubMed PMID: 1673220)
21. Petroni K, Cominelli E, Consonni G, Gusmaroli G, Gavazzi G, Tonelli C (2000) The developmental expression of the maize regulatory gene Hopi determines germination-dependent anthocyanin accumulation. Genetics 155(1):323–336 (PubMed PMID: 10790406)
22. McCarty DR, Carson CB, Stinard PS, Robertson DS (1989) Molecular analysis of viviparous-1: an abscisic acid-insensitive mutant of maize. Plant Cell 1(5):523–532 (PubMed PMID: 12359900)

23. Cocciolone SM, Cone KC (1993) Pl-Bh, an anthocyanin regulatory gene of maize that leads to variegated pigmentation. Genetics 135(2):575–588 (PubMed PMID: 7694886)
24. Grotewold E, Drummond BJ, Bowen B, Peterson T (1994) The myb-homologous P gene controls phlobaphene pigmentation in maize floral organs by directly activating a flavonoid biosynthetic gene subset. Cell 76(3):543–553
25. Ferreyra ML, Rius S, Emiliani J, Pourcel L, Feller A, Morohashi K et al (2010) Cloning and characterization of a UV-B-inducible maize flavonol synthase. Plant J 62(1):77–91 (PubMed PMID: 20059741)
26. Sharma M, Chai C, Morohashi K, Grotewold E, Snook ME, Chopra S (2012) Expression of flavonoid 3'-hydroxylase is controlled by P1, the regulator of 3-deoxyflavonoid biosynthesis in maize. BMC Plant Biol 12:196 (PubMed PMID: 23113982. Pubmed Central PMCID: 3509002)
27. Carey CC, Strahle JT, Selinger DA, Chandler VL (2004) Mutations in the pale aleurone color1 regulatory gene of the Zea mays anthocyanin pathway have distinct phenotypes relative to the functionally similar TRANSPARENT TESTA GLABRA1 gene in *Arabidopsis thaliana*. Plant Cell 16(2):450–464 (PubMed PMID: 14742877)
28. Selinger DA, Chandler VL (1999) A mutation in the pale aleurone color1 gene identifies a novel regulator of the maize anthocyanin pathway. Plant Cell 11(1):5–14 (PubMed PMID: 9878628)
29. Kong Q, Pattanaik S, Feller A, Werkman JR, Chai C, Wang Y et al (2012) Regulatory switch enforced by basic helix-loop-helix and ACT-domain mediated dimerizations of the maize transcription factor R. Proc Natl Acad Sci U S A 109(30):E2091–E2097 (PubMed PMID: 22778424. Pubmed Central PMCID: 3409794)
30. Feller A, Hernandez JM, Grotewold E (2006) An ACT-like domain participates in the dimerization of several plant basic-helix-loop-helix transcription factors. J Biol Chem 281(39):28964–28974 (PubMed PMID: 16867983)
31. Hernandez JM, Feller A, Morohashi K, Frame K, Grotewold E (2007) The basic helix loop helix domain of maize R links transcriptional regulation and histone modifications by recruitment of an EMSY-related factor. Proc Natl Acad Sci U S A 104(43):17222–17227 (PubMed PMID: 17940002)
32. Sainz MB, Grotewold E, Chandler VL (1997) Evidence for direct activation of an anthocyanin promoter by the maize C1 protein and comparison of DNA binding by related Myb domain proteins. Plant Cell 9(4):611–625
33. Kato S, Ishikawa J (1921) On the inheritance of the pigment of red rice. Jpn J Genet 1:1–7
34. Sweeney MT, Thomson MJ, Pfeil BE, McCouch S (2006) Caught red-handed: Rc encodes a basic helix-loop-helix protein conditioning red pericarp in rice. Plant Cell 18(2):283–294 (PubMed PMID: 16399804)
35. Kim CK, Cho MA, Choi YH, Kim JA, Kim YH, Kim YK et al (2011) Identification and characterization of seed-specific transcription factors regulating anthocyanin biosynthesis in black rice. J Appl Genet 52(2):161–169 (PubMed PMID: 21286901. Pubmed Central PMCID: 3088801)
36. Furukawa T, Maekawa M, Oki T, Suda I, Iida S, Shimada H et al (2007) The Rc and Rd genes are involved in proanthocyanidin synthesis in rice pericarp. Plant J 49(1):91–102 (PubMed PMID: 17163879)
37. Reddy AM, Reddy VS, Scheffler BE, Wienand U, Reddy AR (2007) Novel transgenic rice overexpressing anthocyanidin synthase accumulates a mixture of flavonoids leading to an increased antioxidant potential. Metab Eng 9(1):95–111 (PubMed PMID: 17157544)
38. Abdel-Aal el SM, Young JC, Rabalski I (2006) Anthocyanin composition in black, blue, pink, purple, and red cereal grains. J Agric Food Chem 54(13):4696–4704 (PubMed PMID: 16787017)
39. Miyamoto T, Everson E (1958) Biochemical and physiological studies of wheat seed pigmentation. Agron J 50:733–734
40. McIntosh R, Devos K, Dubcovsky J, Rogers J, Morris CF, Appels R, Somers DJ, Anderson OA (2014) Catalogue of gene symbols for wheat: 2007 Supplement. http://wheat.pw.usda.gov/ggpages/wgc/2007upd.html. Accessed 3 April 2014

41. Waddington C (1942) Canalization of development and the inheritance of acquired characters. Nature 150:563–565
42. McClintock B (1961) Some parallels between gene control systems in maize and in bacteria. Am Nat 95:265–277
43. Comfort NC (2001) From controlling elements to transposons: Barbara McClintock and the Nobel Prize. Endeavour 25(3):127–130 (PubMed PMID: 11725307)
44. Cold Spring Harbor Laboratory Press (ed) (1996) Epigenetic mechanisms of gene regulation. Cold Spring Harbor Laboratory, Woodbury
45. Martienssen R (1996) Epigenetic phenomena: paramutation and gene silencing in plants. Curr Biol 6(7):810–813 (PubMed PMID: 8805298)
46. Wolffe AP, Matzke MA (1999) Epigenetics: regulation through repression. Science 286(5439):481–486 (PubMed PMID: 10521337)
47. Feng S, Jacobsen SE, Reik W (2010) Epigenetic reprogramming in plant and animal development. Science 330(6004):622–627 (PubMed PMID: 21030646. Pubmed Central PMCID: 2989926)
48. Chandler VL (2007) Paramutation: from maize to mice. Cell 128(4):641–645 (PubMed PMID: 17320501)
49. Lewis E (1954) The theory and application of a new method of detecting chromosomal rearrangements in *Drosophila melanogaster*. Am Nat 88:225–239
50. Napoli C, Lemieux C, Jorgensen R (1990) Introduction of a chimeric chalcone synthase gene into petunia results in reversible co-suppression of homologous genes in trans. Plant Cell 2(4):279–289 (PubMed PMID: 12354959. Pubmed Central PMCID: 159885)
51. Romano N, Macino G (1992) Quelling: transient inactivation of gene expression in Neurospora crassa by transformation with homologous sequences. Mol Microbiol 6(22):3343–3353 (PubMed PMID: 1484489)
52. Fire A, Xu S, Montgomery MK, Kostas SA, Driver SE, Mello CC (1998) Potent and specific genetic interference by double-stranded RNA in Caenorhabditis elegans. Nature 391(6669):806–811 (PubMed PMID: 9486653)
53. Arteaga-Vazquez M, Sidorenko L, Rabanal FA, Shrivistava R, Nobuta K, Green PJ et al (2010) RNA-mediated trans-communication can establish paramutation at the b1 locus in maize. Proc Natl Acad Sci U S A 107(29):12986–12991 (PubMed PMID: 20616013. Pubmed Central PMCID: 2919911)
54. Hagemann R (1969) Somatic conversion (paramutation) at the sulfurea locus of *Lycopersicon esculentum* Mill. III. Studies with trisomics. Can J Genet Cytol 11:346–358
55. Chandler VL, Eggleston WB, Dorweiler JE (2000) Paramutation in maize. Plant Mol Biol 43(2–3):121–145 (PubMed PMID: 10999400)
56. Shi J, Wang H, Schellin K, Li B, Faller M, Stoop JM et al (2007) Embryo-specific silencing of a transporter reduces phytic acid content of maize and soybean seeds. Nat Biotechnol 25(8):930–937 (PubMed PMID: 17676037)
57. Pilu R, Panzeri D, Cassani E, Cerino Badone F, Landoni M, Nielsen E (2009) A paramutation phenomenon is involved in the genetics of maize low phytic acid1–241 (lpa1–241) trait. Heredity (Edinb) 102(3):236–245 (PubMed PMID: 18781168)
58. Brink RA (1956) A genetic change associated with the R Locus in maize which is directed and potentially reversible. Genetics 41(6):872–889 (PubMed PMID: 17247669. Pubmed Central PMCID: 1224369)
59. Brown DF, Brink RA (1960) Paramutagenic action of paramutant R and R alleles in maize. Genetics 45(10):1313–1316 (PubMed PMID: 17248001. Pubmed Central PMCID: 1210007)
60. Coe EH (1959) A regular and continuing conversion-type phenomenon at the B Locus in maize. Proc Natl Acad Sci U S A 45(6):828–832 (PubMed PMID: 16590451. Pubmed Central PMCID: 222644)
61. Coe EH (1966) The properties, origin, and mechanism of conversion-type inheritance at the B locus in maize. Genetics 53(6):1035–1063 (PubMed PMID: 17248307. Pubmed Central PMCID: 1211079)

62. Hollick JB, Patterson GI, Coe EH, Jr., Cone KC, Chandler VL (1995) Allelic interactions heritably alter the activity of a metastable maize pl allele. Genetics 141(2):709–719 (PubMed PMID: 8647404. Pubmed Central PMCID: 1206767)
63. Das OP, Messing J (1994) Variegated phenotype and developmental methylation changes of a maize allele originating from epimutation. Genetics 136(3):1121–1141 (PubMed PMID: 8005419. Pubmed Central PMCID: 1205868)
64. Sidorenko LV, Peterson T (2001) Transgene-induced silencing identifies sequences involved in the establishment of paramutation of the maize p1 gene. Plant Cell 13(2):319–335 (PubMed PMID: 11226188. Pubmed Central PMCID: 102245)
65. Pilu R, Panzeri D, Gavazzi G, Rasmussen SK, Consonni G, Nielsen E (2003) Phenotypic, genetic and molecular characterization of a maize low phytic acid mutant (lpa241). Theor Appl Genet 107(6):980–987 (PubMed PMID: 14523526)
66. Pilu R, Landoni M, Cassani E, Doria E, Nielsen E (2005) The maize lpa241 mutation causes a remarkable variability of expression and some pleiotropic effects. Crop Sci 45:2096–2105
67. Hagemann R (1958) Somatic conversion in *Lycopersicon esculentum* Mill. Z Vererbungsl 89(4):587–613 (PubMed PMID: 13604837. Somatische Konversion bei *Lycopersicon esculentum* Mill)
68. Ehlert B, Schottler MA, Tischendorf G, Ludwig-Muller J, Bock R (2008) The paramutated SULFUREA locus of tomato is involved in auxin biosynthesis. J Exp Bot 59(13):3635–3647 (PubMed PMID: 18757490. Pubmed Central PMCID: 2561159)
69. Kermicle JL, Eggleston WB, Alleman M (1995) Organization of paramutagenicity in R-stippled maize. Genetics 141(1):361–72 (PubMed PMID: 8536983. Pubmed Central PMCID: 1206733)
70. Sidorenko L, Chandler V (2008) RNA-dependent RNA polymerase is required for enhancer-mediated transcriptional silencing associated with paramutation at the maize p1 gene. Genetics 180(4):1983–1993 (PubMed PMID: 18845841. Pubmed Central PMCID: 2600936)
71. Lund G, Das OP, Messing J (1995) Tissue-specific DNase-I-sensitive sites of the maize P-gene and their changes upon epimutation. Plant J 7(5):797–807 (PubMed PMID: WOS:A1995QY31400008. English)
72. Chandler VL, Stam M (2004) Chromatin conversations: mechanisms and implications of paramutation. Nat Rev Genet 5(7):532–544 (PubMed PMID: 15211355)
73. Stam M, Scheid OM (2005) Paramutation: an encounter leaving a lasting impression. Trends Plant Sci 10(6):283–90 (PubMed PMID: WOS:000230051100007. English)
74. Alleman M, Sidorenko L, McGinnis K, Seshadri V, Dorweiler JE, White J et al (2006) An RNA-dependent RNA polymerase is required for paramutation in maize. Nature 442(7100):295–298 (PubMed PMID: WOS:000239122100037. English)
75. Talbert PB, Henikoff S (2010) Histone variants—ancient wrap artists of the epigenome. Nat Rev Mol Cell Biol 11(4):264–275 (PubMed PMID: WOS:000275850400011. English)
76. Bonasio R, Tu SJ, Reinberg D (2010) Molecular signals of epigenetic states. Science 330(6004):612–616 (PubMed PMID: WOS:000283580600032. English)
77. Moazed D (2009) Molecular biology. Rejoice–RNAi for yeast. Science 326(5952):533–534 (PubMed PMID: 19900888. Pubmed Central PMCID: 3245747)
78. Herman H, Lu M, Anggraini M, Sikora A, Chang Y, Yoon BJ et al (2003) Trans allele methylation and paramutation-like effects in mice. Nat Genet 34(2):199–202 (PubMed PMID: 12740578. Pubmed Central PMCID: 2744043)
79. Hollick JB (2010) Paramutation and development. Annu Rev Cell Dev Biol 26:557–579 (PubMed PMID: 19575656)
80. Rassoulzadegan M, Magliano M, Cuzin F (2002) Transvection effects involving DNA methylation during meiosis in the mouse. EMBO J 21(3):440–450 (PubMed PMID: 11823436. Pubmed Central PMCID: 125843)
81. Walker EL (1998) Paramutation of the r1 locus of maize is associated with increased cytosine methylation. Genetics 148(4):1973–1981 (PubMed PMID: 9560410. Pubmed Central PMCID: 1460097)

82. Mittelsten Scheid O, Afsar K, Paszkowski J (2003) Formation of stable epialleles and their paramutation-like interaction in tetraploid *Arabidopsis thaliana*. Nat Genet 34:450–454
83. Zaratiegui M, Irvine DV, Martienssen RA (2007) Noncoding RNAs and gene silencing. Cell 128(4):763–776 (PubMed PMID: 17320512)
84. Stam M, Belele C, Dorweiler JE, Chandler VL (2002) Differential chromatin structure within a tandem array 100 kb upstream of the maize b1 locus is associated with paramutation. Genes Dev 16(15):1906–1918 (PubMed PMID: 12154122. Pubmed Central PMCID: 186425)
85. Pilu R, Piazza P, Petroni K, Ronchi A, Martin C, Tonelli C (2003) pl-bol3, a complex allele of the anthocyanin regulatory pl1 locus that arose in a naturally occurring maize population. Plant J 36(4):510–521
86. Hagemann R, Berg W (1978) Paramutation at the sulfurea locus of Lycopersicon esculentum Mill. Theoretical Appl Genet 53:113–123
87. Dorweiler JE, Carey CC, Kubo KM, Hollick JB, Kermicle JL, Chandler VL (2000) Mediator of paramutation1 is required for establishment and maintenance of paramutation at multiple maize loci. Plant Cell 12(11):2101–2118 (PubMed PMID: 11090212. Pubmed Central PMCID: 150161)
88. Sidorenko L, Dorweiler JE, Cigan AM, Arteaga-Vazquez M, Vyas M, Kermicle J et al (2009) A dominant mutation in mediator of paramutation2, one of three second-largest subunits of a plant-specific RNA polymerase, disrupts multiple siRNA silencing processes. PLoS Genet 5(11):e1000725 (PubMed PMID: 19936058. Pubmed Central PMCID: 2774164)
89. Hale CJ, Stonaker JL, Gross SM, Hollick JB (2007) A novel Snf2 protein maintains transgenerational regulatory states established by paramutation in maize. PLoS Biol 5(10):e275 (PubMed PMID: 17941719. Pubmed Central PMCID: 2020503)
90. Barbour JE, Liao IT, Stonaker JL, Lim JP, Lee CC, Parkinson SE et al (2012) Required to maintain repression2 is a novel protein that facilitates locus-specific paramutation in maize. Plant Cell 24(5):1761–1775 (PubMed PMID: 22562610. Pubmed Central PMCID: 3442568)
91. Mach J (2012) Identification of a novel maize protein important for paramutation at the purple plant1 locus. The Plant Cell 24:1709
92. Erhard KF, Jr., Stonaker JL, Parkinson SE, Lim JP, Hale CJ, Hollick JB (2009) RNA polymerase IV functions in paramutation in Zea mays. Science 323(5918):1201–1205 (PubMed PMID: 19251626)
93. Nobuta K, Lu C, Shrivastava R, Pillay M, De Paoli E, Accerbi M et al (2008) Distinct size distribution of endogeneous siRNAs in maize: evidence from deep sequencing in the mop1–1 mutant. Proc Natl Acad Sci U S A 105(39):14958–14963 (PubMed PMID: 18815367. Pubmed Central PMCID: 2567475)
94. Lu C, Kulkarni K, Souret FF, MuthuValliappan R, Tej SS, Poethig RS et al (2006) MicroRNAs and other small RNAs enriched in the *Arabidopsis* RNA-dependent RNA polymerase-2 mutant. Genome Res 16(10):1276–1288 (PubMed PMID: 16954541. Pubmed Central PMCID: 1581437)
95. Sekhon RS, Wang PH, Sidorenko L, Chandler VL, Chopra S (2012) Maize unstable factor for orange1 is required for maintaining silencing associated with paramutation at the pericarp color1 and booster1 loci. Plos Genet 8(10):e1002980 (PubMed PMID: WOS:000310528400016. English)
96. Sun W, Julie Li YS, Huang HD, Shyy JY, Chien S (2010) microRNA: a master regulator of cellular processes for bioengineering systems. Annu Rev Biomed Eng 12:1–27 (PubMed PMID: 20415587)
97. Arteaga-Vazquez MA, Chandler VL (2010) Paramutation in maize: RNA mediated trans-generational gene silencing. Curr Opin Genet Dev 20(2):156–163 (PubMed PMID: 20153628. Pubmed Central PMCID: 2859986)
98. Brzeska K, Brzeski J, Smith J, Chandler VL (2010) Transgenic expression of CBBP, a CXC domain protein, establishes paramutation in maize. Proc Natl Acad Sci U S A 107(12):5516–5521 (PubMed PMID: 20212123. Pubmed Central PMCID: 2851796)

99. Jiang J, Benson E, Bausek N, Doggett K, White-Cooper H (2007) Tombola, a tesmin/TSO1-family protein, regulates transcriptional activation in the Drosophila male germline and physically interacts with always early. Development 134(8):1549–1559 (PubMed PMID: 17360778. Pubmed Central PMCID: 2229809)
100. Hollick JB (2012) Paramutation: a trans-homolog interaction affecting heritable gene regulation. Curr Opin Plant Biol 15(5):536–543 (PubMed PMID: 23017240)
101. Pilu R (2011) Paramutation: just a curiosity or fine tuning of gene expression in the next generation? Curr Genomics 12(4):298–306 (PubMed PMID: 22131875. Pubmed Central PMCID: 3131737)
102. Matzke M, Kanno T, Daxinger L, Huettel B, Matzke AJ (2009) RNA-mediated chromatin-based silencing in plants. Curr Opin Cell Biol 21(3):367–376 (PubMed PMID: 19243928)
103. Grewal SI (2010) RNAi-dependent formation of heterochromatin and its diverse functions. Curr Opin Genet Dev 20(2):134–141 (PubMed PMID: 20207534. Pubmed Central PMCID: 3005588)
104. Rassoulzadegan M, Grandjean V, Gounon P, Vincent S, Gillot I, Cuzin F (2006) RNA-mediated non-mendelian inheritance of an epigenetic change in the mouse. Nature 441(7092):469–474 (PubMed PMID: 16724059)
105. Renaud S, de Lorgeril M (1992) Wine, alcohol, platelets, and the French paradox for coronary heart disease. Lancet 339(8808):1523–1526 (PubMed PMID: 1351198)
106. McCullough ML, Feskanich D, Stampfer MJ, Giovannucci EL, Rimm EB, Hu FB et al (2002) Diet quality and major chronic disease risk in men and women: moving toward improved dietary guidance. Am J Clin Nutr 76(6):1261–1271 (PubMed PMID: 12450892)
107. Humble CG, Malarcher AM, Tyroler HA (1993) Dietary fiber and coronary heart disease in middle-aged hypercholesterolemic men. Am J Prev Med 9(4):197–202 (PubMed PMID: 8398218)
108. Liu S, Stampfer MJ, Hu FB, Giovannucci E, Rimm E, Manson JE et al (1999) Whole-grain consumption and risk of coronary heart disease: results from the Nurses' Health Study. Am J Clin Nutr 70(3):412–419 (PubMed PMID: 10479204)
109. Meyer KA, Kushi LH, Jacobs DR, Jr., Slavin J, Sellers TA, Folsom AR (2000) Carbohydrates, dietary fiber, and incident type 2 diabetes in older women. Am J Clin Nutr 71(4):921–930 (PubMed PMID: 10731498)
110. Muller M, Kersten S (2003) Nutrigenomics: goals and strategies. Nat Rev Genet 4(4):315–322 (PubMed PMID: 12671662)
111. Toufektsian MC, Salen P, Laporte F, Tonelli C, de Lorgeril M (2011) Dietary flavonoids increase plasma very long-chain (n-3) fatty acids in rats. J Nutr 141(1):37–41 (PubMed PMID: 21068183)
112. Burdge GC, Calder PC (2005) Conversion of alpha-linolenic acid to longer-chain polyunsaturated fatty acids in human adults. Reprod Nutr Dev 45(5):581–597 (PubMed PMID: 16188209)
113. de Lorgeril M, Salen P, Martin JL, Boucher F, de Leiris J (2008) Interactions of wine drinking with omega-3 fatty acids in patients with coronary heart disease: a fish-like effect of moderate wine drinking. Am Heart J 155(1):175–181 (PubMed PMID: 18082510)
114. di Giuseppe R, de Lorgeril M, Salen P, Laporte F, Di Castelnuovo A, Krogh V et al (2009) Alcohol consumption and n-3 polyunsaturated fatty acids in healthy men and women from 3 European populations. Am J Clin Nutr 89(1):354–362 (PubMed PMID: 19056552)
115. Vauzour D, Vafeiadou K, Rodriguez-Mateos A, Rendeiro C, Spencer JP (2008) The neuroprotective potential of flavonoids: a multiplicity of effects. Genes Nutr 3(3–4):115–126 (PubMed PMID: 18937002. Pubmed Central PMCID: 2593006)
116. Spencer JP (2009) Flavonoids and brain health: multiple effects underpinned by common mechanisms. Genes Nutr 4(4):243–250 (PubMed PMID: 19685255. Pubmed Central PMCID: 2775888)
117. Wu X, Beecher GR, Holden JM, Haytowitz DB, Gebhardt SE, Prior RL (2006) Concentrations of anthocyanins in common foods in the United States and estimation of normal consumption. J Agric Food Chem 54(11):4069–4075 (PubMed PMID: 16719536)

118. Tsuda T, Horio F, Uchida K, Aoki H, Osawa T (2003) Dietary cyanidin 3-O-beta-D-glucoside-rich purple corn color prevents obesity and ameliorates hyperglycemia in mice. J Nutr 133(7):2125–2130 (PubMed PMID: 12840166)
119. Tsuda T, Horio F, Uchida K, Aoki H, Osawa T (2003) Dietary cyanidin 3-O-beta-D-glucoside-rich purple corn color prevents obesity and ameliorates hyperglycemia in mice. J Nutr 133(7):2125–2130
120. Juhan-Vague I, Alessi MC, Mavri A, Morange PE (2003) Plasminogen activator inhibitor-1, inflammation, obesity, insulin resistance and vascular risk. J Thromb Haemost 1(7):1575–1579 (PubMed PMID: 12871293)
121. Pradhan AD, Manson JE, Rifai N, Buring JE, Ridker PM (2001) C-reactive protein, interleukin 6, and risk of developing type 2 diabetes mellitus. JAMA 286(3):327–334 (PubMed PMID: 11466099)
122. Tsuda T, Ueno Y, Yoshikawa T, Kojo H, Osawa T (2006) Microarray profiling of gene expression in human adipocytes in response to anthocyanins. Biochem Pharmacol 71(8):1184–1197
123. Tsuda T, Horio F, Kitoh J, Osawa T (1999) Protective effects of dietary cyanidin 3-O-beta-D-glucoside on liver ischemia-reperfusion injury in rats. Arch Biochem Biophys 368(2):361–366 (PubMed PMID: 10441388)
124. Tsuda T, Horio F, Kato Y, Osawa T (2002) Cyanidin 3-O-beta-D-glucoside attenuates the hepatic ischemia-reperfusion injury through a decrease in the neutrophil chemoattractant production in rats. J Nutr Sci Vitaminol (Tokyo) 48(2):134–141 (PubMed PMID: 12171434)
125. Tsuda T (2008). Regulation of adipocyte function by anthocyanins; possibility of preventing the metabolic syndrome. J Agric Food Chem 56(3):642–646 (PubMed PMID: 18211021)
126. Brown MS, Goldstein JL (1999) A proteolytic pathway that controls the cholesterol content of membranes, cells, and blood. Proc Natl Acad Sci U S A 96(20):11041–11048 (PubMed PMID: 10500120. Pubmed Central PMCID: 34238)
127. Kahn BB, Alquier T, Carling D, Hardie DG (2005) AMP-activated protein kinase: ancient energy gauge provides clues to modern understanding of metabolism. Cell Metab 1(1):15–25 (PubMed PMID: 16054041)
128. Zhou G, Myers R, Li Y, Chen Y, Shen X, Fenyk-Melody J et al (2001) Role of AMP-activated protein kinase in mechanism of metformin action. J Clin Invest 108(8):1167–1174 (PubMed PMID: 11602624. Pubmed Central PMCID: 209533)
129. Takikawa M, Inoue S, Horio F, Tsuda T (2010) Dietary anthocyanin-rich bilberry extract ameliorates hyperglycemia and insulin sensitivity via activation of AMP-activated protein kinase in diabetic mice. J Nutr 140(3):527–533 (PubMed PMID: 20089785)
130. Lopes AA (2009) End-stage renal disease due to diabetes in racial/ethnic minorities and disadvantaged populations. Ethn Dis 19(1):47–51 (PubMed PMID: WOS:000265519000013. English)
131. Ohga S, Shikata K, Yozai K, Okada S, Ogawa D, Usui H et al (2007) Thiazolidinedione ameliorates renal injury in experimental diabetic rats through anti-inflammatory effects mediated by inhibition of NF-kappa B activation. Am J Physiol-Renal Physiol 292(4):F1141-F1150 (PubMed PMID: WOS:000245469200004. English)
132. Li J, Lim SS, Lee JY, Kim JK, Kang SW, Kim JL et al (2012) Purple corn anthocyanins dampened high-glucose-induced mesangial fibrosis and inflammation: possible renoprotective role in diabetic nephropathy. J Nutr Biochem 23(4):320–331 (PubMed PMID: WOS:000302189000002. English)
133. Li J, Kang MK, Kim JK, Kim JL, Kang SW, Lim SS et al (2012) Purple corn anthocyanins retard diabetes-associated glomerulosclerosis in mesangial cells and db/db mice. Euro J Nutr 51(8):961–973 (PubMed PMID: WOS:000310995200008. English)
134. Kang HK, Kim D (2012) Efficient bioconversion of rice straw to ethanol with TiO2/UV pretreatment. Bioprocess Biosyst Eng 35(1–2):43–48 (PubMed PMID: 21909680)

135. Ling WH, Cheng QX, Ma J, Wang T (2001) Red and black rice decrease atherosclerotic plaque formation and increase antioxidant status in rabbits. J Nutr 131(5):1421–1426 (PubMed PMID: WOS:000168666400005. English)
136. Ling WH, Wang LL, Ma J (2002) Supplementation of the black rice outer layer fraction to rabbits decreases atherosclerotic plaque formation and increases antioxidant status. J Nutr 132(1):20–26 (PubMed PMID: WOS:000173088600003. English)
137. Xia XD, Ling WH, Ma J, Xia M, Hou MJ, Wang Q et al (2006) An anthocyanin-rich extract from black rice enhances atherosclerotic plaque stabilization in apolipoprotein e-deficient mice. J Nutr 136(8):2220–2225 (PubMed PMID: WOS:000239366300017. English)
138. Guo HH, Ling WH, Wang Q, Liu C, Hu Y, Xia M et al (2007) Effect of anthocyanin-rich extract from black rice (*Oryza sativa* L. indica) on hyperlipidemia and insulin resistance in fructose-fed rats. Plant Foods Hum Nutr 62(1):1–6 (PubMed PMID: WOS:000244451600001. English)
139. Salgado JM, Oliveira AG, Mansi DN, Donado-Pestana CM, Bastos CR, Marcondes FK (2010) The role of black rice (*Oryza sativa* L.) in the control of hypercholesterolemia in rats. J Med Food 13(6):1355–1362 (PubMed PMID: 21091249)
140. Yang Y, Andrews MC, Hu Y, Wang DL, Qin Y, Zhu YN et al (2011) Anthocyanin extract from black rice significantly ameliorates platelet hyperactivity and hypertriglyceridemia in dyslipidemic rats induced by high fat diets. J Agric Food Chem 59(12):6759–6764 (PubMed PMID: WOS:000291568300049. English)
141. Wang Q, Han PH, Zhang MW, Xia M, Zhu HL, Ma J et al (2007) Supplementation of black rice pigment fraction improves antioxidant and anti-inflammatory status in patients with coronary heart disease. Asia Pac J Clin Nutr 16:295–301 (PubMed PMID: WOS:000245926500056. English)
142. Hou Z, Qin P, Ren G (2010) Effect of anthocyanin-rich extract from black rice (*Oryza sativa* L. Japonica) on chronically alcohol-induced liver damage in rats. J Agric Food Chem 58(5):3191–3196 (PubMed PMID: 20143824)
143. Pannangrong W, Wattanathorn J, Muchimapura S, Tiamkao S, Tong-Un T (2011) Purple rice berry is neuroprotective and enhances cognition in a rat model of Alzheimer's disease. J Med Food 14(7–8):688–694 (PubMed PMID: 21510741)

Index

C. Chen (ed.), *Pigments in Fruits and Vegetables*, DOI 10.1007/978-1-4939-2356-4

Printed by Printforce, the Netherlands